# *Drosophila melanogaster* - MODEL FOR RECENT ADVANCES IN GENETICS AND THERAPEUTICS

Edited by **Farzana Khan Perveen**

***Drosophila melanogaster*** **- Model for Recent Advances in Genetics and Therapeutics**
http://dx.doi.org/10.5772/66545
Edited by Farzana Khan Perveen

**Contributors**

Jose F. De Celis, Cristina Martínez-Ostalé, Ana Ruiz-Gómez, Patricia Vega, Mireya Ruiz-Losada, Carlos Estella, Zazali Alias, Vennobaashini Venu, Thao Dang, Kimberly Carlson, Wilfredo Lopez, Alexis Page, Brad Ericson, Darby Carlson, Andrew Swan, Mohammed Bourouh, Sarat Chandra Yenisetti, Carolina Garcia, Vera L. S. Valente, Farzana Khan Khan Perveen, Brian Xia, Steven Steven De Belle, Paola Bellosta, Mariateresa Allocca, Sheri Zola, Svetlana Sarantseva, Elena Ryabova, Nataliya Matiytsiv, Olena Trush, Iryna Mohylyak, Galina Kislik, Pavel Melentev

**Published by InTech**
Janeza Trdine 9, 51000 Rijeka, Croatia



**Publishing Process Manager** Romina Skomersic
**Technical Editor** SPi Global
**Cover** InTech Design team

First published February, 2018
Printed in Croatia
Legal deposit, Croatia: National and University Library in Zagreb

Additional hard copies can be obtained from orders@intechopen.com

*Drosophila melanogaster* - Model for Recent Advances in Genetics and Therapeutics, Edited by Farzana Khan Perveen
p. cm.
Print ISBN 978-953-51-3853-2
Online ISBN 978-953-51-3854-9

For EU product safety concerns:
IN TECH d.o.o., Prolaz Marije Krucifikse Kozulić 3, 51000 Rijeka, Croatia,
info@intechopen.com or visit our website at intechopen.com.

# Meet the editor

Dr. Farzana Khan Perveen (FLS; gold medallist) obtained her BSc (Hons) and MSc (Zoology, Entomology) degrees from the University of Karachi; MAS (Monbusho Scholar, Agriculture, Agronomy) degree from the Nagoya University, Japan; and PhD (Research and Course work from Nagoya University; Toxicology) degree from the University of Karachi. She is the founder/chairperson of the Department of Zoology (DOZ) and Ex-Controller of Examinations at Shaheed Benazir Bhutto University (SBBU) and ex-founder/ex-chairperson of DOZ, Hazara University and Kohat University of Science and Technology. She is the author of 150 high-impact research papers, 135 abstracts, 40 authored books and 7 chapters and 5 edited books and supervised BS (4), MSc (50), MPhil (40) and PhD (1) students. She has organized and participated in numerous international and national conferences and received multiple awards and fellowships. She is a member of research societies, editorial boards of Journals and World Commission on Protected Areas, International Union for Conservation of Nature. Her fields of interest are entomology, toxicology, forensic entomology and zoology.

# Contents

# Preface

*Drosophila* is derived from the Greek word drósos, which means dew loving. The fruit fly, *Drosophila melanogaster* (Meigen, 1830), belongs to Drosophilidae family and is most frequently known as fruit flies. They are destructive agricultural pests. They are capable of colonizing live fruits that are still in the process of ripening, causing massive agricultural damage. There are more than 1,500 species, which are diverse in appearance, behaviour and breeding habitat. However, *Drosophila* spp. are distributed all over the world and found in every type of environment. Furthermore, hibernation takes place in many northern species. Human activities such as transporting of fruits and other fresh food items are responsible for introducing many species throughout the world. Most *Drosophila* spp. are small, about 2–4 mm long, but some are larger than a house fly. They are typically pale yellow to reddish brown or black and have transverse black rings across the abdomen with brick red eyes. Many species have distinct black patterns on the wings with the plumose (feathery) and arista antennae, bristling on the head and thorax. The characters of wing venations are used to diagnose the family. Chromosomal polymorphism is remarkably abundant in *Drosophila,* highlighting the paracentric inversions. Chromosomal inversions suppress meiotic recombination, and thus, natural selection can act to preserve favourable gene complexes. Analyses of natural and laboratory populations show that these polymorphisms provide adaptive advantages to their carriers in relation to diverse factors, such as niche exploration and climatic factors. In *Drosophila* female, the roles of known cell cycle regulators and specific factors in meiosis focus on three important meiotic events: nuclear envelope breakdown or maturation, establishment of the meiosis I spindle and release from metaphase I arrest at ovulation. Many meiotic processes are controlled by the mitotic kinase, Cdk1, with its cyclin partners, cyclins A, B and B3. Nongenetic transgenerational inheritance in *Drosophila* highlights transgenerational programming of metabolic status and longevity, one particular histone modification as an evolutionarily conserved underlying mechanism and important implications of such studies in understanding health and diseases.

Ecological genetics are conducted both at the level of individuals of the *Drosophila* and at the level of strains and natural populations. Cytostatic drugs have been tested for frequencies of early and late embryonic lethals (EEL and LEL), lethals of larva and pupa, fertility, morphometric analysis and somatic recombination test. *Drosophila* imaginal discs are an epithelial tissue perfectly suited to use them as a playground to define the functional contribution of genes to epithelial development and organ morphogenesis. Three interconnected aspects related to the use of *Drosophila* imaginal discs as an experimental system to analyse gene function are (i) imaginal discs biology, with a focus in the genetic mechanisms involved in pattern formation, (ii) concepts and available experimental tools for the analyses of gene function and (iii) uses of *Drosophila* and the imaginal discs for addressing biomedical problems. *Drosophila* has considerable biological similarity to mammalian systems. It has been solidified as a key model organism for elucidating many aspects of human disease. Parkinson's disease (PD) is a medical condition that has been known since ancient times. It is the second most common neurodegenerative disorder affecting approximately 1% of the population over age 50. It is characterized by both motor and non-motor symptoms. *Drosophila* offers many advantages for studying human neurodegenerative diseases and their underlying molecular and cellular pathology. *Drosophila* is a potential model system to understand the pathology associated with PD.

Ubiquitin carboxyl-terminal hydrolase L1 (UCH-L1) is a protein, which plays an important role in ubiquitin-proteasome system. *Drosophila* model was utilized to examine the role of UCH-L1. Besides, knock-down dUCH in dopaminergic neurons resulted in some PD-like phenotypes in fly. *Drosophila* model contributed a significant of dUCH in tissue development and function. There have been many studies dedicated to NTE/*sws*. A new allele is identified in *sws* using genome screening. Functional studies were carried out for *sws* in vivo, using larval neuromuscular junctions (NMJ) of *D. melanogaster* as a good system of HSP modelling. The *sws* was widely expressed in the larval nervous system, especially in glial cells. Mutations were established in the *sws* gene and alter NMJ morphology, the distribution of synaptic markers, microtubule network and synaptic microtubule organization. *Drosophila melanogaster* glutathione transferases E6 and E7 (DmGSTE6 and DmGSTE7) were successfully cloned, purified and biochemically characterized. The recombinant proteins were readily purified using the combination of both anionic and BSP/GSH-agarose affinity chromatographies. Thin-layer chromatography analysis showed that both isoforms were not able to conjugate several tested insecticides. The inhibition kinetics of natural products and dyes towards GSTs in vitro revealed that phenol red possessed inhibition effects only on GSTE6, while rose bengal and cardiogreen inhibit significantly on both GSTE6 and GSTE7. In contrast, methylene blue dye and *trans*-chalcone have been shown to stimulate GSTE7 activity towards CDNB. *Drosophila melanogaster* is an extremely useful model to study innate immunity mechanisms. A fundamental understanding of these mechanisms, as they relate to various pathogens, has come to light over the past 30 years. The discovery of small-interfering RNAs (RNAi) provided a mean, by which antiviral immunity was accomplished in invertebrates. At least three of these pathways (Imd, JAK-Stat and RNAi) show signal integration in response to viral infection, demonstrating a coordinated immune response against viral infection. The viral pathogens that infect invertebrates have developed countermeasures to some of these pathways, in particular to RNAi. The evolutionary arms race of pathogen vs host is ever ongoing.

**Dr. Farzana Khan Perveen FLS**
Founder Chairperson and Associate Professor
Department of Zoology
Ex-Controller of Examination
Shaheed Benazir Bhutto University (SBBU)
Khyber Pakhtunkhwa, Pakistan

Section 1

# *Drosophila* - Model for Genetics

Chapter 1

# Introduction to *Drosophila*

Farzana Khan Perveen

Additional information is available at the end of the chapter

http://dx.doi.org/10.5772/67731

## 1. Introduction

*Drosophila* derived from the Greek word drósos means dew loving. They belong to the Droso–philidae family; and are most frequently known as fruit flies or often called vinegar, wine or pomace flies. Their main distinguishing character is to stay on fruits, which are ripped or rotten. There is another related family Tephritidae, their members are also called as true fruit flies or fruit flies. *Drosophilae* are different from them. They feed primarily on unripe or ripe fruits. Many species of *Drosophila* are agricultural pests, especially the Mediterranean fruit flies. They oviposit through ovipositor and capable of colonizing in live fruits that are still in the process of ripening, causing massive agricultural damage (**Table 1**; **Figure 1a** and **f**) [1, 2]. Currently, the genus *Drosophila* is considered as paraphyletic group. The entire genus, however, contains more than 1500 species [3], which are very diverse in their appearance, behavior, and breeding habitat [4]. However, many members of the family Drosophilidae are categorized into two subgenera, in which around 1100 species belong to *Drosophila* subgenera, moreover, about 330 species belong to *Sophophora* subgenera including *D.* (*S.*) *melanogaster*. Furthermore, another *Drosophila* species, i.e., Hawaiian spp. have more than 500 species in which only 380 species are described. Furthermore, they are occasionally documented as a separate subgenus or genus, i.e., *Idiomyia grimshawi* [5], but this is not widely accepted. About 250 species are part of the genus *Scaptomyza*, which arose from the Hawaiian *Drosophila* and later recolonized in continental areas [6]. However, *Drosophila* spp. are distributed all over the earth; moreover, many species are found in the tropical regions. Furthermore, the alpine zones, cities, deserts, swamps, and tropical rainforest also confine them. Furthermore, hibernation takes place in many northern species [7].

Their breeding takes place in numerous types of rotten vegetation and mycological materials, comprising barks, flowers, fruits, mushrooms, and slime fluxes. However, the maggots of *D. suzukii* act as the pest and feed on fresh fruits. Moreover, some species of *Drosophila* have achieved the status of parasites and predators. Furthermore, several species attract to lure of mushrooms and fermented bananas, but others deny attracting to every type of bait. Furthermore, females and males are assembled for mating on appropriate propagating

**Kingdom:** Animalia

**Subkingdom:** Invertebrata

**Division:** Eumetazoa

**Subdivision:** Ecdysozoa

**Superphylum:** Tactopoda

**Phylum:** Arthropoda

**Subphylum:** Atelocerata

**Superclass:** Hexapoda

**Class:** Insecta

**Infraclass:** Neoptera

**Subclass:** Pterygota

**Superorder:** Endopterygota

**Order:** Diptera

**Family:** Drosophilidae

**Subfamily:** Drosophilinae

**Tribe:** Drosophilini

**Genus:** *Drosophila* Fallén, 1823

**Type species:** *Musca funebris* Fabricius, 1787

**Synonyms:** *Oinopota* Kirby & Spence, 1815

**Subgenera:**

- *Drosophila*
- *Sophophora*
- *Chusqueophila*
- *Dorsilopha*
- *Dudaica*
- *Phloridosa*
- *Psilodorha*
- *Siphlodora*

**Table 1.** Taxonomic rank of the fruit fly, *Drosophila* spp. Fallén, 1823.

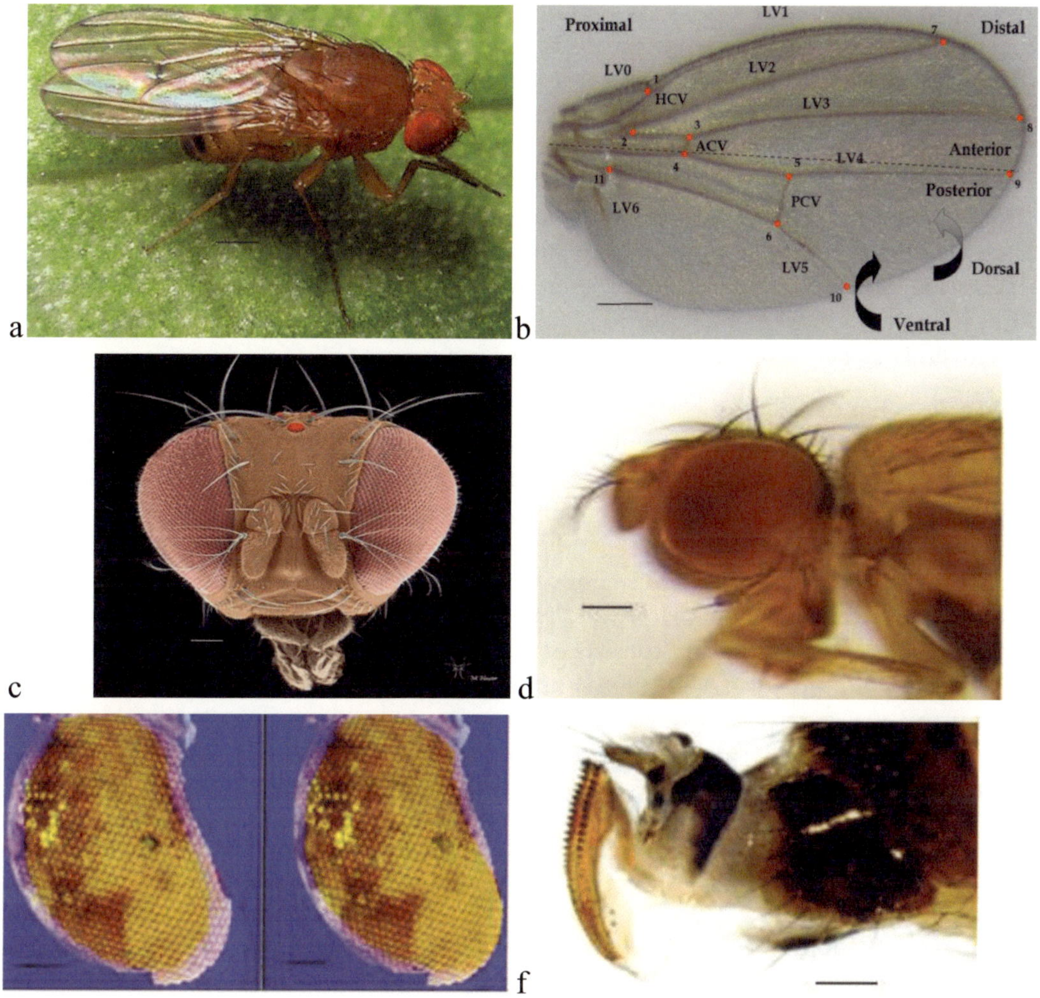


**Figure 1.** Adult of the fruit fly, *Drosophila* spp. Fallén, 1823 (Diptera: Drosophilidae) (a) [9]; wing morphology: ventral view of left wing and landmark positioning; LV: longitudinal vein; HCV: humeral cross-vein; ACV: anterior cross-vein; PCV: posterior cross-vein; the proximal, distal, anterior-posterior, and dorsal-ventral axes are shown (b) [10]; head, frontal view (c); lateral view of head shows characteristic bristles above the eye (d); stero images of *Drosophila* (e); ovipositor of an adult female spotted wing drosophila, *Drosophilia suzukii* (Matsumura), lateral view (f); photographs (c)–(f) by courtesy of Martin Hauser, Department of Food and Agriculture, California, USA; bar on photograph: 25 cm.

materials separate from breeding sites to form leeks. Also, many *Drosophila* spp., comprising *D. immigrans*, *D. melanogaster*, and *D. simulans*, are found neighboring and accompanying with humans and are called domestic species. Also, human activities, such as transporting of fruits and other fresh food items, are responsible for introducing many species throughout the world, including *D. immigrans*, *D. melanogaster*, *D. simulans*, *D. subobscura*, and *Zaprionus indianus* [8].

### 1.1. Morphology

Most *Drosophila* spp. are small, about 2–4 mm long, but some are larger than a house fly. They are typically pale yellow to reddish brown or black and transverse black rings across the abdomen with brick red eyes. Many species have distinct black patterns on the wings (**Figure 1b**) with plumose (feathery) and arista antennae, bristling on the head and thorax (**Figure 1c** and **d**) [11]. The characteristics of wing venations are used to diagnose the family. *Drosophila* flight path of straight sequencing with rapid and jerky turns of the wings with intersperse between positions of rest is known as saccades movement. However, when it turns in saccades movement, it can be revolved at the angle of 90° in about 50 milliseconds. Moreover, *Drosophila*'s wings can beat 220 times per second [12].

*Drosophila* contains one of the most advanced forms of eye among insects, i.e., compound eye. The unit structure of it is ommatidia; however, there are 760 ommatidia per compound eye, moreover, a cornea, eight photoreceptor cells (R1–8), many pigment cells, and some support cells are also found in each ommatidium. Reddish pigment cells are found in wild-type *Drosophila*, excess blue light is absorbed by them; therefore, ambient light is not made the fly blind [13]. As far as photoreceptor cells are concerned, they have two main parts, the rhabdomere and the cell body. However, the nucleus is an active part of the cell body, while rhabdomere is 100-μm long and consists of toothbrush-like masses of membrane, which are called microvilli. Moreover, the length and diameter (dm) of each microvillus are around 1–2 μm and 60 nm, respectively. Further, the rhodopsin is the visual protein; their approximately 100 million molecules are wrapped in rhabdomere's membrane. Accordingly, the function of rhodopsin is absorption of light. On the other hand, there are many other visual proteins that are also present in rhabdomere, which are tightly bound in the spaces among microvilli, hence, there is very little spaces for cytoplasm [14].

In *Drosophila*, there are many types of proteins that are present in photoreceptor cells, which are expressed in rhodopsin isoform, for example, blue light (480 nm) is absorbed by rhodopsin1 (Rh1), which is present in the R1–R6 photoreceptor cells. Similarly, UV light (345 and 375 nm) is absorbed by an expression of a combination of Rh3 or Rh4, which is present in the R7 and R8 photoreceptor cells. In the same way, blue (437 nm) and green (508 nm) lights are absorbed by Rh5 or Rh6, respectively. More likely, a protein opsin is also present in each photoreceptor cells, which is covalently linked to a carotenoid chromophore, i.e., 11-cis-3-hydroxyretinal. This protein is found in each rhodopsin molecule (**Figure 1c–e**) [15].

### 1.2. Lifecycle

Sexual dimorphism is characteristic of *Drosophila* spp. Therefore, males can be easily differentiated from females having differences in size and color. However, the length of female is ca. 2.5 mm, moreover, male is somewhat smaller than female with dorsal sites of male's body being darker due to a distinct black patch at the abdomen. Furthermore, in newly emerged flies and sex comb, sexual dimorphism is less noticeable (**Figure 2**) [16]. They also vary widely in their reproductive capacity. Females lay some 400 eggs (embryos), about 5 at a time, into rotting fruits or other suitable materials such as decaying mushrooms and sap fluxes. *Drosophila melanogaster*

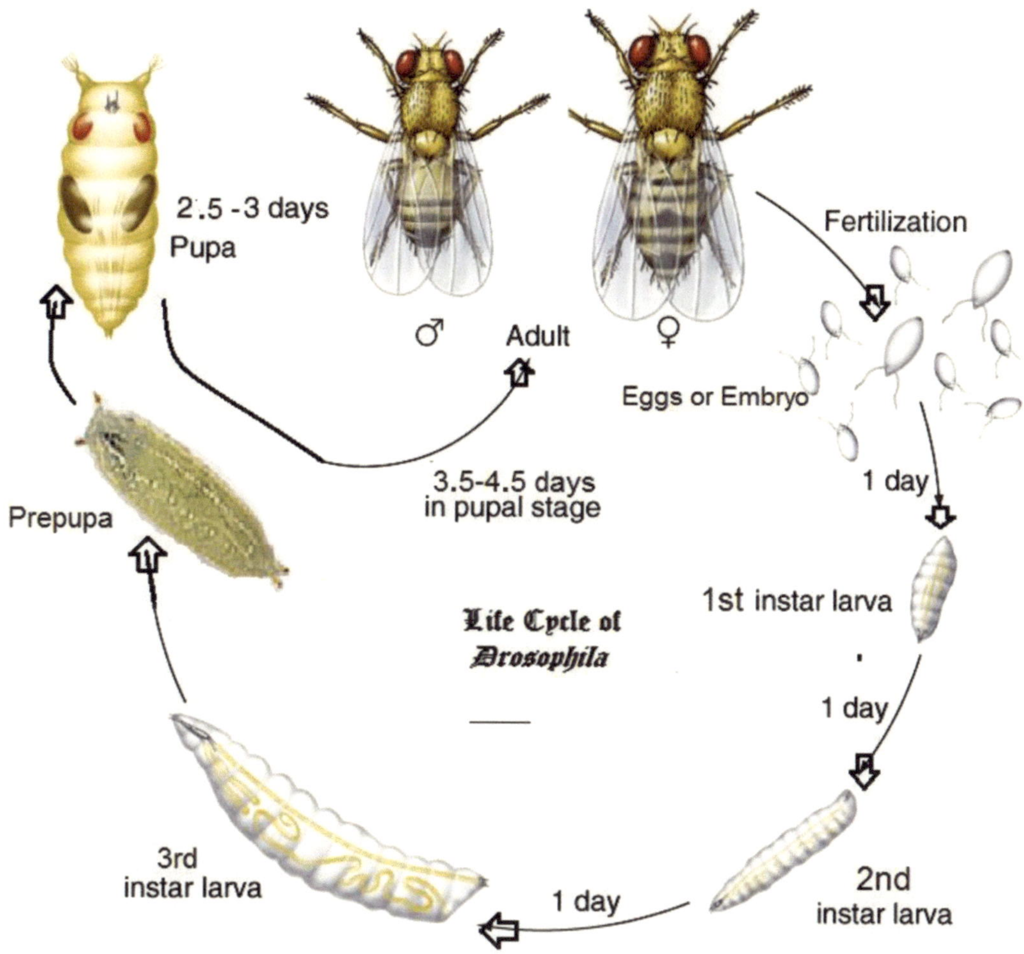


**Figure 2.** Life-cycle of *Drosophila* (♂: male; ♀: female)with three larval instar and a pupal stages; bar on photograph: 25 cm [27].

breeds in bulky and comparatively scarce substrates. About 10–20 eggs are matured at the same time; therefore, female lays them together in one place. However, others species those oviposit only one egg in a day, breed in additional-rich but a smaller amount of nourishing resources, such as fresh leaves and grasses [17]. Eggs of *Drosophila* are ca. 0.5 mm in length, silvery, oblong, ovoid, and somewhat compressed when view laterally. Internally, an indistinguishable skinny vitelline covering warps the egg together with an exterior extracellular covering is called a chorion. At the front end, two minor respiratory filaments prolong from the dorsal surface close to the front termination; however, the tips of these extend above the surface and allow oxygen ($O_2$) to reach the embryo. The anterior end can be recognized by the micropyle, a structure on the external coating surrounding the egg [18].

In usual environmental conditions, hatching of eggs takes place after 12–15 hours at 25°C (77°F) into small, white first instar maggots (larvae). Then food is taken by the resulting maggots and their growth takes place for nearly 4 days (at 25°C). After that, they molt two times into second and third instar maggots during 24 and 48 hours after hatching, respectively. During the period of the larval stage, they are actively feed on bacteria, microbes, germs, and detritus, which are present on the rotten and decaying breeding resources that decompose the fruits, as well as on the sugar of the fruits itself, vegetable matters, and yeasts. The mother puts feces on the egg sacs to establish the same microbial composition in the larvae's guts, which works positively for them [19].

In specific conditions, their development time varies widely (between 7 and more than 60 days) between species to species and depends on the environmental factors, such as temperature, breeding substrates, and crowding. Numerous studies have shown that eggs oviposited by *Drosophila* spp. pass through an ecological cycle [20]. Nocturnally, oviposited eggs pass through favorable environmental circumstances, thus, such eggs are not vulnerable to withering from parasites [21]. Consequently, the maggots hatched from those eggs are healthy and have higher appropriateness in contrast to the maggots that are hatched from diurnal oviposited eggs. In eggs of *D. melanogaster*, a biological clock has been observed and their maggots greatly adapt to their ecological cycles, therefore, their survival becomes easy and they gain the highest benefits in their environment [22].

Then, the larvae encapsulate (pupation) in the puparium and undergo a 4-day long metamorphosis (at 25°C), after which the adults emerge [23]. The developmental period for *D. melanogaster* varies with temperature, as with many ectothermic species. They are Endopterygota, also known as Holometabola, is a superorder of insects within the infraclass Neoptera, therefore, they go through distinctive larval, pupal, and adult stages and their wings are developed internally. Duration of lifecycle of *Drosophila* spp. is variable and depends on various factors. However, it increases with the increase of temperature due to the stress of heat. Moreover, the shortest lifecycle from egg to adult, 7 days, is achieved at 28°C (82°F). Further, the same 11 days is at 30°C or 86°F. Furthermore, under the best environmental conditions, duration of lifecycle is 8.5 days at 25°C (77°F), under moderate conditions, 19 days is at 18°C (64°F), under low conditions, it takes over 50 days at 12°C (54°F) [23]. While under crowded condition, lifecycle is prolonged with emergence of smaller flies. However, their average life span is 35–45 days. (**Figure 2**; **Table 2**) [6].

| S. no | LC in d | Temperatures | | Type of life cycle |
|---|---|---|---|---|
| | | °C | °F | |
| 1. | 7 | 28 | 82 | Shortest life cycle |
| 2. | 8.5 | 25 | 77 | Under ideal conditions |
| 3. | 11 | 30 | 86 | – |
| 4. | 19 | 18 | 64 | – |
| 5. | 50 | 12 | 54 | Longest life cycle |
| 6. | 35–45 | 15–35 | 59–95 | Average life cycle |

LC, life cycle of *Drosophila* spp.; d, days; °C, degree celsius; °F, fahrenheit.

**Table 2.** Duration of life cycle of *Drosophila* spp. on different temperatures.

According to the studies of living organisms related to sperm, the longest sperm cells, i.e., 58 mm (2.3 in) in length, are found in males of *Drosophila bifurca* compared to any animal that exists on this universe [24]. Structurally, sperms consist of a head with nucleus and a long tail with tangled coils. Other species of genus *Drosophila* also form somewhat gigantic sperm cells; however, the longest sperms are produced by *D. bifurca* [19]. The length of the sperm cells of *D. melanogaster* is 1.8 mm, which has an adequate length, while, a human sperm is still around 35 times shorter than *D. bifurca*'s sperm. A number of *Drosophila* spp. are known to mate by traumatic insemination [25]. In *Drosophila* female, sperm are stored in a tubular and two mushroom-shaped receptacle structures, which are called spermathecae. In polyandrous female, strong sperm competition takes place for fertilization of eggs. It is observed that the sperm precedence of last male mated is utilized for production of descendent. The female sires about 80% of her offspring with the sperm inseminated by the previous male to mate. Both displacement and incapacitation are responsible for this precedence [26].

## 2. *Drosophila* genetics

*Drosophila* was the first organisms to be studied genetically; perhaps it is the best understood animal in genetic systems. It has a small size, short lifecycle, high reproductive rate, easy to culture, and easy genetic manipulation. It is one of the most valuable organisms in biology, developmental biology, genetics, medicine, human disease, and stem cell research. One species of *Drosophila* in particular, *D. melanogaster,* has been comprehensively used in genetics as a common model and toolbox organism in modern biology. People have been working on it, since ancient time, consequently, very much is already known about it, therefore, it is easy to handle and well understood. It is cheap and easy to keep in large numbers. From the past century as well as at present, *Drosophila* has been used for advancement in learning, studies, education, and research. Many different aspects of *Drosophila* have been studied by thousands of scientists. Its significance for human health was documented by the award of the Nobel Prize in medicine/physiology to Lewis, Volhard and Wieschaus in 1995 [28].

### 2.1. Sex distinguish

In *Drosophila,* female's abdomen consists of seven segments with many dark transverse stripes and pointed tip. However, male's abdomen consists of five segments with two dark stripes and more curved with heavily pigmented tip. Moreover, in a newly immerged adult male, the pigmentation is not observed. Further, the gender of *Drosophila* can be differentiated by the structure of the external genitalia and their color. Furthermore, the abdomen is pale and relatively smooth in mature female, in comparison with dark genitalia that are found in mature male. Additionally, a secondary sexual character is also present in the male flies, which is called sex comb, a structure that consists of a minor cluster of about 10 black hairs in front of the last large segment (third segment counting from the end of the body). The same is also present even in immature males. Likewise, another secondary sexual character in male, the presence of a cluster of spiky hairs (claspers) surrounding the reproducing parts used to attach to the female during mating (**Figure 3**) [16].

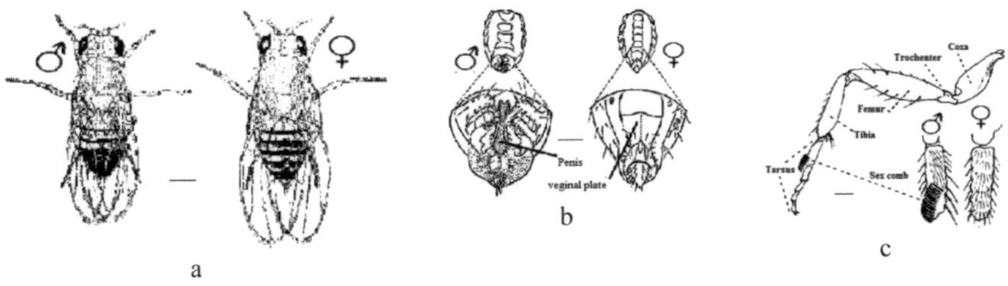


**Figure 3.** Dorsal view of male (left) and female (right) *Drosophila*: five segments and rounded abdomen in male and seven segments and pointed abdomen in female (a); ventral view of genitalia of male (left) and female (right) (b); left foreleg of male with sex comb on tarsus of male fly (left) and female sex comb absent in female fly (c); drawn by W. Hewitt; bar on photograph: 25 cm.

## 2.2. Virgin females

All female flies, which are used in genetic experiments for making control crosses, should be virgin. As a concern, after 8 hours of emerging from pupal stage, *Drosophila* females are able to mate with males. Likewise, they are capable of mating with many or multiple males; therefore, they are called as polyandrous. On the other hand, when they mate, males inseminate millions of viable sperms, which are stored in spermathecae of females for several days and this will puzzle the outcomes of a following orderly mating. Therefore, to prevent multiple mating, all adult *Drosophila*, specially, females are isolated 7 hours earlier to utilize them for experiment, consequently, that all freshly produced *Drosophila* will remain virgin [29].

## 2.3. Sex chromosomes

Different organisms have diverse sex determination mechanisms. Mostly, females are defined as homogametic (all gametes will carry the X chromosome: XX); however, males are known as heterogametic (half the gametes carry the X and half the Y chromosome: XY). The examples of the same are species such as humans and *Drosophila*, etc. The genes carried on the X chromosome and those carried on the Y chromosome, consequently, distinction has made between them. Since sex chromosomes as well as autosomes have been correlated the law of segregation. The genes on the X chromosome are distributed independently from genes on the Y chromosome (**Figure 4**) [28].

**Figure 4.** Metaphase chromosomes from dividing cells of larval ganglion (brain) of *Drosophila melanogaster*: chromosomes of female (a) consist of three pairs of autosomes and one pair of rod-shaped X chromosomes; chromosomes of male (b) consist of three pairs of autosomes, one X chromosome, and one Y chromosome; bar on photograph: 10 μm [30].

### 2.4. *Drosophila* as model systems

The boundaries of human genetics, however, sort it essential to practice prototypical systems to evaluate precious genes and passageways in more detail. During the past 20 years, investigation utilizing the genetically acquiescent fruit fly has known *D. melanogaster* as an appreciated model system in the learning of human neurodegeneration. These studies offer reliable models for Parkinson's, motor neuron, and Alzheimer's diseases, as well as models for trinucleotide repeat expansion diseases, including Huntington's and ataxias disease. As a consequence of these studies, many signaling pathways comprising target of rapamycin (TOR), c-Jun N-terminal kinase (JNK), bone morphogenetic protein (BMP) signaling, and phosphatidylinositol 3-kinase (PI3K)/Akt have been revealed to be decontrolled in models of proteinopathies suggesting that two or more starting actions may activate disease formation in an age-related manner [31]. Moreover, these studies also determine that *Drosophila* can be utilized to monitor chemical compounds for their prospective to inhibit or enhance the disease, which in order can openly monitor medical research and the expansion of original therapeutic approaches for handling of human neurodegenerative diseases. Human neurodegenerative diseases are demoralizing illnesses that principally disturb aging individuals. The bulk of the ailments are related with pathogenic oligomers from misfolded proteins, ultimately producing the development of masses and the advanced damage of neurons in the brain and central nervous system (CNS). Many of these proteinopathies are sporadic and the source of pathogenesis leftovers obscure. Inborn forms are linked with genetic deficiencies, suggesting that the affected protein is causally related to diseases formation and/or progression [32, 33].

## 3. Summary

*Drosophila* belongs to the family Drosophilidae, whose members are most frequently called the fruit flies. Since ancient time, *Drosophila* is used as a model and toolbox for biology, genetics, medicine, human disease, and stem cell research. It was nearly 100 years ago that Thomas H. Morgan reported the identification of the white gene in *Drosophila melanogaster*. Genetic approaches dominated the first 50 years of research in *Drosophila* (1910–1960), concentrating on dissecting the principles of inheritance. In this period, important concepts and tools were developed that allowed the study of many other biological processes during 1960–2010. Certainly, investigators realized in the early 1950s that genetic approaches could be used to study problems other than heredity. The uninterrupted development of research tools during 1960–2010 has driven numerous new discoveries in *Drosophila*. This is an appropriate time to reflect on the past and present contributions of *Drosophila* research in different fields, and therefore, the present book on "*Drosophila*" aims to publish this. *Drosophila* offers many unique advantages that will ensure that it is a premier research organism since ancient time to recent to future and for every field of research. After 100 years, fruit flies continue to be the choice model system for many neuroscientists. Example of recent contribution of research in *Drosophila* relates to numerous aspects of the physiological properties of sleep that is shared between *Drosophila* spp. and humans. Another example relates to Parkinson's disease. The work on parkin and PINK1 mutation, *D. melanogaster* has provided evidence that regulating mitochondrial remodeling and dysfunction is a cause of Parkinson's disease. Recently, many *Drosophila* experts have focused their attention on dissecting the molecular and cellular basis of the behavior. These include phototaxis, chemotaxis, aggression, physical response to mechanical stimuli, escape behavior, and sex. These studies will undoubtedly advance our understanding of how the nervous system of *Drosophila* works and provide us with very valuable paradigms to study mammalian brain function. This book provides window for *Drosophila* research as a toolbox for biology and medicine and its profile acts in top creatures for science experimentation. *Drosophila's* contribution to ageing, basal body, stem cell, nanoparticles, and artificial intelligence research is helping us to open new doors of research. A fluorescent tagging approach can be used in *Drosophila* to pest and vector risk assessment. Cloning and characterization in *Drosophila* can be exposed novel entries in the future of *Drosophila*. This book aims to provide the readers with a comprehensive overview on the historical, modern, and future prospects on this important insect, *Drosophila*, featuring an easy-to-follow, vignette-based format that will be focusing on the most important research-oriented evidences on various advantageous aspects for parasitologists, entomologists, neurologists, evolutionists, researchers, scientists, students, and others.

## Author details

Farzana Khan Perveen

Address all correspondence to: farzana_san@hotmail.com

Department of Zoology, Shaheed Benazir Bhutto University (SBBU), Main Campus, Sheringal, Khyber Pakhtunkhwa (KP), Pakistan

## References

[1] Green MM. It really is not a fruit fly. Genetics. 2002;**162**:1-3

[2] Beers EH, Smith TJ, Walsh DB. Spotted wing drosophila, *Drosophila Suzukii* (Matsumura) (Diptera: Drosophilida). Orchard Pest Management. 2010;47-39. Online: http://jenny.tfrec.wsu.edu/opm/displaySpecies.php?pn=165 (Accessed: 29/4/2011)

[3] Patterson C. Evolution. Cornell University Press, Cornell, United State of America (USA). 1999;45-55

[4] Markow TA, O'Grady PM. *Drosophila*: A guide to species identification and use. Elsevier, London, United Kingdom (UK). 2005;56-76

[5] Brake I, Bächli G. Drosophilidae (Diptera). World Catalogue of Insects. 2008;1-412

[6] O'Grady P, DeSalle R. Out of Hawaii: the origin and biogeography of the genus *Scaptomyza* (Diptera: Drosophilidae). Biology Letters. 2008;**4**(2):195-199. DOI: 10.1098/rsbl.2007.0575

[7] Castrezana S. New records of *Zaprionus indianus* Gupta, 1970 (Diptera: Drosophilidae) in North America and a key to identify some *Zaprionus* species deposited in the *Drosophila* Tucson Stock Center. Drosophila Information Service. 2007;**90**:34-36

[8] Linde KVD, Steck GJ, Hibbard K, Birdsley JS, Alonso LM, Houle D. First records of *Zaprionus indianus* (Diptera: Drosophilidae), a pest species on commercial fruits, from Panama and the United States of America. Florida Entomologist. 2006;**89**(3):402-404. DOI: 10.1653/0015-4040(2006)89[402:FROZID]2.0.CO;2

[9] Online. Fly (Fruit) *Drosophila melanogaster*. Taranaki Educational Resource: Research, Analysis and Information Network (TER: RAIN); 2016. Online: http://www.terrain.net.nz/friends-of-te-henui-group/local-flies/common-fruit-fly-drosophila-melanogaster.html (Accessed: 23/11/2016)

[10] Carreira VP, Soto IM, Mensch J, Fanara JJ. Genetic basis of wing morphogenesis in *Drosophila*: sexual dimorphism and non-allometric effects of shape variation. BMC Developmental Biology. 2011;**11**:32. DOI: 10.1186/1471-213X-11-32

[11] Vilela CR. Is *Zaprionus indianus* Gupta, 1970 (Diptera: Drosophilidae) currently colonizing the Neotropical region? Drosophila Information Service. 1999;**82**:37-39

[12] Fry S, Dickinson M. The aerodynamics of free-flight maneuvers in *Drosophila* (PDF). Science. 2003;**300**(5618):495-498. DOI: 10.1126/science.1081944

[13] Hardie RC, Raghu P. Visual transduction in Drosophila. Nature. 2001;**413**(6852):186-93. DOI: 10.1038/35093002

[14] Raghu P, Colley NJ, Webel R, et al. Normal photo transduction in *Drosophila* photoreceptors lacking an InsP(3) receptor gene. Molecular and Cellular Neuroscience. 2000; **15**(5):429-45. DOI: 10.1006/mcne.2000.0846

[15] Wang T, Xu H, Oberwinkler J, Gu Y, Hardie R, Montell C, et al. Light activation, adaptation and cell survival functions of the Na+/Ca2+ exchanger CalX. Neuron. 2005;**45**(3): 367-378. DOI: 10.1016/j.neuron.2004.12.046

[16] Chiang HC, Hodson AC. An analytical study of population growth in *Drosophila melanogaster*. Ecological Monographs. 1950;**20**(3):173-206. DOI: 10.2307/1948580

[17] Bakker K. An analysis of factors which determine success in competition for food among larvae of *Drosophila melanogaster*. Archives Neerlandaises de Zoologie. 1961;**14**(2):200-281. DOI: 10.1163/036551661X00061

[18] Broughton SJ, Piper MD, Ikeya T, Bass TM, Jacobson J, Driege Y, Martinez P, Hafen E, Withers DJ, Leevers SJ, Partridge L. Longer lifespan, altered metabolism, and stress resistance in *Drosophila* from ablation of cells making insulin-like ligands. Proceedings of the National Academy of Sciences of the United States of America. 2005;**102**(8):3105-3110. DOI: 10.1073/pnas.0405775102

[19] Pitnick S, Spicer GS, Therese A, Markow A. How long is a giant sperm? Nature. 1995;**375**(6527):109. DOI: 10.1038/375109a0

[20] Markow TA, O'Grady PM. *Drosophila*: a guide to species identification and use. Elsevier Publisher, London, United Kingdom (UK). 2005;34-49

[21] Remsen J, O'Grady P. Phylogeny of Drosophilinae (Diptera: Drosophilidae), with comments on combined analysis and character support. Molecular Phylogenetics and Evolution. 2002;**4**(2):249-264. DOI: 10.1016/S1055-7903(02)00226-9

[22] Patterson C. Evolution. Cornell University Press, New York (NY), United State of America (USA). 1999;1-70

[23] Ashburner M, Thompson JN. The laboratory culture of *Drosophila*. In: Ashburner M, Wright TRF (eds). The Genetics and Biology of *Drosophila*. Academic Press, Georgia, United State of America. 1978;**2A**:1-81

[24] Joly D, Luck N, Dejonghe B. Adaptation to long sperm in *Drosophila*: correlated development of the sperm roller and sperm packaging. Journal of Experimental Zoology B: Molecular and Developmental Evolution. 2007;**310B**(2):167-178. DOI: 10.1002/jez.b.2167

[25] Howlader G, Sharma KV. Circadian regulation of egg-laying behavior in fruit flies *Drosophila melanogaster*. Journal of Insect Physiology. 2006;**52**:779-785

[26] Pitnick S. Investment in testes and the cost of making long sperm in *Drosophila*. American Naturalist. 1996;**148**:57-80. DOI: 10.1086/285911

[27] Weigmann K, Klapper R, Strasser T, Rickert C, Technau G, Jäckle H, Janning W, Klämbt C. FlyMove: a new way to look at development of *Drosophila*. Trends in Genetics. 2003;**19**(6):310-311. DOI: 10.1016/S0168-9525(03)00050-7

[28] Wolf MJ, Rockman HA. Drosophila melanogaster as a model system for genetics of postnatal cardiac function. Drug Discovery Today Diseases Models. 2008;**5**(3):117-123. DOI: 10.1016/j.ddmod.2009.02.002

[29] Sokolowski MB. Drosophila: genetics meets behaviour. Nature Review Genetics. 2001;**2**(11):879-890. DOI: 10.1038/35098592

[30] Demerece M, Kaufmann BP. *Drosophila* Guide. 10th Edition; Carnegie Institute of Washington DC, United State of America (USA). 1996;1-46

[31] Kamimura Y. Twin intermittent organs of *Drosophila* for traumatic insemination. Biology Letters, The Royal Society, London, United Kingdom (UK). 2007;**3**(4):401-404. DOI: 10.1098/rsbl.2007.0192

[32] Catherine P, Kelly D, Jerry C. Sperm competition between *Drosophila* males involves both displacement and incapacitation. Nature. 1999;**400**(6743):449-452. DOI: 10.1038/22755

[33] Hirth F. *Drosophila melanogaster* in the study of human neurodegeneration. CNS and Neurological Disorders Drug Targets. 2010;**9**(4):504-523. DOI: 10.2174/187152710791556104

Chapter 2

# *Drosophila* Chromosomal Polymorphism: From Population Aspects to Origin Mechanisms of Inversions

Carolina Garcia and Vera L. S. Valente

Additional information is available at the end of the chapter

http://dx.doi.org/10.5772/intechopen.73246

**Abstract**

High rates of chromosomal rearrangements are remarkably abundant in *Drosophila* Fallén, 1832 (Insecta, Diptera) genus, highlighting the paracentric inversions. Since different species of this genus are paradigms for genetics, evolutionary, and population studies, polymorphism analyses for chromosomal inversions have provided basic knowledge for beautiful biological questions. Chromosomal inversions suppress meiotic recombination and thus, natural selection can act to preserve favorable gene complexes. Analyses of natural and laboratory populations show that these polymorphisms provide adaptive advantages to their carriers in relation to diverse factors, such as niche exploration and climatic factors. In addition, due to their monophyletic origin, they also serve as genetic markers for the construction of unrooted phylogenies. With the increasing domain of molecular techniques and genome sequencing, factors such as the reuse of breakpoints by different inversions and the mechanisms that give rise to these polymorphisms have been exploited with scientific refinement. These analyses show the presence of regions that are hot spots for breakpoints, fitting the fragile breakage chromosomal evolution model, as well as the involvement of transposition elements at the origin of chromosomal inversions.

**Keywords:** chromosomal evolution, chromosomal inversion, polytene chromosomes, staggered breaks, transposable elements

## 1. Introduction

Structural chromosome rearrangements originate from chromosomal breaks at different sites, followed by reconstitution of these breaks in a distinct combination. They involve large quantities of genetic material at the cytological level and can be visualized under light microscopy.

The analysis of different rearrangements in the karyotype of the species of this genus was favored due to the presence of the polytene chromosomes. These polytene chromosomes are formed in interphase nuclei and are the final product of successive replication cycles without the consequent separation of the daughter chromatids, resulting in a huge structure that presents natural banding, formed by the precise synapses of parallel chromomeres of the sisters' chromatids. It is estimated that the polytene chromosomes founded in the salivary glands undergo $2^{10}$ replication events, generating up to 1024 filaments in each chromosomal pair of a diploid cell [1], originating a unique visualization magnitude. Tissues and organs containing cells with polytene chromosomes are, in general, involved in intense short time secretory functions, in a fast-growing context. Another peculiarity of the interphase polytene chromosomes is the non-segregation after replication; the parental chromosomes remain united and paired in the same conformation only seen in meiosis I of most other organisms [2].

The physical structure of the polytene chromosomes enables the accurate analysis of the different chromosomal rearrangements in *Drosophila* focusing on inversions—the most frequent rearrangement of the genus. This rearrangement consists in the simultaneous break of two sites in a chromosome and the reorganization of this area with a 180° inverted order.

Inversions are classified in two types, in diploid organisms: paracentric (do not involve the centromere in its formation, occurring in the same chromosome arm) and pericentric (involve the centromere and more than one chromosome arm). This rearrangement can be visualized as heterozygous during the pairing of the homologous chromosomes in meiosis I when only one of the parental chromosomes carries the inversion, forming an inversion loop for the correct pairing of the homologous chromosomes; or as homozygous when both parental chromosomes carry the inversion. These chromosomal conformations can be visualized on the *Drosophila* interphase polytene chromosomes (**Figure 1**) [3].

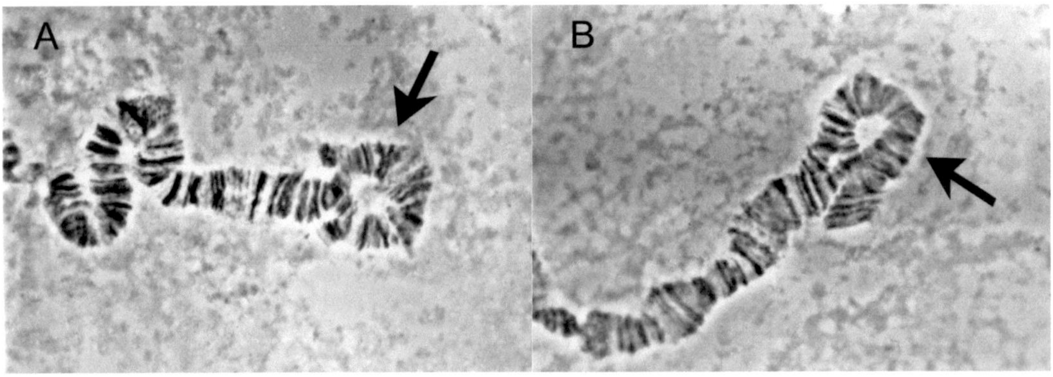


**Figure 1.** Chromosome inversions in heterozygosis in *Drosophila willistoni* Sturtevant, 1916 polytene chromosomes. (A) IIR-E inversion in the IIR chromosome arm. (B) III-J inversion in the chromosome III. Arrows point to the inversion loops formed by the correct pairing of the homologous chromosomes. Both inversions are physically at the distal end of the chromosome. Note that the precise synapse of the sister chromatids, according to the degree of compaction along the chromosome, forms a pattern of dark and light bands (bands and interbands, respectively). Regarding the chromosome polymorphism in *D. willistoni*, see the review in [56]. Source: Collection of images by Professor Vera L. S. Valente, deposited in the Laboratory of *Drosophila* at the Federal University of Rio Grande do Sul, Brazil.

Chromosomal inversions, compared to the other structural chromosomal rearrangements, use to be better tolerated by the organisms that carry them, since do not imply, theoretically, an increase or reduction of the genomic material. An inversion that occurs within a gene, however, can result in mutation, often lethal to the organism. The changing position of the genes, related to each other's and their controlling sequences, which is called the position effect, is another consequence of the inversion, resulting in alterations of gene expression and, consequently, alterations at the phenotypic level.

The behavior of a heterozygous inversion and the consequences it may entail differs during meiosis and mitosis. In meiosis I, the occurrence of crossing over inside of a paracentric inversion loop induces the formation of a dicentric chromosome (with two centromeres) and an acentric fragment (without centromere), resulting in gametes with deletions. In contrast, the occurrence of a meiotic recombination at the pericentric inversion loop results in the normal segregation of the chromosomes during meiosis I, since the centromeres are contained in the inversion, but originates gametes with deletion and duplications at meiosis II ending. During the mitosis, a heterozygous inversion does not imply major difficulties for the course of the cycle, since each chromosome duplicates and the sister chromatids are directed to the resulting daughter cells [4]. Illustrations of this are found in several genetics books, usually in Structural Chromosomal Alterations chapter.

Species of the *Drosophila* genus are model organisms for the study of chromosomal inversions, given the high resolution of the polytene chromosomes analysis, coupled with the fact that more than half of the studied species of *Drosophila* are naturally polymorphic for inversions [5]. However, based on the knowledge of the genomic destabilization and effects on the production of gametes that the inversions can originate, the high occurrence of chromosomal polymorphism is not expected *a priori* in the different living beings. The species of the genus *Drosophila* present a high rate of paracentric inversions, without a major deleterious effect on their reproductive success duo the presence of defense mechanisms in males and females, preventing the production of gametes bearing unbalanced chromosomal rearrangements [6].

There is a mechanism in the meiosis of females of *Drosophila melanogaster* Meigen, 1830, carrying heterozygous inversion that selectively eliminates the recombinant gametes during the formation of the polar corpuscles. In this mechanism, the first polar corpuscle to be excluded is one of the balanced chromatids (standard order, or inverted order). The second polar corpuscle eliminated is the dicentric chromosome. The acentric fragment is not oriented in the meiotic spindle and is later degraded. The last polar corpuscle to be eliminated, which will be effectively fertilized, also presents the standard order, or inverted order [7].

The mechanism of protection against the production of inviable gametes in males of *D. melanogaster* seems to be the suppression of recombination in spermatogenesis [8]. Mutations in genes that affect the segregation of chromosomes that did not undergo meiotic exchange in *Drosophila* females do not have the same effect in males, suggesting that the exchange is not necessary for the correct segregation of homologous chromosomes in meiosis I in males of this genus [6].

Aside from the inferred suppression of recombination in males, reports of its occurrence at the meiotic level are present in the literature, evidencing some peculiarities. Among these, the high occurrence in males showing the phenomenon of the hybrid dysgenesis of different species stands out. This phenomenon is also characterized by the presence of high frequencies of inviable offspring, mutations, structural chromosomal alterations, and distortion of the rate of transmission of alleles by one sex [9].

Another peculiarity is the spontaneous occurrence of recombination in males of species with a high degree of polymorphism for paracentric inversions, such as *D. melanogaster* [10], *Drosophila ananassae* Doleschall, 1858 [11], and *D. willistoni* [12].

Despite the exceptions, the presence of several cases of multiple heterozygosities occurring in many species of *Drosophila* support the great efficiency of these mechanisms and direct us to other biological aspects involving these chromosome rearrangements. The purpose of this chapter is to provide a basic overview of the knowledge of the evolutionary basis of its wide occurrence, and the adaptability conferred by the chromosomal polymorphism to the bearers of paracentric inversions found in this genus, converging in the present day in the analyses at the genomic level of the mechanisms that originate these inversions.

## 2. Population studies of chromosomal inversions in the genus *Drosophila*

The high polymorphism of chromosomal inversions has been used as a model for different adaptative processes, involved in the maintenance of the genetic variation. The concerns of Theodosius Dobzhansky and collaborators, more than 80 years ago, originated the early studies encompassing analyses of chromosomal inversions in natural populations of *Drosophila persimilis* Dobzhansky and Epling, 1944 and *Drosophila pseudoobscura* Frolova, 1929 [13]. Their findings were the stimuli for many of the discoveries that constituted the basis of modern evolutionary synthesis, which intricately combines Charles Darwin theory of evolution of species with Mendelian heritage patterns and population genetics.

The work of Dobzhansky "Genetics and the Origin of Species" [14] was a great incentive to the development of experimentation in evolutionary and population genetics.

Several experiments with *D. pseudoobscura* performed by Dobzhansky and colleagues were the basis to the postulation of the co-adaptation model of the genes contained in inversions [15]. Dobzhansky established that the reduced recombination in the inversions of this species is able to sustain positive combinations of genes in epistasis with other gene arrangement prevailing in the population. Therefore, gene complexes linked in an inversion in the different chromosome types are inherited as blocks and are rarely corrupted by meiotic recombination. Thus, the heterozygosity would be preferable to homozygosity, as predicted by the balancing selection [13, 14]. Thenceforth, the analysis and characterization of the chromosomal inversion polymorphism in natural populations of other species have become extensively explored. Also, indirect evidence of the association of chromosomal inversion with a better adaptation of the carrier individuals based on statistics was reported.

*Drosophila pseudoobscura* presents a broad geographic distribution in North America, being founded since west Canada, USA, and part of Central America, with the presence of a subspecies in Colombia (*D. pseudoobscura bogotana*), and individuals collected in New Zealand (Oceania) [16]. The *ST, AR, CH, PP, SC, OL, EP,* and *TL* arrangements founded on the chromosome 3 of this species are extensively monitored and traditionally present altitudinal cline on their frequencies. Among these, the *TL* inversion presents a frequency increase on the Pacific coast since the 70 decade, which seems to be related to environmental changes [17].

*Drosophila subobscura* Collin, 1936 is a species with high chromosomal inversion polymorphism. Their rearrangements have been traditionally associated with adaptation to environmental variables. This Palearctic species invaded the American continent in the 70/80 decades [18]. Studies encompassing the frequency of inversion in European, North-, and South American populations show an inversely proportional relation of the increase in the frequency of inversions occurring in low latitudes (hot climate areas) and a decrease of frequency of the inversions occurring in high latitudes (cold climate areas) [19]. The chromosomal polymorphism in this species has also been related to environmental heavy metal contamination [20].

*Drosophila buzzatii* Patterson and Wheeler, 1942 belongs to the cactophilic species of the *repleta* group. It is originally from Southern Latin America, and its occurrence has been reported in the 1970s in the Mediterranean region, the Canary Islands, equatorial Africa and Australia, associated with cactus species of the genus *Opuntia*, which have been disseminated by human interference [21]. In this species, latitudinal clines in the frequency of some inversions have been inferred for the populations of the original areas and the colonized areas. The polymorphism described for the second chromosome, for example, the *2j* arrangement has been related to the longer development time, and larvae viability [22].

The Neotropical species *Drosophila mediopunctata* Dobzhansky and Pavan, 1943, belongs to the *Drosophila* subgenus, *tripunctata* group (the second largest group of Neotropical species). The acrocentric chromosome II of this species is highly polymorphic, with 17 inversions described, which are distributed in the distal (inversions *DA, DI, DS, DP, DR, DL,* and *DJ*) and proximal (inversions *C0, PC1, PC2, PC3, PC4, PC5, PB0, PA0,* and *PA8*) regions. Based on the 72 haplotypes already described for this chromosome it is possible to infer that the inversions at the distal and proximal regions practically do not overlap, and there is strong linkage disequilibrium between them. Thus, *DA* inversion is mostly found in association with *PA0* inversion. In the same way, *DI* inversion is associated with *PB0* inversion, *DS* with *PC0, DP* with *PC0,* and *DS* with *PC0.* Thus, it is difficult to find one of these distal inversions not associated with the corresponding proximal inversions. These five haplotypes are the most frequent (>90%) in the natural populations of *D. mediopunctata* from Southeastern Brazil. Since 1980, the inversions of chromosome II of this species have been analyzed as potential bioindicator of genetic responses to environmental changes, under the action of natural selection. Collections conducted from 1986 to 1988 and from 1991 to 2002 in different places of Southeast and Southern Brazil showed that *DA, DP,* and *DS* inversions present seasonal variation of their frequencies, and the inversion *DA* increased in dry and cold periods, and *DP* and *DS* inversions during rainy and hot periods. In addition, this panorama is related to altitudinal clines. Later collections (2007–2010) in one of the sampled sites (Itatiaia National Park, Rio de Janeiro, Brazil) allowed the comparison of the mean frequencies of inversions at the distal region, with the previous frequencies for this site. It was observed that the mean frequencies of *DA* and *DI* inversions increased, while *DS, DP,* and *DV* (associated with higher temperatures) decreased their frequencies; and the *DA* inversion no longer has a significant correlation with altitude. Considering the climatic changes that occurred during these two decades in the region of Itatiaia Park, this suggest that temperature change has little influence on the seasonal changes in the frequencies of inversions in this species. Climate changes may have affected other genetic or morphological features, which may be more directly related to the inversions in chromosome II of *D. mediopunctata* [23–26].

Although several characteristics are indirectly associated with the inversions, little progress has been made in defining the genetic-evolutionary basis of these associations [23]. Direct shreds of evidence associating chromosomal inversions and selective pressures have been presented with the advancement of molecular techniques and genome sequencing.

Increasing amounts of data tend to confirm the inhibition of the recombination within the inversion area and also in adjacent areas, which is fundamental to the maintenance of the adaptive role. The patterns of linkage disequilibrium (LD) located within these regions reflect the inversion history and the gene flow since its origin [27–31].

An example of this case comes from the study of genetic variation and the unbalance of cosmopolitan inversion *In(3R)P* in two *D. melanogaster* populations from Australia, one from a tropical region (subdivided between individuals with inversion and individuals with the standard arrangement) and another from a temperate region (whose individuals carried only the standard arrangement). Since their high frequencies are related to higher temperatures, this inversion is known to be associated with climatic adaptations and the success of an evolutionarily recent migratory event (100 years) of this species in Australia. The results of this analysis support the hypothesis that *In(3R)P* inversion is associated with capture of locally adapted alleles, which interact substantially with loci external to the inversion. However, it was not possible to clarify whether these alleles are either in an additive or epistatic mode. Interestingly, high rates of LD in the region within the inversion are also found in the corresponding genomic region of the individuals that carried the standard arrangement in the tropical population, evidencing selection of such loci. Another result showed a high differentiation of the genomic region that involves the *In(3R)P* inversion between the tropical and the temperate population [30].

Despite the confirmed association of chromosomal inversions with the maintenance of combinations of alleles that lie within this region, gene recombination in the inverted region of a chromosome is possible because viable recombinant gametes arise through double meiotic recombination within the inverted region and also in consequence of gene conversion [31].

The prediction of recombination rates analysis in chromosomes carrying a heterozygous inversion, based on two mathematical models (Poison and Couting), made by Navarro and collaborators [32] infer three main points about this: "(i) the lower the inversion, the greater the effect on the reduction of the double meiotic recombination rate; (ii) in short inversions and in regions around the breakpoint, inversion reduces the rate of recombination but does not have the same capacity to prevent gene conversion; (iii) reduction of the recombination rate is not uniform throughout the chromosome, generally reducing the gene flow between different arrangements to near zero close to the breakpoints, and higher recombination rates are found in the central regions of the inversion." The inversion also influences the events of recombination of regions outside their limits. All these findings have implications for the analyses that use balancer chromosomes [32–34].

It should be noted that a fraction of these chromosomal polymorphisms occurring in the different species is adaptively neutral, and thus suffer less selective pressure (or none), and its fixation, or loss, depends on population size and migration. These inversions can also reach high frequencies through other mechanisms, such as the inversion *In(1)Be* of the X chromosome of *D. melanogaster*. This inversion, considered of recent origin, has its maintenance probably due to the distortion of the transmission ratio through males of the species [35].

Despite the high acceptance and diffusion of the co-adaptation model of the genes contained in the inversions in *Drosophila* [15], alternative hypotheses point to different scenarios for the propagation and distribution of chromosomal inversions in populations of living beings, as a result of the increasing acquisition of knowledge and domain of improved analysis techniques [33, 34, 36].

## 3. Inversions breakpoints in *Drosophila*: chromosomal distribution

Parallel to the evolutionary-population studies of chromosomal inversions in *Drosophila*, the concern about the cause and origin of these polymorphisms in populations was already present.

Krimbas and Powel [37] wrote the best definition of the traditional point of view for the genesis of inversions: "It is that they are the result of two independent breaks, occurring at the same time, followed by the reconnection of the broken parts of the chromosome in an inverted orientation with respect to neighboring regions. Thus, the multiple overlapping inversions found in many *Drosophila* species would have occurred sequentially, not due to the simultaneous occurrence of multiple breaks. Regarding *in tandem* inversions (side-by-side inversions), the coincidence of breakpoints is attributed to chance, in events that occurred at different times. The hypothesis of the unique origin of the inversion is reinforced by the rarity of a chromosomal inversion event. It is even rarer that two events originating the same inversion occur spontaneously at the same time in the same chromosome site [37, 38]."

The monophyletic origin of the inversions implies that different rearrangements in the same chromosome can clarify some aspect of the evolutionary history of the analyzed species (or distinct species, when inter-crossings are possible), establishing the inversions as genetic markers for the reconstruction of unrooted phylogenies [39, 40].

"The first genetic dataset used for phylogenetic construction were the inversions of the chromosome 3 of *D. pseudoobscura* [41]." For this, the karyotype of a given populations of this species was arbitrarily inferred as the standard arrangement, being named *ST*. The crossings of males collected in the wild (as well as male offspring of the collected females), with females of the *ST* lineage, showed the differences of the chromosomal arrangements between the populations due to the formation of inversion loops in the F1 offspring. This comparative methodology of chromosomal inversions allowed relating the different triads of overlapping heterozygous inversions found in an unrooted phylogenetic tree. Based on this, a hypothetical central arrangement in the phylogeny, which has never been found in nature in later works, has been inferred. However, the key point for this analysis was that all copies of a particular inversion would have a unique origin, the arrangement being seen in the individuals of a population as a replica of the single arrangement that arose in the past in a single common ancestor; in other words, its monophyletic origin. Later, molecular phylogenies corroborated the unique origin of the inversions in the chromosome 3 in *D. pseudoobscura*, and the topology of molecular phylogeny is in accordance with the topology obtained from the cytogenetic data [41].

The analysis of the phylogenetic relationships of overlapping inversions [39] considers the most parsimonious route (those with the small amount of inversion) for the evolutionary inference. Phylogenies were constructed for various species groups, such as *melanogaster* [42], *cardini* [43], Hawaiian *Drosophila* [44], *virilis* [45], *fasciola* subgroup [46], *willistoni* subgroup [47], among others.

Considering the traditional point of view of an inversion genesis, the distribution of the inversions along the chromosomes occurs randomly [37]. Sometimes, this characteristic seems to be well suited to the chromosomal distribution of the arrangement of chromosome 3 of *D. pseudoobscura* in natural populations [48], sometimes does not seem [39]. The inversion breakpoints induced by X-ray in *Drosophila* (and many other organisms, in general), seem to cluster preferentially in regions closer to the centromere [37, 49]. Add to this postulate, the evolutionary random breakage model, which gained notoriety with analyses of genomic comparisons, mainly between humans and mouse, later extending to other mammals in the 1980s. This model, in a simplistic way, assumed that the chromosome rearrangements, responsible for the breakdown of the synteny between these organisms, had their breakpoints distributed randomly along the chromosomes [50–52].

However, increasingly consistent studies evidencing the occurrence of repeated breaks in the same site for different inversions in a considerable amount of species have raised doubts regarding the randomness of the breakpoints distribution. These sites were denominated "hot spots," and may involve particular structural instabilities of these regions [37].

The availability of the complete human genome and other mammals showed the effects of the limitations of the random breakage model, since it did not consider countless regions of the genomes of these organisms, because they were not available. The analysis of 281 syntenic blocks up to 1 Mb shared between humans and mouse showed the presence of 190 additional blocks with less than 1 Mb in size, which was very difficult to identify by alignment, and were totally unknown until then. The comparison of the chromosomal rearrangements occurred during the divergence between the two species showed a large number of breakpoints close to each other. This characteristic did not fit the random breakage model theory, so the fragile breakage model was proposed [53, 54].

This model was based on the inference that breakpoints of chromosome rearrangements occur mainly within fragile genome sites (hot spots), in other words, regions prone to breakage. These fragile sites may correspond to regions with lots of transposable elements (TEs), to segmental duplications, or to a palindromic sequence. "The reuse feature does not imply the use of the same genomic position (at the nucleotide level) repeatedly, but rather that, the breakpoint presents multiple genomic regions that originate chromosomal rearrangements [53, 54]."

Pioneering results at cytological level, on the reuse of breakpoints by different inversions, provided challenging data about the randomness of these breaks in *Drosophila*. Cáceres et al. [55] analyzed 86 paracentric (heterozygous and fixed) inversions described for species of the *D. buzzatii* complex and 18 inversions induced in *D. buzzatii* by introgression, through crossings with *Drosophila koepferae* Fontdevila and Wasserman, 1988. The authors found that inversions of intermediate size are the most successful for the fixation in this species. They also observed that the breakpoints distribution of chromosome 2 inversions of these species, taking into account the location of the band involved in the break, is not random. The authors founded up to eight breakpoints at the same band in certain chromosomal segments. Similar results were observed in *D. subobscura* [56], Hawaiian *Drosophila* [57], and *D. willistoni* [58].

Although the reuse intra or interspecific of the inversions breakpoints, at cytological level, is common and well documented in the *Drosophila* genus, the characterization at DNA sequence level is still limited [59–61]. *In silico* comparisons of total genomes of different species available [62], estimate between 1.5 [63] and 2.27 [64] times the reuse of breakpoints throughout the evolutionary history of the species of this genus.

## 4. Characterization of inversion breakpoints in *Drosophila* and origin mechanisms

Delimitation and characterization of the inversion breakpoints are fundamental to determinate the mechanisms that originate them. In *Drosophila*, two main mechanisms have been highlighted in the origin of chromosomal inversions.

The first mechanism is the non-allelic homologous recombination (NAHR, also called ectopic recombination) between repetitive sequences, especially, the TEs [65–67]. The molecular machinery used by this mechanism is the same as allelic recombination, which has direct involvement with the genetic recombination in meiosis I. When ectopic recombination occurs between two copies of a repetitive sequence (very similar or identical), which are located physically at different chromosomal sites and in opposite orientations, the resulting inverted chromosome segment is flanked by two copies of these sequences, which are chimeric due to the exchange between them [66, 67]. The minimum identity between two sequences required for recombination is called minimal effective processing segment (MEPS). This parameter is not yet satisfactorily elucidated, *in vitro* analyses with prokaryotic organisms and mammalian cells infer that efficient MEPS for NAHR is 50 bp and between 270 and 280 bp, respectively [68]. However, the genomic approach of NAHR between copies of *Ty* retrotransposons in *Saccharomyces cerevisiae* Meyen ex E. C. Hansen, 1883 points out that more important than the identity between the copies of TEs is the genomic distance between them [69]. **Figure 2** illustrates a schematic for this mechanism. Based on this, it is important to note that when NAHR involves transposable elements, target site duplications (TSDs) of these can also be changed during recombination, a feature that has been very relevant for the recognition of this mechanism (see Section 4.1). For a long time, the TEs were considered junk DNA, and the involvement of these in the genesis of inversions of *Drosophila* genus provides solid knowledge to support the participation of these sequences in the molding of the genomes of living beings.

The second mechanism is via the erroneous repair of the free extremities, resulting from the chromosomal staggered breaks, by the non-homologous end joining (NHEJ). The physically close breaks in the chromosome cause failures in the correct pairing of the nitrogen bases, and

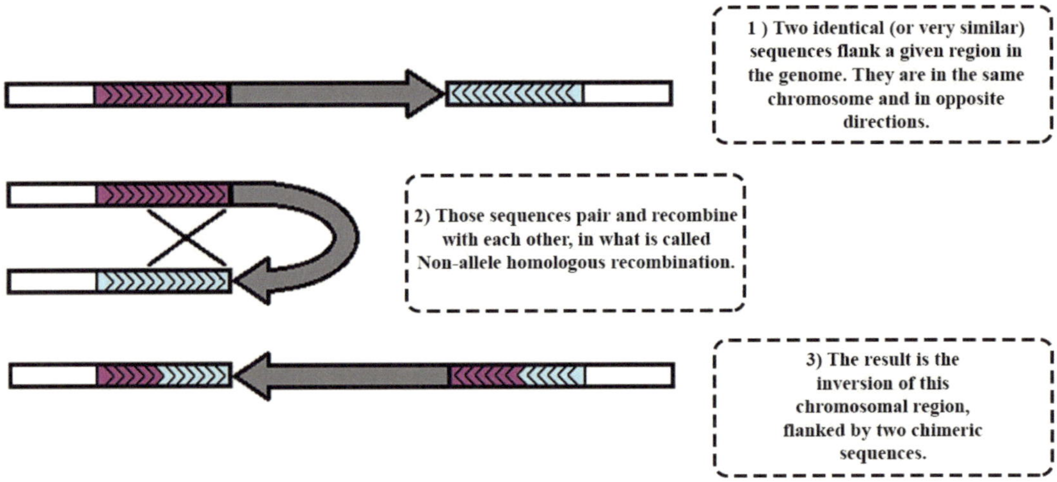


**Figure 2.** Standard schematic representation of the origin of chromosomal inversions via the non-allelic homologous recombination mechanism between identical or similar repetitive sequences. The arrow in dark gray represents the region that undergoes the inversion and its orientation. The repetitive sequences that flank this region in the left and right sides are represented with the tips indicating their orientations.

the chromosomal regions separate. The inversion is due to the junction of the 5′ end with the 3′ end of the other breakpoint [60, 70]. Duplicated DNA segments and in opposite orientations (delimiting the inverted chromosome segment) are the result of the repair and the main recognition mark of this mechanism [59]. In **Figure 3** it is possible to notice that staggered breaks occurred on both sides, duplicating two sequences that were originally single copies. However, based on the same figure, it is possible to extrapolate the occurrence of staggered breaks in only one side, and a simple break in the other side. The result is the duplication of just one originally single copy segment flanking the inversion. These duplicate sequences may involve genes. Gene duplication has been implicated as one of the main sources for the evolution of the genomes. The duplicate copy often does not undergo selective pressure, thus mutating more rapidly than the other essential regions of the genome. This may result in new gene functions, which is considered one of the most important results of these duplication events [71]. Thus, the repair of the free ends of staggered breaks by NHEJ gives rise to two different structural rearrangements: chromosomal inversion and duplication. Although in the case in question, duplications have small chromosomal magnitude compared to inversions, when they involve genes, they can also provide genomic variability in populations, and act on adaptive processes, speciation, and chromosome evolution.

The contribution of these two mechanisms is not completely clarified, and intriguing questions such as "whether these mechanisms are generalized among species of the genus and whether there are functional implications through the chromosomal evolution maintained by these inversions, remain open [59]".

**Table 1** presents a compilation of the different studies that characterized the inversion breakpoints at the molecular level in different species of the *Drosophila* genus. As can be

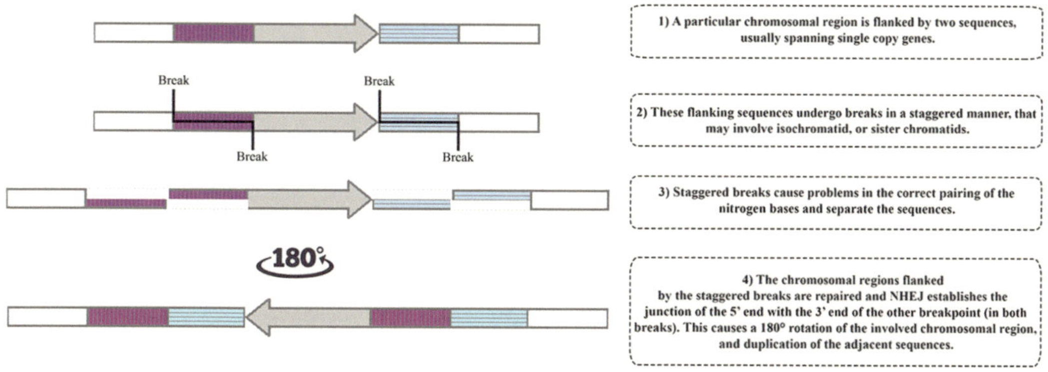


**Figure 3.** Standard schematic representation of the origin of chromosomal inversions via repair of the staggered breaks by non-homologous end joining. The gray arrow represents the region that suffers the inversion and its orientation. Adjacent sequences, originally single copies, are represented in the top of the image by vertical bars (left side) and horizontal bars (right side).

seen, the origin of different inversions, besides being via NAHR between TEs and other repetitive sequences, and staggered breaks followed by NHEJ, is also via simple breaks and repair. The breakpoint analysis does not always allow us to infer the probable origin of the inversions, a point that may be related to the antiquity of the inversion genesis, implying a greater amount of modifications in these regions, and loss of the signals that point to their origin mechanisms.

| **Species** | **Chromosomal inversion** | **Breakpoints description and mechanism of chromosomal inversion genesis** |
|---|---|---|
| *D. melanogaster* | *In(3R)P* | Analysis by microdissection and sequencing of the inversion region in the chromosome. Absence of repetitive sequences at the breakpoints [72]. |
| *D. melanogaster* x *D. subobscura* | Fixed inversion in the X chromosome of *D. subobscura* | Sequences of approximately 30–50 bp rich in thymines flanking the breakpoints [73]. |
| *D. melanogaster* | *In(2L)t* | Analysis of the proximal breakpoint and presence of a TE – LINE [74]. |
| *D. buzzatii* | *2j* | Presence of homologous copies of a TE denominated *Galileo* at the breakpoints, and origin of the inversion by NAHR between inverted copies of this TE [65]. |
| *D. buzzatii* | $2q^7$ | Presence of homologous copies of a TE denominated *Galileo* at the breakpoints, and origin of the inversion by NAHR between inverted copies of this TE [66]. |
| *D. pseudoobscura* | *Arrowhead* | Presence of 128 and 315 bp repetitive motifs in opposite orientation at the breakpoints of the inversion. Origin of the inversion by NAHR between the inverted copies of these repetitions [75]. |
| *D. melanogaster* | *In(3R)Payne* | Small duplications in both breakpoints of the inversion [76]. |

| Species | Chromosomal inversion | Breakpoints description and mechanism of chromosomal inversion genesis |
|---|---|---|
| *D. melanogaster* x *D. simulans* x *D. yakuba* | 29 inversions | 17 (59%) of the inversions presented inverted duplications at the breakpoints, including the *In(3R)84F1;93F6–7* inversion, which traditionally differentiates the karyotype of *D. melanogaster* and *D. simulans*. Origin of these inverions by staggered breaks mechanism [59]. |
| *D. americana* | *In(4)a* | Repetitive sequences in opposite orientation of a *MITE* element in both breakpoints of the inversion [77]. |
| *D. mojavensis* x *D. arizonae* | Inversion in the X chromosome | Absence of repetitive sequences at the breakpoints of the inversion [78]. |
| *D. pseudoobscura* x *D. persimilis* | Inversion in the X and II chromosomes | *In tandem* repetitions of a 319 bp motif at the breakpoints of the inversion in the XR arm of *D. persimilis* [79]. |
| *D. buzzatii* | $2z^3$ | Presence of homologous copies of the TE *GalileoN* at the breakpoints and origin of the inversion by NAHR between the inverted copies of this TE [67]. |
| *D. buzzatii* | *5 g* | Absence of significant repetitive sequences at the breakpoints [80]. |
| *D. mojavensis* | *Xe* | Absence of significant repetitive sequences at the breakpoints. Probable origin by single breaks [70]. |
| *D. americana* x *D. virillis* | Inversions *(In) Xa* and *(In)5a* | Presence of copies of the *MITE DAIBAM* at the breakpoints of the inversions in *D. americana*. Origin of the inversions by NAHR between the inverted copies of this TE [81]. |
| *D. buzzatii* | Inversions *2m* and *2n* | *2m* inversion with 13 Kbp duplications in both breakpoints; origin of the inversion by staggered breaks [60] |
| *D. melanogaster* | Inversions *In(2L)t, In(2R)NS, In(3R)K,*<br>*In(3R)Mo,*<br>*In(3R)P,*<br>*In(3L)P,*<br>*In(1)A,*<br>*In(1)Be* | Presence of inverted duplications at the breakpoints of the *In(2R)NS, In(3R)K, In(3R)P, In(1)A, In(1)Be* inversions [82]. |
| *D. mojavensis* | Inversions *2c, 2f, 2g, 2h, 2q* and *2r* | Presence of copies of the TE *But-5* in both breakpoints of the *2s* inversion by NAHR between the inverted copies of this TE. Presence of inverted duplications at the breakpoints of the *2h* and *2q* inversions; origin of the inversions by staggered breaks [83]. |
| *D. subobscura* | $O_3$ | 300 bp sequence in both breakpoints; origin of the inversion by staggered breaks [84]. |
| *D. subobscura* | Inversions $E_1$ and $E_2$ | Probable origin of the $E_1$ inversion by staggered breaks and duplication of a region with approximately 400 bp, named β motif; origin of the $E_2$ inversion by NAHR between α motifs (~ 700 bp) [61]. |

| Species | Chromosomal inversion | Breakpoints description and mechanism of chromosomal inversion genesis |
|---|---|---|
| *D. subobscura* | Inversions $E_9$ and $E_3$ | Presence of duplicated region (~8 Kbp) at the breakpoints of the $E_9$ inversion and of the duplicated region (~3.5 Kbp) at the breakpoints of $E_3$ inversion. Origin of the inversions by staggered breaks [85]. |
| *D. subobscura* | $E_{12}$ | Presence of the *Ugt58Fa* gene in both breakpoints. Origin of the inversions by staggered breaks [86]. |
| *D. subobscura* | Inversions $O_4$ and $O_8$ | Duplications in both breakpoints of the inversions; origin by staggered breaks [87]. |

**Table 1.** Molecular characterization studies of the inversion breakpoints in species of the *Drosophila* genus.

### 4.1. Involvement of the transposable elements at the origin of the inversions: non-allelic homologous recombination

Transposable elements are interesting and dominant components of the prokaryote and eukaryote genomes, meaning that the comprehension of their biology is a fundamental subject in genetics. Since their discovered by McClintock [88], much has been learned regarding the molecular properties of the TEs and their contribution to genome configuration of living beings.

These elements are classified according to their characteristics and transposition mode. Class I elements, also called retrotransposons, replicate through a "copy and paste" method and involve the production of an RNAm intermediary, processed by reverse transcription to DNA and re-inserted in the genome. The retrotransposons subdivide into elements with Long Terminal Repeats (LTRs), for example, *copia* and *Gipsy* elements in *Drosophila*, that are similar to retroviruses; and the retrotransposons without LTRs, as Long Interspersed Elements (LINEs) and Short Interspersed Nuclear Elements (SINEs), which do not encode their reverse transcriptase and are also called retroposons [89, 90].

Class II elements, or DNA transposons, replicate, generically by a "cut and paste" mechanism, where the elements are physically excised from the genome and inserted into another site. In this case, there is an increase in the number of copies during the repairing of the excision sites of the DNA transposon by the host during DNA synthesis, or by the insertion of the TE in a genome site which has not been replicated [90, 91]. Still, among Class II elements, there is a non-autonomous element group denominated MITEs (Miniature Inverted-repeat Transposable Elements). These elements are short sequences with several copies in the genome and without coding capacity, as suggested by *Mar* element, which seems to be restricted to the *D. willistoni* subgroup [92].

The TEs of both classes are also classified in Subclass, Order, Superfamily, Family, and Subfamily based on their sharing of certain structures and sequence similarities [91].

The studies associating TEs with chromosomal rearrangements breakpoints in *Drosophila* genus begin mostly with the analysis of lineages presenting hybrid dysgenesis syndrome. This syndrome is caused by crossing certain lineages of *Drosophila* and is characterized by high mutation rates in germinative cells, causing a high frequency of inviable offspring, recombination in males, mutation and structural chromosomal abnormalities [9]. The cause of hybrid dysgenesis has been reported to the activation of several TEs families, including *P, I,* and *hobo* elements in *D. melanogaster* [93] and *Penelope, Ulysses, Helena,* and *Telemac* in *Drosophila virilis* Sturtevant, 1916 [94]. Subsequently, studies involving programed crossings and also cytogenetic and molecular analyses of the offspring followed the movement of the involved TEs and the appearance of chromosomal rearrangements associated with this movement [95].

The association of TEs insertions at cytological level with inversions breakpoints in natural populations of *Drosophila* has also been reported. Among them, stands out the analysis of the transposon *hobo* in *D. melanogaster* [96], the *P* element in *D. willistoni* [97], and the retroelements *Penelope* and *Ulysses* in *D. virilis* species group [98, 99].

The first analysis that directly evidenced the involvement of a TE at the origin of an inversion in a natural *Drosophila* population was made by Cáceres et al. [65]. This study analyzed the breakpoints of the polymorphic inversion *2j* (of the second chromosome) of the species *D. buzzatii* (subgenus *Drosophila, repleta* group), which originated from the *2st* (standard) chromosome arrangement. For the analysis of the breakpoints of the *2j* inversion, these were delimited by chromosome walk, cloned, and sequenced in two lineages of *D. buzzatti,* which presented the *2st* (lineage *st-1*) and *2j* (lineage (*j-1*) arrangements in homozygosis. For organization purposes, the breakpoints were designated AB and AC (distal breakpoint), CD and BD (proximal breakpoint) in the *2st* and *2j* lines, respectively. Sequencing and alignment of these regions in both lineages showed large insertions at the two inversion breakpoints, which were not present in the *2st* standard arrangement. The insertion between A and C had 392 bp with long inverted repeats terminals (ITRs) of 106 bp. The insertion between B and D had 4319 bp, with ITRs as those of the 106 and 47 bp AC inserts. The central 180 bp of the AC insert and the BD sequence had 95% homology but was in opposite orientations. Sequences of 7 bp separated and inverted flanked each insert and resembled TSDs, which are the result of the *TE* insertion event. These characteristics pointed out that inversion *2j* was generated by intrachromosomal pairing and recombination between the two homologous sequences inserted at distant sites and opposite orientations. The original structure of these inserts was homologous at approximately 274 bp and sustained a NAHR in *Drosophila.* These same insertions of the inversion breakpoints *2j* were characterized as copies of a Class II TE, which was named *Galileo* [65].

Subsequently, the *Galileo* element was classified as a member of the *P* Superfamily of Class II Transposons [100] and subdivided into three subfamilies: *GalileoG* (*Galileo*), *GalileoN* (*Newton*), and *GalileoK* (*Kepler*) [67]. The involvement of this family was also pointed on the origin of two more polymorphic inversions of the chromosome 2 of *D. buzzatii*: $2q^7$ [66] e $2z^3$ [67]. These analyses showed, through cytological, molecular, and *in silico* analyses, that the origin of these inversions was due to the occurrence of NAHR between two copies of the TE *Galileo,* present at the breakpoints of these inversions.

Still, with respect to inversion *2j* of *D. buzzatii* chromosome 2, its effect on the *CG13617* gene was analyzed. This gene was chosen because it is very close to the proximal breakpoint of this inversion (12 bp), and the embryos of homozygous lineage for the *2j* arrangement have the expression five times lower compared to the standard lineages, without the presence of the inversion. Based on the characterization of this region in the *D. buzzatti* genome and analysis of the mRNA levels, the authors pointed that the TE denominated *Kepler* is responsible for originating an antisense RNA, which forms a complex with the mRNA of the *CG13617* gene, performing a post-transcriptional regulation, making it inactive. *Kepler* TE is inserted adjacent to the proximal breakpoint in the lineages that carry the inversion *2j* and is not found in this same region of the breakpoint in the lineages without the inversion. The results of this study show a scenario of the interaction of antisense RNA with the *CG13617* gene via position effect. "Thus, the silencing of the *CG13617* gene is not due to the influence of the inversion *2j* itself, but rather due to the performance of sequences associated with them [101]."

There are also analyses of fixed inversions *2m* and *2n* in *D. buzzatti*, which are distributed *in tandem* and share the central breakpoint at the cytological level [60]. The delimitation and molecular characterization of the breakpoints were based on the genomic library of bacterial artificial chromosomes (BACs), and physical map of this species [102], and in the genome of the related species *Drosophila mojavensis* Patterson, 1940 [62], which did not exhibit such inversions. It was possible to establish which clones contained the regions of the three breakpoints in *D. buzzatii* (breakpoints denominated AC, BE, DF, whose direction from the left to the right is from the telomere to the centromere), by means of chromosomal walk by *in situ* hybridization, using BACs as probes. These positive BACs had their terminal portions sequenced, and these sequences served as a basis for delimiting the breakpoints (denominated AB, CD, EF, although not fully representative of the ancestral karyotype) in the genome of *D. mojavensis*. Subsequently, probes based on this genome were physically mapped on the polytene chromosomes of *D. buzzatii*, thus allowing the gene delimitation of the three breakpoints of *2m* and *2n* inversions. The comparison of these regions at the molecular level presented a very complex scenario. Small fragments of the *BuT-5* TE were found at both breakpoints of the *2n* inversion (breakpoints BE and DF), which may indicate their probable origin by ectopic recombination between these copies. However, due to the age of inversion, this assumption cannot be strongly based since these regions have already undergone many modifications and the TSDs have not been found. On the other hand, the *2m* inversion (AC and BE breakpoints) is flanked by ~13 Kbp duplications, which contain the *CG4673* gene. Thus, its most probable origin is via staggered breaks followed by NHEJ (See Section 4.2).

There is an extensive analysis of the mechanisms of origin of fixed inversions in *Drosophila mojavensis*, another representative of the *repleta* group. This species is the only representative of the *mulleri* complex that inhabits the Sonora desert, one of the aridest known environments, with fauna and flora quite peculiar [83]. The analysis of the chromosome evolution of *D. mojavensis* shows 10 fixed inversions in relation to the primitive arrangement I of the *repleta* group, along the evolution: one on the chromosome X (*Xe*), seven on chromosome 2 (*2c, 2f, 2g, 2h, 2q, 2r,* and *2s*) and two on chromosome 3 (*3a* and *3d*) [83, 103]. The molecular characterization of the breakpoints of the seven inversions of chromosome 2 of this species occurred by means of end sequencing of clones of chromosome 2 of the genomic library of BACs of *D. buzzatii* [102]. Subsequently, these sequences were mapped in the genome of *D. mojavensis* and compared with the genome of *D. virilis* (external species with the karyotype without inversions). The breakpoints of *2c, 2r,* and *2s* inversions showed copies of TEs flanking both sides of the inversion. However, the *2s* inversion stood out, due to the presence of the *BuT-5* transposon at its breakpoints. The distal copy had 981 bp delimited by 9 bp AAGGCAAGT and CTGTATAAT sequences. At the proximal breakpoint, the copy of *BuT-5* TE was a 27 bp fragment, with 12 bp identical to one end and the remaining 15 bp were identical to the other end of this TE, delimited by sequences of 9 bp ACTTGCCTT and ATTATACAG. The sequences ACTTGCCTT and CTGTATAAT are the inverted complementary sequences of AAGGCAAGT and ATTATACAG, respectively, and constitute the TSDs derived from the insertion of the element. These characteristics indicate that the origin of the *2s* inversion of *D. mojavensis* is due to ectopic recombination between the two copies of the *BuT-5* TE. Functional inference of this inversion in the *D. mojavensis* genome indicates that the proximal copy of *BuT-5* TE acts on the *Dmoj\CG10375* gene promoter (which probably relates to the *Hsp40* gene family). *In silico* analyses show that *2s* inversion and the proximal copy of *BuT-5* TE increase the expression of this gene and may have direct implication with the thermotolerance regulation in this species [83].

Another species that clearly presents the involvement of TEs in the genesis of their inversions is the *Drosophila americana* Spencer 1938 (subgenus *Drosophila, virilis* group). The neo-X chromosome of this species is derived from a centromeric fusion segregating between the X-chromosome (Muller element A) and chromosome 4 (Muller Element B) in this species. This chromosomal fusion is positively correlated with latitude and has a polymorphic *In(4)a* inversion [77]. In addition, the arrangement of *D. americana* chromosome 4 is homosequential to the arrangement of the same chromosome in *D. virilis*, a related species that has its genome sequenced. Thus, its genome served as the basis for the design of the analysis, associated with the construction of a genomic library of BACs of *D. americana*. The analysis of *In(4)a* inversion of neo-X indicated its probable origin by means of ectopic recombination between two copies of a repetitive MITE element, which was widely dispersed in the genome of *D. virilis*. These same sequences were not present in the corresponding region in strains of the species without inversion (analysis made by PCR) and in *D. virilis*. The characteristics of this repeating sequence that support its identity as TE is the presence of 240 bp TIRs flanking an internal region of 869 bp. Comparisons of the multiple copies present in the genome of *D. virilis* with the sequences found at breakpoints in *D. americana* indicate that the copy present at the proximal breakpoint is a canonical element, whereas the copy present at the distal breakpoint is a rearranged element. From the functional point of view, the proximal breakpoint of this inversion presents allelic associations consistent with co-adaptation [77].

Subsequently, sequencing with low genome coverage of two strains of *D. americana* allowed the analysis of the *Xa* inversion fixed in *D. americana* and absent in *D. virilis* and the polymorphic *5a* inversion in *D. americana* [81]. The alignment of the breakpoints of both inversions between the two species indicated that in the regions where the alignment was corrupted, there was always a sequence varying between 500 and 1130 bp, present only in the lineages carrying the inversions. These sequences showed by BLASTN high similarity to an incomplete MITE sequence, with TIRs of 240 bp. In this study, the authors named it *DAIBAM* (*Drosophila americana* Inversion Breakpoints Associated MITE). In *Xa* inversion, it was possible to find clear TSDs and defective copies of TE *DAIBAM* flanking the inversion. In *5a* inversion, copies of the *DAIBAM* element flanking the inversion had more than 70% nucleotide similarity. Considering that TE *DAIBAM* copies are defective and that the analyzed inversions are old, the authors infer that the data found supported the origin of inversions *Xa* and *5a* by ectopic recombination between the *DAIBAM* elements present at the breakpoints of these inversions. It was also found that this element was the same as that present at the breakpoints of inversion *In(4)a* [77]. Thus, the *DAIBAM* element is involved in the origin of at least 20% of the inversions occurring in the *virilis* group [77, 81].

### 4.2. Inversion origin via staggered breaks and repair by non-homologous end joining.

It has now been characterized that the origin of the inversions via staggered breaks followed by repair by NHEJ, is prevalent in two chromosomal systems: between the fixed chromosomal inversions that differentiate the *D. melanogaster* karyotype from those of *D. simulans* Sturtevant, 1919 and *D.* yakuba Burla, 1954 [59]; and the chromosomal polymorphism of the E and O chromosomes of *D. subobuscura* [61, 84–87].

*Drosophila melanogaster, D. simulans,* and *D. yakuba* are members of the *D. melanogaster* subgroup (*Sophophora* subgenus). The main karyotypic difference between *D. melanogaster* and its cryptic *D. simulans* is the occurrence of inversion in the right arm of the 3 chromosome, denominated *In(3R)84F1;93F6-7* [104]. *Drosophila yakuba,* on the other hand, has at least 28 paracentric inversions differentiating its chromosomes from those of *D. melanogaster.*

The study of Ranz et al. [59] analyzed the breakpoints of 29 interspecific inversions in these species through experimental and computational methods.

The analysis of the breakpoints of the *In(3R)84F1;93F6-7* inversion highlighted that the breakpoints were proximally flanked by the *CG2708* and *CG7918* genes, and distally by *CG31176* and *CG34034* in *D. melanogaster*. Among these regions, there are occurrences of expressed sequences, and three of these sequences (*HDC14862, pfd800* e *HDC12400*) are duplicated and in opposite directions, in both breakpoints of the inversion in *D. melanogaster*, with 95% of identity between them. These sequences are single copies in *D. simulans* and *D. yakuba,* indicating that these duplications are a derived state with respect to the chromosomal arrangement of these species. Comparisons of the 3R chromosomal arm of *D. melanogaster, D. simulans,* and *D. yakuba* at the molecular level, highlighted a fixed inversion in the latter species (*In3R(7)*), that reuses the breakpoints of the *CG7918-CG34034* region, also used by the *In(3R)84F1;93F6–7* inversion. In both breakpoints of the *In3R(7)* inversion, there were two duplicated sequences (*CG34034* e *CG31286*) and in opposite orientation [59].

Due to the presence of inverted duplications associated with the *In(3R)84F1;93F6-7* and *In3R(7)* inversion breakpoints, the most parsimonious mechanism involved on its origins is through staggered breaks, proposed and schematized for the first time in this analysis. These staggered breaks can be isochromatid, occurring during the premeiotic mitosis and involving staggered single-strand breaks; or chromatid, occurring during the meiotic prophase involving staggered double-strand breaks [59].

The same *in silico* study analyzed the breakpoints of 28 paracentric inversions that differentiate the *D. melanogaster* chromosomes from those of *D. yakuba,* as well as a pericentric inversion in the chromosome 2. The genomic and phylogenetic evidences suggest that among these 29 inversions, 28 originated in the *D. yakuba* lineage. The analysis of the inversions breakpoints showed that in approximately 62% of the cases (18 of 29 inversions), occurred the presence of duplicate sequences, which were presents with just a single copy in the *D. melanogaster* genome. Sequences of both breakpoints were inverted and duplicated in six of these inversions (as in **Figure 3**), and the sequence of just one of the breakpoints was duplicated in 12 inversions, which can be explained by several factors, for example, modifications occurred along the time. Most of these duplications (except three) did not prove to be functional. The comparative analysis of these breakpoints among *D. yakuba, D. melanogaster*, and other species, regarding the occurrence of TEs and its involvement in the origin of these inversions via NAHR¸ showed little support for this mechanism. It is clear in this analysis that most of the inversions that differentiate the *D. melanogaster* chromosomes from those of *D. yakuba* originated by staggered breaks in the latter species (17 of 29 analyzed inversions) and point to a rapid chromosomal evolution in the lineage that leads to *D. yakuba* [59].

The polymorphism of the Palearctic species *D. subobscura* (*Sophophora* subgenus, *obscura* group) has been extensively characterized and monitored for more than seven decades, which allows associating its variation with climate changes [18–20, 56]. Its karyotype is composed of six pairs of chromosomes, with the highest level of polymorphisms for inversions in all of them (except in the dot chromosome). This polymorphism is well characterized for the presence of complex chromosomal arrangements, formed by the occurrence of overlapping inversions, being the E and O chromosomes the ones with the highest occurrence of these arrangements in natural populations.

One of the pioneering analyzes in this species involved the characterization of the breakpoints of the $O_3$ inversion, which can be found in the $O_{st}$ lineages (corresponding to the current standard arrangement of the species). This inversion originated from the extinct ancestral $O_3$ arrangement, that also gave rise to the $O_{3+4}$ arrangement, which segregates the $O_4$ inversions in the different populations. For this analysis, breakpoints of the $O_3$ inversion in the extinct arrangement and without the $O_3$ inversion were denominated AB (proximal breakpoint) and CD (distal breakpoint). The $O_{3+4}$ chromosome arrangement differs from the $O_3$ arrangement due to a small inversion of its distal breakpoint (called DC), presenting the same order of the proximal breakpoint (AB). In turn, the $O_{st}$ chromosomal arrangement differs from $O_3$ by inversion $O_3$ (note that the extinct $O_3$ arrangement does not involve the $O_3$ inversion, which occurs in the $O_{st}$ chromosome), involving *B* and *C* regions (their breakpoints being then called AC and BD). The analysis was a strategy that mixed *in situ* hybridization and *in silico* tests, together with the knowledge of the location of previously established probes [84].

New probes were established via comparisons with the available genomes of *D. melanogaster* and *D. pseudoobscura*, which made it possible to delimit the genomic region containing the breakpoints of the $O_3$ inversion in the chromosomes of the $O_{3+4}$ arrangement. The posterior sequencing of this region in the $O_{st}$ e $O_{3+4}$ lineages allowed the comparisons between the breakpoints of the $O_3$ e $O_{3+4}$ inversions, respectively. As a result, it was found that the breakpoints AB and DC in the $O_{3+4}$ inversion comprised two small regions of 309 and 63 bp, respectively. The 63 bp sequence was the same 309 bp sequence, which was deleted at the origin of the $O_{3+4}$ inversion. In turn, the same 309 bp sequence was present at both $O_3$ inversion breakpoints, indicating that at the origin of this inversion such region was duplicated, being present in regions B and C of the breakpoints. The AB and DC regions in the lineage that carries the $O_{3+4}$ inversion showed no similarity to any known TE. This meticulous analysis, based on the absence of TEs and the duplication of the 309 bp fragment, infers that the origin of the inversion $O_3$, present in the chromosomal arrangement $O_{st}$, was by means of staggered double-strand breaks [84].

Still, in the O chromosome of *D. subobscura*, the breakpoints of the $O_4$ and $O_8$ inversions were delimited, sequenced and analyzed. Just as the inversion $O_4$ segregates only with the $O_3$ arrangement (giving rise to the complex chromosome arrangement $O_{3+4}$), the inversion $O_8$ segregates only with the arrangement $O_{3+4}$ (giving rise to the chromosomal arrangement $O_{3+4+8}$). Comparisons of the $O_4$ inversion breakpoints with the respective regions in the $O_{st}$ arrangement (without the inversion) pointed the occurrence of *Pxd*, *CG5225*, *Acf*, and *Set8* gene fragments at the proximal breakpoint, and the *CG5225*, *Pxd*, and *Acf* gene fragments at the distal breakpoint [87].

In the regions corresponding to the breakpoints of the $O_{st}$ arrangements, fragments of the *Pxd*, *CG5225*, and *CG4009* genes were found at the proximal breakpoint. The distal breakpoint of the $O_{st}$ arrangement encompasses fragments of the *Set8* and *Acf* genes, It is evident that at the origin of inversion $O_4$ fragments of the *Set8* and *Acf* genes were duplicated at the proximal breakpoint, and fragments of the *CG5225* and *Pxd* genes were duplicated at the distal breakpoint. This scenario fits the origin of inversion $O_4$ by the staggered double-strand break mechanism. The $O_8$ inversion breakpoints in the $O_{3+4}$ arrangement (without the inversion) and $O_{3+4+8}$ arrangements presented a similar picture to that of the $O_4$ inversion. The presence of the *Prosβ2R2* gene at both $O_8$ inversion breakpoints shows that this was doubled and fits the origin of this inversion $O_8$ by the staggered double-strand break mechanism. This analysis also found that genes *CG5225* and *Prosβ2R2* are involved in multiple rearrangements (duplications and transpositions, in addition to inversions) occurring along the chromosomal evolution of the species of the genus *Drosophila* [87].

The *D. subobscura* species also had the breakpoints of the $E_1$ e $E_2$, $E_9$, $E_3$, and $E_{12}$ inversions of the acrocentric chromosome E delimited, sequenced and analyzed. These inversions give rise to the complex arrangements $E_{1+2}$, $E_{1+2+9}$, $E_{1+2+9+3}$, and $E_{1+2+9+12}$. These chromosome constitutions, besides providing a great system for the analysis of the mechanisms of origin of inversions, also provide a basis for studying the reuse of the inversion breakpoints at the molecular level [61, 85–87].

The $E_1$ and $E_2$ inversions share, cytologically, one of the breakpoints. The comparison of the breakpoints of the standard lineage $E_{st}$ (AB, EF, GH breakpoints) with the $E_{1+2}$ lineage (AG, FB, EH breakpoints) showed two motifs, denominated $\alpha$ and $\beta$, which share the terminal portion named $\delta$, in opposite orientations. The $\alpha$ motif was present at the AB and AG breakpoints with the same orientation, but with inverted orientation in the GH breakpoint (two copies with inverted orientation in the $E_{st}$ chromosome and a single copy in the $E_{1+2}$ chromosome). The $\beta$ motif was present with the same orientation at the EF and EH breakpoints, and with inverted orientation at the *FB* breakpoint (a single copy in the *Est* chromosome and two copies in the $E_{1+2}$ chromosome). The $\alpha$ motif exhibits small fragments similar to the *SGM* element, whereas the $\beta$ motif is not similar to any described TE. Based on this scenario, the probable origin of the $E_1$ inversion was inferred by staggered breakpoints, that lead to the duplication of the $\beta$ motif present at the FB and EH breakpoints. The origin of the $E_2$ inversion, on the other hand, was inferred due to the ectopic recombination between two $\alpha$ motifs, present in both AB and GH breakpoints. The reuse was inferred by the presence of 400–700 bp repetitions at the breakpoints; however, it was impossible to elucidate which of the two inversions originated first [61].

The extensive analysis done in the classical rearrangements of the E and O chromosomes, mentioned above, showed that, with the exception of the *E2* inversion, the other chromosomal arrangements originated via staggered double-strand break mechanism. Thus, *D. subobscura* resembles *D. melanogaster*, and both emphasize a possible predominance of this mechanism in the origin of the inversions of the species belonging to the subgenus *Sophophora*. In addition, duplicate regions in these events range from a few hundred base pairs to about 8 Kbp (see **Table 1**), encompassing whole and partial genes in some of these duplications. However, no dose effect or generation of new transcripts was detected in the analyses [61, 84–87].

Still considering staggered break mechanism followed by erroneous repair by NHEJ, the molecular characterization of the inversion breakpoints in *D. mojavensis* indicates that its inversions *2h* and *2q* originated by this route. The *2h* inversion would have originated by staggered single-break at the distal breakpoint in the parental chromosome, resulting in a duplicated region of approximately 7 Kb, encompassing *CG1792, Dmoj\GI23402*, and *pasha* genes. This event resulted in the origin of the gene *Dmoj\GI23123*, located at the proximal breakpoint of inversion *2h*. This gene, by similarity, showed a relationship with the *pasha* gene, and according to the prediction of the modENCODE software, it is also functional. Thus the *Dmoj\GI23123* gene originated from the duplication of the *pasha* gene (the extra copy of the gene giving rise to a new gene) in the event that resulted in the *2h* inversion [83].

Staggered single-break occurred in the two breakpoints of the parental chromosome in the *2q* inversion. This event resulted in a duplication of an approximately 4 Kb region containing a partial fragment of the *CG1208* gene. The duplication of this gene resulted in the origin of a new gene, called *Dmoj\GI22075*, at the distal breakpoint of the *2q* inversion. The new gene maintained the MFS domain (Major Facilitator Superfamily) as an important feature of the *CG1208* gene [83].

The *2h* and *2q* inversions of *D. mojavensis* are pioneer examples of the origin of new genes with possible new functions, via duplication, based on the origin of an inversion by the staggered break mechanism followed by NHEJ [83].

## 5. Concluding remarks

Inversions are structural chromosomal alterations that, most of the time, neither imply genetic unbalance, nor phenotypic modifications in its carriers. However, one of its characteristics is to be a source of genetic variability, in which natural selection acts. Thus, the inversions participate in the chromosomal evolution of numerous species, including *Homo sapiens*. The basic knowledge about the biological influence of inversions is largely based on the analysis of the polytene chromosomes of the *Drosophila* model organism, which extends to other living beings.

The first works, with descriptive approaches to the frequency of chromosome polymorphism in different natural populations, while indirectly pointing out that the inversions provided advantages to its users, raised questions that until now guide the analysis on this theme: How does natural selection work in inversions? How do inversions offer greater adaptability to living beings? What is the role of the inversions in the speciation processes? What are the functional consequences of inversions in living beings? Are inversions randomly distributed on chromosomes? How do inversions originate?

The *Drosophila* model organism provides knowledge and answers to these questions nowadays, with the availability of complete genomes of different species, improved cytomolecular techniques, as well as a solid knowledge about the cytogenetics of polytene chromosomes.

The molecular characterization of the inversion breakpoints tells us about the mechanisms that originate these rearrangements, the genomic composition of the region involved in the inversion—which allows to analyze the nucleotide variation and to show which genes are under selection—the reuse of certain regions for the breakage of different inversions that occurred at different times, the age of the inversion, its monophyletic origin, possible positional effect and its influence on the genes that are inside and outside the inversion, among others. Valuable understandings emerge, but are still incipient.

These analyses go far from being simplistic, but, with the current resources, we have never had so much opportunity to acquire knowledge. Let us live the new time in science, and avail the most of the knowledge already established, with the certainty that many other questions will arise.

As the eminent geneticist Michael Ashburner of the University of Cambridge, United Kingdom, compiles: "What a wonderful time to be a biologist [105]."

## Acknowledgements

We are grateful to Mr. Leonardo Lindenmeyer for all the assistance provided with the images of this manuscript. Grants and fellowships of Coordenação de Aperfeiçoamento de Pessoal de Nível Superior (CAPES), Conselho Nacional de Desenvolvimento Científico e Tecnológico (CNPq, process 455101/2014-0) and Fundação de Amparo à Pesquisa do Estado do Rio Grande do Sul (FAPERGS) from Brazil.

## Author details

Carolina Garcia* and Vera L. S. Valente

*Address all correspondence to: cafagibio@hotmail.com

Department of Genetics, Institute of Biosciences, Federal University of Rio Grande do Sul, Porto Alegre, Brazil

## References

[1] Swift H. Nucleic acids and cell morphology in dipteran salivary glands. In: Allen JM, editor. The Molecular Control of Cellular Activity. New York: McGraw Hill; 1962. pp. 73-125

[2] Zhimulev IF, Koryakov DE. Polytene chromosomes. Encyclopedia of Life Sciences (ELS). Chichester: John Wiley & Sons, Ltd; 2009. DOI: 10.1002/9780470015902.a0001183.pub2

[3] Hartl DL, Jones EW. Human karyotypes and chromosome behavior. In: Genetics: Analysis of Genes and Genomes. 7th ed. London: Jones and Bartlett; 2009. pp. 250-294

[4] Griffiths AJF, Wessler SR, Carroll SB, John Doebley J. Large-scale chromosomal changes. In: Introduction to Genetic Analysis. 9th ed. New York: WH Freeman & Company; 2015. pp. 481-520

[5] Powell JR. Population genetics—Inversions. In: Progress and Prospects in Evolutionary Biology—The *Drosophila* Model. New York: Oxford University; 1997. pp. 48-80

[6] Orr-Weaver TL. Meiosis in *Drosophila*: seeing is believing. Proceedings of the National Academy of Sciences of the United States of America. 1995;**92**(23):10443-10449. PMCID: PMC40628

[7] Hinton CW, Lucchesi JC. A Cytogenetic study of crossing over in inversion heterozygotes of *Drosophila melanogaster*. Genetics. 1960;**45**(1):87-94. PMCID: PMC1210005

[8] Morgan TH. Complete linkage in the second chromosome of the male of *Drosophila*. Science. 1912;**36**(934):719-720. DOI: 10.1126/science.36.934.719

[9] Kidwell MG, Kidwell JF, Sved JA. Hybrid dysgenesis in *Drosophila melanogaster*: A syndrome of aberrant traits including mutation, sterility, and male recombination. Genetics. 1977;**86**(4):813-833. PMCID: PMC1213713

[10] Henderson SA, Woodruff RC, Thompson JN. Spontaneous chromosome breakage at male meiosis associated with male recombination in *Drosophila melanogaster*. Genetics. 1978;**88**(1):93-107. PMCID: PMC1213793

[11] Goñi B, Matsuda M, Yamamoto MT, Vilela CR, Tobari YN. Crossing over does occur in males of *Drosophila ananassae* from natural populations. Genome. 2012;**55**(7):505-511. DOI: 10.1139/g2012-037

[12] Santos-Colares MC, Degrand TH, Valente VLS. Cytological detection of male recombination in *Drosophila willistoni*. Cytologia (Tokyo). 2004;**69**(4):359-365. DOI: 10.1508/cytologia.69.359

[13] Dobzhansky T. Genetics of the Evolutionary Process. New York: Columbia University Press; 1970

[14] Dobzhansky T. Genetics and the Origin of Species. 1st, 2nd, 3rd ed. New York: Columbia University Press, 1937, 1941, 1951

[15] Dobzhansky T. On some of the problems of population genetics and evolution. Symposium sui Fattori Ecologici e Genetici della Speciazione Degli Animali. La Reserca Scientifica. 1949;**19**(Suppl):11

[16] Lambert DM, McLea MC. *Drosophila pseudoobscura* from New Zealand. Drosophila Information Service. 1983;**59**:72-73

[17] Anderson WW, Arnold J, Baldwin DG, Beckenbach AT, Brown CJ, Bryant SH, Coyne JA, Harshman LG, Heed WB, Jeffery DE, et al. Four decades of inversion polymorphism in *Drosophila pseudoobscura*. Proceedings of the National Academy of Sciences of the United States of America. 1991;**88**(22):10367-10371. PMCID: PMC52929

[18] Brncic D, Prevosti A, Budnik M, Monclús M, Ocaña J. Colonization of *Drosophila subobscura* in Chile I. First population and cytogenetic studies. Genetica. 1981;**56**(1):3-9. DOI: 10.1007/BF00126923

[19] Balanyà J, Huey RB, Gilchrist GW, Serra L. The chromosomal polymorphism of *Drosophila subobscura*: A microevolutionary weapon to monitor global change. Heredity (Edinburgh). 2009;**103**(5):364-367. DOI: 10.1038/hdy.2009.86

[20] Kenig B, Kurbalija Novičić Z, Patenković A, Stamenković-Radak M, Anđelković M. Adaptive role of inversion polymorphism of *Drosophila subobscura* in lead stressed environment. PLoS One. 2015;**10**(6):e0131270. DOI: 10.1371/journal.pone.0131270

[21] Fontdevila A, Ruiz A, Ocaña J, Alonso G. Evolutionary history of *Drosophila buzzatii*. II. How much has chromosomal polymorphism changed in colonization? Evolution. 1982; **36**(4):843-851. DOI: 10.1111/j.1558-5646.1982.tb05450.x

[22] Hoffmann AA, Sgrò CM, Weeks AR. Chromosomal inversion polymorphisms and adaptation. Trends in Ecology & Evolution. 2004;**19**(9):482-488. DOI: 10.1016/j.tree.2004.06.013

[23] Schaeffer SW, Goetting-Minesky MP, Kovacevic M, Peoples JR, Graybill JL, et al. Evolutionary genomics of inversions in *Drosophila pseudoobscura*: Evidence for epistasis. Proceedings of the National Academy of Sciences of the United States of America. 2003;**100**(14):8319-8324. DOI: 10.1073/pnas.1432900100

[24] Galina A, Peixoto AA, Bitner-Mathé CB, Souza WN, Silva LB, Valente VLS, Klaczko LB. Chromosomal inversion polymorphism in *Drosophila mediopunctata*: Seasonal, altitudinal, and latitudinal variation. Genetics and Molecular Biology. 2004;**27**(1):61-69. DOI: 10.1590/S1415-47572004000100011

[25] Klaczko LB. Evolutionary genetics of *Drosophila mediopunctata*. Genetica. 2006;**126**(1):43-55. DOI: 10.1007/s10709-005-1431-6

[26] Batista MR, Ananina G, Klaczko LB. Unexpected long-term changes in chromosome inversion frequencies in a Neotropical *Drosophila* species. Climate Research. 2012;**53**(2):131-140. DOI: 10.3354/cr01088

[27] Pegueroles C, Ordóñez V, Mestres F, Pascual M. Recombination and selection in the maintenance of the adaptive value of inversions. Journal of Evolutionary Biology. 2010;**23**(12):2709-2717. DOI: 10.1111/j.1420-9101.2010.02136.x

[28] McGaugh SE, Noor MA. Genomic impacts of chromosomal inversions in parapatric *Drosophila* species. Philosophical Transactions of the Royal Society B. 2012;**367**(1587):422-429. DOI: 10.1098/rstb.2011.0250

[29] Fuller ZL, Haynes GD, Zhu D, Batterton M, Chao H, et al. Evidence for stabilizing selection on codon usage in chromosomal rearrangements of *Drosophila pseudoobscura*. G3 (Bethesda). 2014;**4**(12):2433-2449. DOI: 10.1534/g3.114.014860

[30] Rane RV, Rako L, Kapun M, Lee SF, Hoffmann AA. Genomic evidence for role of inversion *3RP* of *Drosophila melanogaster* in facilitating climate change adaptation. Molecular Ecology. 2015;**24**(10):2423-2432. DOI: 10.1111/mec.13161

[31] Chovnick A. Gene conversion and transfer of genetic information within the inverted region of inversion heterozygotes. Genetics. 1973;**75**(1):123-131. PMCID: PMC1212990

[32] Navarro A, Betran E, Barbadilla A, Ruiz A. Recombination and gene flux caused by gene conversion and crossing over in inversion heterokaryotypes. Genetics. 1997;**146**:695-709. PMCID: PMC1208008

[33] Andolfatto P, Depaulis F, Navarro A. Inversion polymorphisms and nucleotide variability in *Drosophila*. Genetical Research. 2001;**77**(1):1-8. DOI: 10.1017/S0016672301004955

[34] Miller DE, Cook KR, Yeganeh Kazemi N, Smith CB, Cockrell AJ, Hawley RS, Bergman CM. Rare recombination events generate sequence diversity among balancer chromosomes in *Drosophila melanogaster*. Proceedings of the National Academy of Sciences of the United States of America. 2016;**113**(10):E1352-E1361. DOI: 10.1073/pnas.1601232113

[35] Corbett-Detig RB, Hartl DL. Population genomics of inversion polymorphisms in *Drosophila melanogaster*. PLoS Genetics. 2012;**8**(12):e1003056. DOI: 10.1371/journal.pgen.1003056

[36] Kirkpatrick M, Barton N. Chromosome inversions, local adaptation and speciation. Genetics. 2006;**173**(1):419-434. DOI: 10.1534/genetics.105.047985

[37] Krimbas CB, Powell JR. Chromosomal polymorphism in natural and experimental populations. In: Krimbas CB, Powell JR, editors. *Drosophila* Inversion Polymorphism. Florida, USA: CRC; 1982. pp. 2-52

[38] Sperlich D, Pfriem P. Chromosomal polymorphism in natural and experimental population. In: Ashburner M, Carson HL, Thompson JN Jr, editors. The Genetics and Biology of *Drosophila*. Vol. 3e. London: Academic Inc.; 1986. pp. 257-309

[39] Sturtevant AH, Dobzhansky T. Inversions in the third chromosome of wild races of *Drosophila pseudoobscura*, and their use in the study of the history of the species. Proceedings of the National Academy of Sciences of the United States of America. 1936;**22**(7):448-450. PMCID: PMC1076803

[40] Dobzhansky T, Sturtevant AH. Inversions in the chromosomes of *Drosophila pseudoobscura*. Genetics. 1938;**23**:28-64. PMCID: PMC1209001

[41] Wallace AG, Detweiler D, Schaeffer SW. Evolutionary history of the third chromosome gene arrangements of *Drosophila pseudoobscura* inferred from inversion breakpoints. Molecular Biology and Evolution. 2011;**28**(8):2219-2229. DOI: 10.1093/molbev/msr039

[42] Lemeunier F, Aulard S. Inversion polymorphism in *Drosophila melanogaster*. In: Krimbas CB, Powell JR, editors. Drosophila Inversion Polymorphism. Florida, USA: CRC; 1982. pp. 339-405

[43] Heed WB, Russell JS. Phylogeny and population structure in island and continental species of the *cardini* group of *Drosophila* studied by inversion analysis. University of Texas Publishing. 1971;**7103**:91-130

[44] Carson HL, Kaneshiro KY. *Drosophila* of Hawaii: Systematics and ecological genetics. Annual Review of Ecological Systems. 1976;**7**:311-345. DOI: 10.1146/annurev.es.07.110176.001523

[45] Throckmorton LH. The *virilis* species group. In: Ashburner M, Carson HL, Thompson JN, editors. The Genetics and Biology of *Drosophila*. Vol. 3b. London: Academic Press; 1982. pp. 227-296

[46] Diniz NM, Sene FM. Chromosomal phylogeny of the *Drosophila fasciola* species subgroup revisited (Diptera, Drosophilidae). Genetics and Molecular Biology. 2004;**27**(4):561-556. DOI: 10.1590/S1415-47572004000400016

[47] Rohde C, Garcia AC, Valiati VH, Valente VL. Chromosomal evolution of sibling species of the *Drosophila willistoni* group. I. Chromosomal arm IIR (Muller's element B). Genetica. 2006;**126**(1-2):77-88. DOI: 10.1007/s10709-005-1433-4

[48] Olvera O, Powell JR, de la Rosa ME, Salceda VM, et al. Populations genetics of Mexican *Drosophila*. VI. Cytogenetics aspects of the inversion polymorphism in *Drosophila pseudoobscura*. Evolution. 1979;**33**:381-395. DOI: 10.2307/2407628

[49] Jackson WD, Barber HN. Patterns of chromosome breakage after irradiation and ageing. Heredity. 1958;**12**:1-25. DOI: 10.1038/hdy.1958.1

[50] Ohno S. Ancient linkage groups and frozen accidents. Nature. 1973;**244**(5414):259-262. DOI: 10.1038/244259a0

[51] Sachs RK, Levy D, Chen AM, Simpson PJ, Cornforth MN, et al. Random breakage and reunion chromosome aberration formation model; an interaction-distance version based on chromatin geometry. International Journal of Radiation Biology. 2000;**76**(12):1579-1588. DOI: 10.1080/09553000050201064

[52] Nadeau JH, Taylor BA. Lengths of chromosomal segments conserved since divergence of man and mouse. Proceedings of the National Academy of Sciences of the United States of America. 1984;**81**:814-818. PMCID: PMC344928

[53] Pevzner P, Tesler G. Human and mouse genomic sequences reveal extensive breakpoint reuse in mammalian evolution. Proceedings of the National Academy of Sciences of the United States of America. 2003;**100**(13):7672-7677. DOI: 10.1073/pnas.1330369100

[54] González J, Casals F, Ruiz A. Testing chromosomal phylogenies and inversion breakpoint reuse in *Drosophila*. Genetics. 2007;**175**(1):167-177. DOI: 10.1534/genetics.106.062612

[55] Cáceres M, Barbadilla A, Ruiz A. Inversion length and breakpoint distribution in the *Drosophila buzzatii* species complex: Is inversion length a selected trait? Evolution. 1997;**51**(4):1149-1155. DOI: 10.2307/2411044

[56] Krimbas CB, Loukas M. The inversion polymorphism of *Drosophila subobscura*. In: Hecht MK, Steere WC, Wallace B, editors. Evo biol. Vol. 12. New York: Plenum Press; 1980. pp. 163-234

[57] Tonzetich J, Lyttle TW, Carson HL. Induced and natural break sites in the chromosomes of Hawaiian *Drosophila*. Proceedings of the National Academy of Sciences of the United States of America. 1988;**85**(5):1717-1721. PMCID: PMC279846

[58] Rohde C, Valente VLS. Three decades of studies on chromosomal polymorphism of *Drosophila willistoni* and description of fifty different rearrangements. Genetics and Molecular Biology. 2012;**35**(4):966-979. DOI: 10.1590/S1415-47572012000600012

[59] Ranz JM, Maurin D, Chan YS, von Grotthuss M, Hillier LW, Roote J, et al. Principles of genome evolution in the *Drosophila melanogaster* species group. PLoS Biology. 2007;**5**: 1366-1381. DOI: 10.1371/journal.pbio.0050152

[60] Calvete O, González J, Betrán E, Ruiz A. Segmental duplication, microinversion, and gene loss associated with a complex inversion breakpoint region in *Drosophila*. Molecular Biology and Evolution. 2012;**29**(7):1875-1889. DOI: 10.1093/molbev/mss067

[61] Puerma E, Orengo DJ, Salguero D, Papaceit M, Segarra C, Aguadé M. Characterization of the breakpoints of a polymorphic inversion complex detects strict and broad breakpoint reuse at the molecular level. Molecular Biology and Evolution. 2014;**31**(9):2331-2341. DOI: 10.1093/molbev/msu177

[62] Drosophila 12 Genomes Consortium. Evolution of genes and genomes on the *Drosophila* phylogeny. Nature. 2007;**480**:203-218. DOI: 10.1038/nature06341

[63] Bhutkar A, Schaeffer SW, Russo SM, et al. Chromosomal rearrangement inferred from comparisons of 12 *Drosophila* genomes. Genetics. 2008;**179**(3):1657-1680. DOI: 10.1534/genetics.107.086108

[64] von Grotthuss M, Ashburner M, Ranz JM. Fragile regions and not functional constraints predominate in shaping gene organization in the genus *Drosophila*. Genome Research. 2010;**20**(8):1084-1096. DOI: 10.1101/gr.103713.109

[65] Cáceres M, Ranz JM, Barbadilla A, Long M, Ruiz A. Generation of a widespread *Drosophila* inversion by a transposable element. Science. 1999;**285**:415-418. DOI: 10.1126/science.285.5426.415

[66] Casals F, Caceres M, Ruiz A. The foldback-like transposon *Galileo* is involved in the generation of two different natural chromosomal inversions of *Drosophila buzzatii*. Molecular Biology and Evolution. 2003;**20**:674-685. DOI: 10.1093/molbev/msg070

[67] Delprat A, Negre B, Puig M, Ruiz A. The transposon Galileo generates natural chromosomal inversions in *Drosophila* by ectopic recombination. PLoS One. 2009;**4**:e7883. DOI: 10.1093/molbev/msg070

[68] Jinks-Robertson S, Michelitch M, Ramcharan S. Substrate length requirements for efficient mitotic recombination in *Saccharomyces cerevisiae*. Molecular and Cellular Biology. 1993;**13**:3937-3950. DOI: 10.1128/MCB

[69] Hoang ML, Tan FJ, Lai DC, Celniker SE, Hoskins RA, et al. Competitive repair by naturally dispersed repetitive DNA during non-allelic homologous recombination. PLoS Genetics. 2010;**6**(12):e1001228. DOI: 10.1371/journal.pgen.1001228

[70] Runcie DE, Noor MA. Sequence signatures of a recent chromosomal rearrangement in *Drosophila mojavensis*. Genetica. 2009;**136**(1):5-11. DOI: 10.1007/s10709-008-9296-0

[71] Ohno S. Evolution by Gene Duplication. New York Inc.: Springer-Verlag; 1970. 160p

[72] Wesley CS, Eanes WF. Isolation and analysis of the breakpoint sequences of chromosome inversion *In(3L)Payne* in *Drosophila melanogaster*. Proceedings of the National Academy of Sciences of the United States of America. 1994;**91**(8):3132-3136. PMCID: PMC43529

[73] Cirera S, Martin-Campos JM, Segarra C, Aguade M. Molecular characterization of the breakpoints of an inversion fixed between *Drosophila melanogaster* and *D. subobscura*. Genetics. 1995;**139**:321-326. PMCID: PM7705632

[74] Andolfatto P, Wall JD, Kreitman M. Unusual haplotype structure at the proximal breakpoint of *In(2L)t* in a natural population of *Drosophila melanogaster*. Genetics. 1999;**153**:1297-1311. PMCID: PM10545460

[75] Richards S, Liu Y, Bettencourt BR, Hradecky P, Letovsky PS, et al. Comparative genome sequencing of *Drosophila pseudoobscura*: Chromosomal, gene, and cis-element evolution. Genome Research. 2005;**15**:1-18. DOI: 10.1101/gr.3059305

[76] Matzkin LM, Merritt TJ, Zhu CT, Eanes WF. The structure and population genetics of the breakpoints associated with the cosmopolitan chromosomal inversion *In(3R)Payne* in *Drosophila melanogaster*. Genetics. 2005;**170**:1143-1152. DOI: 10.1534/genetics.104.038810

[77] Evans AL, Mena PA, McAllister BF. Positive selection near an inversion breakpoint on the neo-X chromosome of *Drosophila americana*. Genetics. 2007;**177**(3):1303-1319. DOI: 10.1534/genetics.107.073932

[78] Cirulli ET, Noor MA. Localization and characterization of X chromosome inversion breakpoints separating *Drosophila mojavensis* and *Drosophila arizonae*. The Journal of Heredity. 2007;**98**(2):111-114. DOI: 10.1093/jhered/esl065

[79] Machado CA, Haselkorn TS, Noor MA. Evaluation of the genomic extent of effects of fixed inversion differences on intraspecific variation and interspecific gene flow in *Drosophila pseudoobscura* and *D. persimilis*. Genetics. 2007;**175**(3):1289-1306. DOI: 10.1534/genetics.106.064758

[80] Prazeres da Costa O, González J, Ruiz A. Cloning and sequencing of the breakpoint regions of inversion *5g* fixed in *Drosophila buzzatii*. Chromosoma. 2009;**118**(3):349-360. DOI: 10.1007/s00412-008-0201-5

[81] Fonseca NA, Vieira CP, Schlötterer C, Vieira J. The *DAIBAM MITE* element is involved in the origin of one fixed and two polymorphic *Drosophila virilis* phylad inversions. Fly (Austin). 2012;**6**(2):71-74. DOI: 10.4161/fly.19423

[82] Corbett-Detig RB, Cardeno C, Langley CH. Sequence-based detection and breakpoint assembly of polymorphic inversions. Genetics. 2012;**192**:131-137. DOI: 10.1534/genetics.112.141622

[83] Guillén Y, Ruiz A. Gene alterations at *Drosophila* inversion breakpoints provide prima facie evidence for natural selection as an explanation for rapid chromosomal evolution. BMC Genomics. 2012;**13**(1):53. DOI: 10.1186/1471-2164-13-53

[84] Papaceit M, Segarra C, Aguadé M. Structure and population genetics of the breakpoints of a polymorphic inversion in *Drosophila subobscura*. Evolution. 2012;**67**:66-79. DOI: 10.1111/j.1558-5646.2012.01731.x

[85] Orengo DJ, Puerma E, Papaceit M, Segarra C, Aguadé MA. Molecular perspective on a complex polymorphic inversion system with cytological evidence of multiply reused breakpoints. Heredity (Edinburgh). 2015;**114**(6):610-618. DOI: 10.1038/hdy.2015

[86] Puerma E, Orengo DJ, Aguadé M. The origin of chromosomal inversions as a source of segmental duplications in the *Sophophora* subgenus of *Drosophila*. Scientific Reports. 2016;**6**:30715. DOI: 10.1038/srep30715

[87] Puerma E, Orengo DJ, Aguadé M. Multiple and diverse structural changes affect the breakpoint regions of polymorphic inversions across the *Drosophila* genus. Scientific Reports. 2016;**6**:36248. DOI: 10.1038/srep36248

[88] McClintock B. Maize genetics. Carnegie Institution of Washington Year Book. 1946;**45**: 176-186

[89] Capy P. Classification and nomenclature of retrotransposable elements. Cytogenetic and Genome Research. 2005;**110**(1-4):457-461. DOI: 10.1159/000084978

[90] Finnegan DJ. Eukaryotic transposable elements and genome evolution. Trends in Genetics. 1989;**5**:103-107. DOI: 10.1016/0168-9525(89)90039-5

[91] Wicker T, Sabot F, Hua-Van A, Bennetzen JL, Capy P, et al. A unified classification system for eukaryotic transposable elements. Nature Reviews. Genetics. 2007;**8**(12):973-982. DOI: 10.1038/nrg2165

[92] Deprá M, Ludwig A, Valente VLS, Loreto ELS. Mar, a MITE family of hAT transposons in *Drosophila*. Mobile DNA. 2012;**3**:13. DOI: 10.1186/1759-8753-3-13

[93] Kidwell MG. Transposable elements and the evolution of dipteran genomes. In: Yeates DK, Wiegmann BM, editors. The Evolutionary Biology of Flies. New York: Columbia University; 2005. pp. 145-173

[94] Zelentsova H, Shostak N, Kozitsina M, Barskyi V, et al. *Penelope*, a new family of transposable elements and its possible role in hybrid dysgenesis in *Drosophila virilis*. Proceedings of the National Academy of Sciences of the United States of America. 1997;**94**(1):196-201. PMCID: PMC19282

[95] Engels WR, Preston CR. Formation of chromosome rearrangements by *P* factors in *Drosophila*. Genetics. 1984;**107**:657-678. PMCID: PMC1202383

[96] Lyttle TW, Haymer DS. The role of the transposable element *hobo* in the origin of endemic inversions in wild populations of *Drosophila melanogaster*. Genetica. 1992;**86**:113-126. DOI: 10.1007/BF00133715

[97] Regner LP, Pereira MSO, Alonso CEV, Abdelhay E, Valente VLS. Genomic distribution of *P* elements in *Drosophila willistoni* and a search for their relationship with chromosomal inversions. The Journal of Heredity. 1996;**87**:191-198. DOI: 10.1093/oxfordjournals.jhered.a022984

[98] Zelentsova H, Poluectova H, Mnjoian L, Lyozin G, Veleikodvorskaja V, et al. Distribution and evolution of mobile elements in the *virilis* species group of *Drosophila*. Chromosoma. 1999;**108**(7):443-456. DOI: 10.1007/s004120050396

[99] Evgen'ev M, Zelentsova H, Mnjoian L, Poluectova H, Kidwell MG. Invasion of *Drosophila virilis* by the Penelope transposable element. Chromosoma. 2000;**109**(5):350-357. DOI: 10.1007/s004120000086

[100] Marzo M, Puig M, Ruiz A. The *Foldback-like* element *Galileo* belongs to the *P* superfamily of DNA transposons and is widespread within the *Drosophila* genus. Proceedings of the National Academy of Sciences of the United States of America. 2008;**105**(8):2957-2962. DOI: 10.1073/pnas.0712110105

[101] Puig M, Cáceres M, Ruiz A. Silencing of a gene adjacent to the breakpoint of a widespread *Drosophila* inversion by a transposon-induced antisense RNA. Proceedings of the National Academy of Sciences of the United States of America. 2004;**101**(24):9013-9018. DOI: 10.1073/pnas.0403090101

[102] González J, Nefedov M, Bosdet I, Casals F, Calvete O, et al. A BAC-based physical map of the *Drosophila buzzatii* genome. Genome Research. 2005;**15**:885-892. DOI: 10.1101/gr.3263105

[103] Ruiz A, Heed WB, Wasserman M. Evolution of the *mojavensis* cluster of cactophilic *Drosophila* with descriptions of two new species. The Journal of Heredity. 1990;**81**(1):30-42. DOI: 10.1093/oxfordjournals.jhered.a110922

[104] Sturtevant AH, Plunkett CR. Sequence of corresponding third chromosome genes in *Drosophila melanogaster* and *Drosophila simulans*. The Biological Bulletin. 1926;**50**:56-60. DOI: 10.2307/1536631

[105] Ashburner M. *Drosophila* Genomes by the Baker's Dozen. Genetics. 2007;**177**(3):1263-1268. PMCID: PMC2147995

# Chapter 3

# Cell Cycle Regulators in Female Meiosis of *Drosophila melanogaster*

Mohammed Bourouh and Andrew Swan

Additional information is available at the end of the chapter

http://dx.doi.org/10.5772/intechopen.70671

**Abstract**

Meiosis is a highly regulated and complex variation on the canonical cell cycle. It depends on the activity of most of the known mitotic cell cycle regulators, as well as many meiosis-specific factors that interact with and modify the activities of this core cell cycle machinery. This review will examine the roles of known mitotic cell cycle regulators and meiosis-specific factors in *Drosophila* female meiosis, focusing on three important meiotic events: nuclear envelope breakdown or maturation, establishment of the meiosis I spindle, and release from metaphase I arrest at ovulation. Many meiotic processes are controlled by the mitotic kinase, Cdk1 with its cyclin partners, cyclins A, B, and B3. Other major mitotic kinases, including Polo and Aurora B have been found to play multiple roles in *Drosophila* meiosis. The Anaphase Promoting Complex or Cyclosome (APC/C) controls many meiotic processes through regulation of Cdk1, the sister chromatid cohesion regulator, Separase and other targets. This review will focus on these and other meiotic regulators, emphasizing some of the technical advances that have driven the field forward in recent years, and highlighting gaps that need to be filled to achieve a more complete picture of how meiosis is regulated in *Drosophila*.

**Keywords:** APC/C, Aurora B, Cdk1, cohesin, Cort, cyclin, *Drosophila*, meiosis, oogenesis, Polo, Separase

## 1. Introduction

The major events of meiosis are conserved throughout eukaryotes, and as with all cell biology, knowledge gained in one model system informs our understanding of meiosis in other organisms. On the other hand, as researchers gain a better understanding of how meiosis is controlled at the molecular level, it becomes clear that there are major differences between model systems and even between males and females in the same organism. This review discusses the regulation of meiosis specifically in one model system, females of *Drosophila melanogaster*. Mitosis and meiosis in other model systems will be brought into the discussion, both to highlight areas of divergence, and to suggest explanations for events that are not yet well understood in this model system.

## 2. Oogenesis and meiosis in *Drosophila* females

Female meiosis takes place in the context of oogenesis. *Drosophila* females have two ovaries, each containing approximately 18 ovarioles, chains of progressively developing egg chambers, each consisting of an oocyte and associated support cells. Ovarioles are organized from anterior to posterior. The anterior contains the germarium, the location of germline stem cells and somatic stem cells that give rise to the egg and the follicle cells respectively. The posterior end contains mature developed eggs. Egg development has been divided into 14 distinct stages, based on major morphological events [1, 2].

In the anterior tip of the germarium, germline stem cells divide asymmetrically to give rise to a stem cell and a daughter cell. The daughter cell undergoes 4 incomplete divisions to generate a cyst of 16 cells that remain connected via cytoplasmic bridges called ring canals [1, 2].

From this cyst of 16 cells, one of the two cells with the most cytoplasmic bridges will differentiate into the oocyte [1]. Oocyte determination occurs while the oocyte is still within the germarium. The oocyte enters meiotic prophase, assembling synaptonemal complexes between homologs and undergoing crossing over. Throughout prophase, which lasts until stage 13 of oogenesis, the chromatin is compacted within the nucleus in a structure referred to as the karyosome. The other 15 germline cells of the cyst enter the endocycle concurrent with entry of the oocyte into meiotic prophase. These polyploid cells, called nurse cells, generate proteins and mRNAs important for meiosis progression, egg maturation and early embryonic development. The nurse cells use the cytoplasmic bridges to transfer their contents into the egg, prior to undergoing apoptosis in late oogenesis [1, 2].

As mentioned, the oocyte is arrested in prophase I until stage 13, at which point oocyte maturation occurs, highlighted by nuclear envelope breakdown (NEB). Spindle formation occurs in the absence of centrosomes, via microtubule polymerization around the karyosome, and a bipolar spindle assembles. Meiosis arrests at stage 14, the final stage of oogenesis, and the arrest is maintained until ovulation triggers egg activation, marked by the resumption of meiosis [2, 3]. Upon egg activation, the meiosis I spindle rotates and undergoes anaphase I. At the completion of the first anaphase, the two meiosis II spindles form around the separated homologs. These spindles are arranged perpendicular to the egg length and are held together by an aster of microtubules. At the completion of the 2nd meiotic division the 4 meiotic products enter a post-meiotic interphase. They then appear to undergo DNA replication in synchrony with the male pronucleus that entered the egg during fertilization. One of the 4 female meiotic products, usually the most interior, migrates towards the male pronucleus, apparently along microtubules that radiate out from the male aster. The male and female pronuclei enter the first mitosis together. The remaining female haploid products come together, undergo nuclear envelope breakdown, and arrest in a mitotic-like state with condensed chromatin arranged on an aster-like array of microtubules called the polar body [4].

## 3. Cell cycle regulation and oocyte maturation

Oocyte development in most metazoans has two arrest points, a primary arrest at prophase, and a secondary arrest in metaphase. In insects, the secondary arrest is at metaphase I, while in many vertebrates it is at metaphase II. The long prophase arrest allows synapsis and crossing over to occur and at the same time, allows for oocyte growth. The secondary arrest facilitates the coordination of completion of meiosis with fertilization and the transition from oogenesis to embryogenesis [5].

The primary arrest in prophase is broken by nuclear envelope breakdown, a process that in most, if not all, eukaryotes is dependent on the mitotic cyclin-dependent kinase (Cdk), Cdk1. The Cdks are the core regulators of the cell cycle. They are activated by phosphorylation on their T-loop via a Cdk activating kinase or CAK, and by dephosphorylation of a Thr and Tyr at 14 and 15, respectively, by the Cdc25 phosphatase. Cdks also require association with cyclin partners, which themselves are subject to both transcriptional control and ubiquitin/ mediated destruction.

The importance of Cdk1 in *Drosophila* female meiosis was first established through analysis of Twine, a germline-specific Cdc25 homologue. *Twine* mutants are viable but females produce eggs that do not hatch. In a study focusing on Polo and its regulator Matrimony (Mtrm, discussed below), it was found that *twine* mutant oocytes undergo NEB in stage 14 instead of in mid-stage 13 [6]. This finding was supported by the finding that a temperature-sensitive allele of *Cdk1* also produces a delay in NEB [7].

These studies illustrate a major challenge in studying meiosis in a genetic system such as *Drosophila*: classical genetics is limited to the study of genes that are non-essential for viability, such as *twine*; or the study of hypomorphic or conditional mutants as with *Cdk1*. The development of transgenic RNAi for the female germline has allowed researchers in the last few years to overcome these limitations and study the meiosis-specific requirements for otherwise essential genes. The Transgenic RNAi Project (TRiP) out of Harvard University has generated a genome-wide collection of transgenic RNAi lines that are driven by the *UAS/Gal4* system [8]. While the earlier collections were not effective in the germline, a micro-RNA based collection is now available that can be very effective. In most studies of meiosis RNAi is expressed using the female germline-specific mat-$\alpha$-Tubulin Gal4 driver which expresses just after the premeiotic divisions in the germarium, thus not affecting mitotic divisions that are necessary for oocyte formation [9]. Using transgenic RNAi, it was found that loss of Cdk1 in meiosis leads to a complete failure of NEB in most oocytes, indicating that Cdk1 is indeed essential for NEB [10].

*Drosophila*, like other metazoans has three mitotic cyclins, Cyclin A, B and B3, though unlike vertebrates and other animals, *Drosophila* has only a single representative of each subtype. Cyclin A is the only one that is essential for viability. *Cyclin B* and *B3* mutants are viable but female sterile, though *Cyclin B/B3* double mutants are lethal [11]. The identity of the cyclin partner for Cdk1 in oocyte maturation was investigated using transgenic RNAi, as well as a conditional mutant for *Cyclin B* and female sterile alleles of *Cyclin B3* [10]. While classic studies of meiosis in *Xenopus* and other vertebrate models have revealed Cyclin B to be the major Cdk1 partner in meiotic maturation, RNAi knockdown and conditional mutants of *Cyclin B* had no effect on NEB timing in *Drosophila* female meiosis [10]. Loss of Cyclin B3 also had no effect on NEB timing, either alone or when combined with *Cyclin B* knockdown. Knockdown of *Cyclin A* resulted in a slight delay in NEB timing. This delay was not enhanced by simultaneous loss of *Cyclin B* or *B3*. However, simultaneous knockdown of all three mitotic cyclins produced a prolonged delay or complete block in NEB, similar to *Cdk1* knockdown. Therefore, in *Drosophila* female meiosis, all three mitotic cyclins function in NEB, with Cyclin A apparently playing the most important role [10] (**Figure 1**).

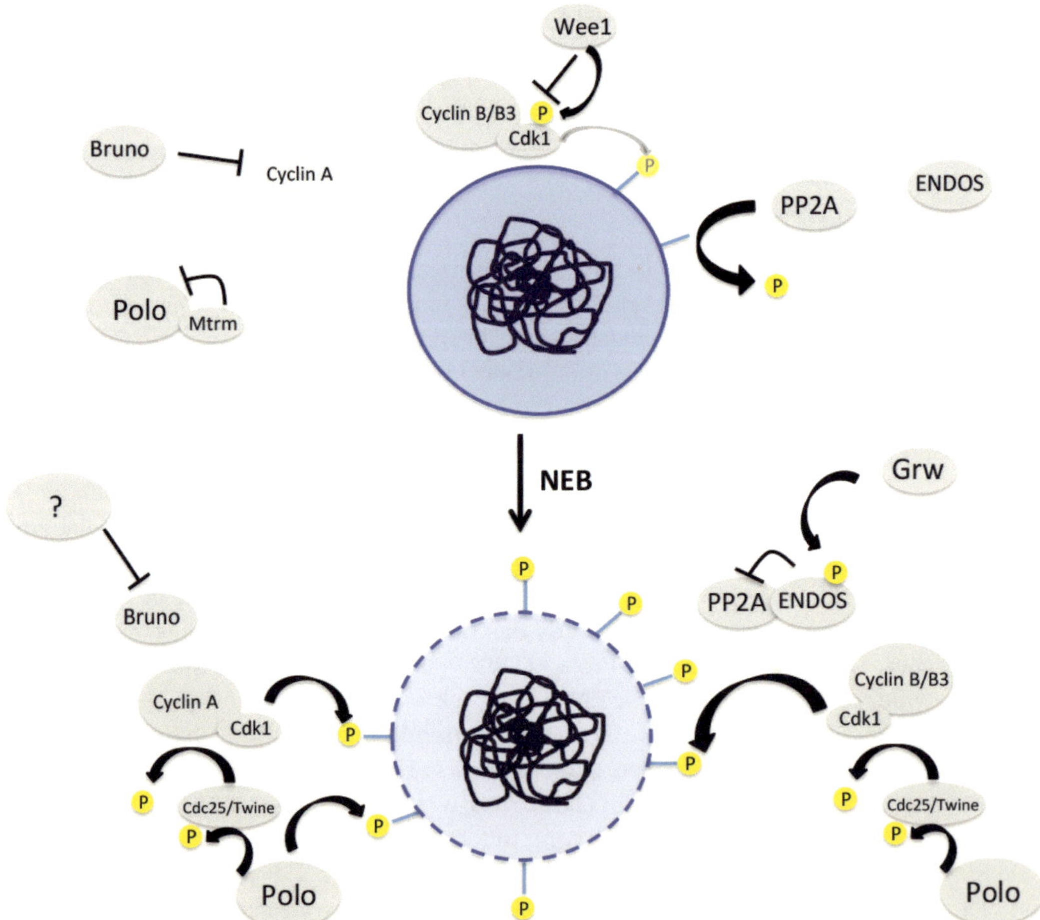


**Figure 1.** Model for nuclear envelope breakdown in *Drosophila* female meiosis. Phosphorylation events that drive nuclear envelope breakdown are here depicted as occurring on the nuclear envelope. The nuclear envelope is depicted in the lower panel as a dashed line to indicate NEB. See text for details and references.

The regulation of Cyclin A expression may contribute to the timing of NEB in *Drosophila*. Cyclin A is translationally repressed during meiotic prophase by the mRNA binding protein Bruno [12]. Bruno levels fall dramatically in stages 12 and 13 of oogenesis by an as yet unknown mechanism. This coincides with a dramatic increase in Cyclin A levels, possibly driving oocyte maturation [13] (**Figure 1**). Therefore, the regulation of Cyclin A translation appears to be critical for NEB timing in *Drosophila*. If this is the case, the forced expression of Cyclin A prior to stage 12 of oogenesis may be expected to result in premature NEB.

In addition to cyclin binding, Cdk1 activity appears to be regulated in many other ways that may contribute to the timing of NEB. The *Endos* gene was found to be required for proper NEB timing [7]. In vertebrates, Endos binds to the B55/Twins subunit of Protein phosphatase 2A (PP2A) to inhibit PP2A activity [14, 15]. This activity appears to be conserved for *Drosophila* Endos [16]. PP2A has been shown to recognize and dephosphorylate Cdk phosphorylated proteins, and as such is a major negative regulator of Cdk-dependent processes.

The ability of Endos to bind and inhibit PP2A appears to in turn be regulated by the phosphorylation of Endos by Greatwall kinase (Gwl) [16, 17]. Gwl was initially discovered in *Drosophila* by the dominant Scant allele, and was found to play multiple roles in meiosis beyond NEB [18]. Both genetic and biochemical evidence points to a simple linear pathway in which Gwl phosphorylation of Endos allows Endos to bind and inhibit PP2A [16, 17, 19, 20]. While Gwl has not yet been directly implicated in oocyte maturation, transgenic Endos with a S68A mutation, abolishing the Gwl phosphorylation site, is unable to rescue the delayed NEB phenotype of a *Endos* null mutant [17]. Thus Gwl-mediated phosphorylation likely plays a role in the Endos-mediated inhibition of PP2A to promote NEB (**Figure 1**).

The activity of Gwl itself is subject to regulation via phosphorylation via both Cdk1 and Polo kinases [21, 22]. Both kinases phosphorylate Gwl in the central region of the protein, disrupting the function of two nuclear localization sequences, thus promoting the cytoplasmic accumulation of Gwl. This may allow it to efficiently inhibit PP2A, which is predominantly cytoplasmic [21, 22].

In mitotic cells, Polo kinase plays multiple roles throughout the cell cycle, regulating centrosome dynamics, chromosome cohesion, and events at cytokinesis [23]. Polo promotes Cdk1 activity through the activation of Cdc25, while also targeting common substrates of Cdk1, such as the APC/C component Cdc27. Polo also recognizes many of its substrates depending on their prior phosphorylation by Cdk1 [23]. Therefore, the activities of Polo and Cdk1 are closely coordinated and often synergistic.

Studies of the Polo-binding protein, Matrimony (Mtrm), suggest a critical role for Polo in the timing of NEB in *Drosophila* female meiosis [6]. Matrimony is a maternally expressed protein that acts as a physical inhibitor of Polo. *Mtrm* mutants, even when heterozygous, display precocious NEB. This is likely due to precocious activity of Polo, since the simultaneous reduction of *Polo* gene dose leads to suppression of this phenotype [6]. Furthermore, a mutation in the Polo-interacting domain of Mtm results in a loss-of-function phenotype, suggesting that the antagonistic relationship between Polo and Mtm reflects inhibition of Polo by Mtm and not the other way around.

The timing of Mtrm expression is consistent with it having a role in NEB timing. Matrimony expression in the oocyte starts to increase in stage 10 and peak levels are reached at stage 11–12, when it localizes to the nucleus and cytoplasm [6]. Levels of Polo start to rise above Matrimony levels in stage 12, possibly allowing Polo to escape inhibition and help to promote NEB in stage 13 [6] (**Figure 1**). Recently, a strong RNAi knockdown allele of *Polo* has been characterized. Surprisingly, NEB still occurs in these oocytes [24], but a possible effect on the timing of NEB remains to be determined.

## 4. Meiosis I spindle assembly and chromosome orientation on the spindle

Meiotic spindles in many organisms, including humans, frogs and *Drosophila*, differ from mitotic spindles in that they are acentrosomal. Meiotic spindle microtubules do not originate from centrosomes but instead appear to nucleate from the chromosomes, and become tapered into bipolar spindles. Another major difference is in the behavior of kinetochores. In metaphase of meiosis I, the kinetochores of sister chromatids contact microtubules from the same pole, referred to as co-orientation. Homologous chromosomes, on the other hand, contact microtubules from opposite poles. This is referred to as biorientation of homologs. Tension created by pulling forces of kinetochore microtubules results in a stereotypic arrangement of chromosomes in metaphase I in which centromeres for each homolog are oriented on either side of the spindle midzone [25]. Assembly of this meiotic spindle and proper orientation of chromosomes requires the actions of many cell cycle regulators.

The problem of how to build a spindle without centrosomes appears to be dealt with differently in different organisms. In Xenopus egg extracts the small GTPase, Ran accumulates in its active GTP-bound form in the vicinity of chromosomes as a result of its chromatin-bound activating GEF, Rcc1. This Ran gradient promotes acentrosomal spindle assembly [26]. Rcc1 associates with the karyosome in *Drosophila* female meiosis, suggesting a similar role in flies [27]. The importance of the Ran gradient was investigated by over-expression of a Ran mutant that is unable to exchange GDP for GTP. This dominant negative allele produced only mild defects in spindle pole formation [27]. Therefore it appears that Ran does not play a central role in spindle assembly in *Drosophila* female meiosis. As discussed below, it appears that Aurora B and the chromosome passenger complex takes on this role.

The chromosome passenger complex (CPC), composed of Aurora B kinase, Incenp, Survivin and Borealin has multiple functions in mitotic cells [28]. The CPC promotes chromatin condensation in prophase. It accumulates at kinetochores in prometaphase, where Aurora B promotes the breakage of kinetochore-microtubule contacts. Importantly, the CPC is sensitive to tension. Attachment of spindle microtubules to a kinetochore results in a pulling of that chromatid toward the spindle pole. When the kinetochores of a pair of attached sister chromatids each make attachments to opposite poles (known as an amphitelic attachment), pulling forces from either pole generate tension across the kinetochores. This is sensed by the CPC, leading to inactivation of Aurora B. As a result, bi-polar attachments are stabilized. Aurora B and other CPC components are also required at kinetochores for activation of the spindle assembly checkpoint (SAC). At anaphase onset the CPC relocates to the spindle midzone where it then plays a role in establishing the cleavage furrow [28].

In *Drosophila* meiosis the CPC has multiple roles, though surprisingly, these are quite different than CPC roles in mitosis. CPC proteins appear to associate with the karyosome immediately after NEB, coincident with the first detectable microtubules forming around the chromatin. CPC localization is independent of microtubules, suggesting that the CPC directly associates with chromatin [29]. Hypomorphic mutations in *Incenp* were found to produce a delay in microtubule polymerization around the karyosome following NEB [30]. Subsequent analysis of strong RNAi knockdown alleles of *Incenp* and *Aurora B* revealed a near complete failure to nucleate microtubules around the chromatin [29]. Therefore the CPC localizes to chromatin following NEB and is required for the first step in meiotic spindle formation, the recruitment of microtubules to the karyosome.

Unlike the situation in mitotic cells, the CPC does not appear to associate with kinetochores in *Drosophila* female meiosis. Nonetheless, core kinetochore proteins fail to assemble in strong Aurora B knockdown oocytes, suggesting that the CPC is necessary for kinetochore assembly [31]. Instead of accumulating on kinetochores in prometaphase, the CPC accumulates on the interpolar microtubules of the spindle midzone early in prometaphase [32]. Partial loss of function mutations in CPC genes permit meiotic spindle formation but these spindles often have a poorly formed spindle midzone. Many spindles appear to have more than two poles, reflecting the importance of interpolar microtubules in maintaining spindle bipolarity. In terms of protein localization and mutant phenotype, CPC hypomorphs resemble *Subito* mutants. *Subito* encodes the *Drosophila* MKLP2 homolog, and is a major organizer of the meiotic spindle midzone [32]. The CPC is required for Subito localization to the spindle midzone [29, 32].

Weaker CPC alleles that permit bipolar spindle formation display defects in chromosome orientation on the meiotic spindle. This failure of biorientation is also observed upon knockdown of *Subito*, suggesting that interpolar microtubules are important for chromosome movements that establish biorientation. Knockdown of the essential kinetochore component *Spc105R* also results in biorientation defects [31]. Interestingly, Ndc80, a kinetochore component required for head-on attachment of spindle microtubules to kinetochores, has only a subtle role in biorientation of homologs [31]. Taken together the results support the idea that side-on interactions between kinetochores and interpolar microtubules lead to chromosome movements that then lead to establishment of end-on attachments. The CPC appears to be important for both the formation of interpolar microtubules and for assembly of the kinetochore [31].

In mitotic cells, the Centralspindlin complex that includes the kinesin-like protein, MKLP1 (Pavarotti in *Drosophila*), is a major regulator of cleavage furrow formation at cytokinesis. MKLP1 at the central spindle recruits the GEF for RhoA [33]. This leads to the local activation of RhoA on overlying cell membrane, leading to actin/myosin recruitment, contractility and cleavage furrow formation [33]. Recently it was found that Centralspindlin components, and surprisingly, Rho and its GEF, play a role in biorientation of homologs in *Drosophila* female meiosis. This implies a novel role for this cytokinesis pathway in organizing the acentrosomal meiotic spindle [24].

In many cell types, Polo kinase plays an important role at the central spindle. In *Drosophila* male meiosis for example, Polo is required for CPC localization to the central spindle, and for localization of Shugoshin to centromeric chromatin where it protects centromeric cohesin complexes during the first meiotic division [34–36]. In *Drosophila* female meiosis, Polo does not specifically accumulate on the central spindle. It is not required for central spindle formation or for the localization of the CPC to the central spindle [24, 32]. *Polo* knockdown results in bi-orientation failure and instead of a compact karyosome in metaphase I, each homolog pair is randomly dispersed on the meiosis I spindle, often appearing as discrete chromatin masses [24]. Interestingly, a similar phenotype is observed in mutants or RNAi knockdown of the major CENP-E homolog in *Drosophila* (*cmet*), suggesting a possible function for Polo in regulation of this plus-end directed microtubule motor that may function to oppose poleward forces during prometaphase [31].

Polo accumulates in a punctate pattern on chromosomes in metaphase I oocytes, likely at kinetochores. It may function at kinetochores to stabilize kinetochore microtubules, as loss of Polo leads to the apparent hollowing out of the meiotic spindle, possibly due to a reduction in the number of kinetochore microtubules [24].

Given the close functional relationship between Polo and Cdk1, it is interesting that biorientation of homologs in meiosis is also dependent on Cyclin A-Cdk1 [10]. Unlike Polo, Cyclin A does not appear to be required for maintaining a compact chromatin mass or for proper spindle morphology, though *twine* mutant oocytes and *Cyclin A*, *Cyclin B3* double knockdown oocytes display a scattered chromatin phenotype that may be similar to that seen in Polo [7, 10]. As yet it is not known how Cyclin A-Cdk1 promotes biorientation. In other systems Cdk1 can phosphorylate Incenp, thereby activating the CPC [37]. Cyclin A-Cdk1 has also been found to promote proper head-on attachment of kinetochore microtubules in prometaphase of mitosis [38].

## 5. The APC/C and control of meiotic anaphase

Mature eggs are maintained in a metaphase I arrest that can be stable for long periods of time. This arrest is maintained by multiple forces, many of which are focused on the inhibition of Anaphase Promoting Complex/Cyclosome (APC/C) activity. The APC/C is a multi-subunit E3 ubiquitin ligase that catalyzes the addition of ubiquitin chains to target proteins, marking them for degradation by the proteasome. This section will first introduce the APC/C in general and in *Drosophila* oocytes more specifically. This will be followed by a discussion of how the APC/C is kept inactive in metaphase I arrested mature oocytes, how it is activated upon egg activation, and how its activity leads to the completion of meiosis and other processes that occur upon egg activation.

Ubiquitination by the APC/C, as with other E3 Ubiquitin ligases, depends on the activity of E1 and E2 enzymes. The E1 activates Ubiquitin and transfers it to an E2. In the case of the APC/C, the E2 appears to mediate Ubiquitin transfer to the substrate, with the APC/C serving to bring E2 and substrate together [39, 40]. The core of the APC/C is composed of APC2, APC11, DOC1 and an activator, CDC20 or Cdh1. Doc1 and CDC20/Cdh1 facilitate the substrate recognition component of the APC/C, whereas APC2 allows for binding of the E2 [39]

The APC/C activators Cdc20 and Cdh1 (known as Fzy and Fzr respectively in *Drosophila*) recognize well-conserved degron motifs on target proteins. Cdc20 can recognize the destruction box (D-Box), which has the consensus sequence RXXLXXXXN. Cdh1 recognizes the D-Box, the KEN box and a small number of other degron motifs [40].

In vertebrate mitosis, the two APC/C activators function in tandem to control cyclin levels. APC/C$^{Cdc20}$ is activated by Cdk1 phosphorylation in mitosis, and drives anaphase by targeting mitotic cyclins and Securin for destruction. Destruction of Securin results in activation of Separase, a protease that cleaves the klesin subunit of the cohesin complexes, thereby releasing sister chromatids. APC/C$^{Cdh1}$ is inhibited by Cdk-mediated phosphorylation and is thus activated following cyclin destruction in anaphase. APC/C$^{Cdh1}$ remains active through G1 and maintains low Cdk activity. It is then inactivated as Cdk activity rises at S-phase, and it remains inactive through G2. Both APC/C activators are important in vertebrate meiosis. In the mouse APC/C$^{Cdh1}$ is active in G2 and prophase to maintain low cyclin levels to prevent precocious NEB [41]. In Xenopus, by contrast, it promotes NEB, targeting Protein Phosphatase 6 for destruction, thereby maintaining Cdk1 phosphorylations that promote NEB [42]. APC/C$^{Fzy}$ is the primary driver of anaphase in vertebrate meiosis, as it is in mitosis [40].

In *Drosophila*, Fzr (Cdh1) is expressed at very low levels in early stage embryos, suggesting that it may not be present during late stages of meiosis [43]. As yet, no role has been described for Fzr in meiosis. Meanwhile, two APC/C complexes cooperate in meiotic anaphase, APC/C$^{Fzy}$ (Cdc20) and the female germline-specific APC/C$^{Cort}$ [40, 44].

## 6. Inhibition of APC/C during meiosis I arrest

Classic studies in Xenopus oocytes and other vertebrate models identified Cytostatic Factor (CSF) as the key to inhibition of APC/C activity in the meiosis II arrest. While the molecular identity of CSF remained unknown for many years, it now seems that the APC/C$^{Fzy}$ inhibitor, Emi2 is responsible for CSF activity. Emi2 functions by competing with APC/C$^{Fzy}$ for interaction with the E2, Ube2S [45]. Emi2 is related to another APC/C inhibitor, Emi1, that functions in meiotic prophase and in mitotic cells. *Drosophila* has a single Emi homolog, Rca1. Rca1, like Emi1, has a clear mitotic role. Its role, if any, in meiosis has not yet been determined. It does not appear capable of interacting with and inhibiting Fzy [46], and no interaction with APC/C$^{Cort}$ has been described.

Emi2 stability and its ability to interact with APC/C depend on phosphorylation by the Rsk kinase, acting downstream of a Map kinase pathway that has Mos as the upstream kinase. This phosphorylation recruits PP2A which in turn reverses the Cdk1-mediated phosphorylation of Emi2 that leads to its inactivation and destruction [47]. The role of Mos and Mapk have been investigated in *Drosophila* and neither appears to be necessary for maintaining the meiosis I arrest [48].

The spindle assembly checkpoint (SAC) plays a key role in assuring faithful chromosome segregation in mitosis by inhibiting anaphase initiation until all chromosomes have made bipolar attachments to the mitotic spindle. During prometaphase, kinetochores that are not yet connected to spindle microtubules act as sites for recruitment of SAC proteins including the core SAC component, Mad2. Mad2 is converted to an active form that can diffuse away to assemble the mitotic checkpoint complex (MCC), which binds and inhibits APC/C$^{Cdc20}$. In some organisms such as yeast and *Drosophila*, the SAC is not essential for normal mitosis, but becomes essential under conditions in which mitotic spindle assembly or chromosome attachment to the spindle is disrupted. In vertebrates, the SAC plays an essential role even in the absence of spindle disruption [49].

A role for the SAC in meiosis has been most clearly established in the mouse where it is important for delaying anaphase I under normal conditions and under conditions in which the spindle is disrupted [50]. A role for the SAC in *Drosophila* is suggested by the protein localization and mutant phenotype for the essential SAC component, Mps1 (also known as Ald) [51]. Mps1 is a kinase that is recruited to unoccupied kinetochores where it functions to recruit other SAC components. In *Drosophila* female meiosis, Mps1 accumulates on kinetochores starting in prometaphase of meiosis. Mutants display a precocious anaphase phenotype, consistent with a SAC function in meiosis I [51]. Another essential SAC component, BubR1, also displays a precocious sister chromatid separation and/or missegregation phenotype in meiosis that could be attributed to a meiotic SAC role [52]. On the other hand, both Mps1 and BubR1 clearly have non-SAC roles in meiosis that could underlie these phenotypes [51, 52].

Unlike many SAC proteins, Mad2 appears to have no function outside of the SAC [53] and, importantly, null alleles of *Mad2*, do not result in precocious anaphase in meiosis I [54]. Furthermore, loss of Mad2 as well as BubR1, Mps1 and another SAC gene, Zwilch, do not result in reduced levels of the APC/C$^{Fzy}$ target, Cyclin B, either globally or locally on the meiosis I spindle, as would be expected if the APC/C were activated [54]. Genetic evidence also argues against a role for the SAC in inhibiting APC/C activity in the 2nd meiotic division [54]. This apparent lack of a requirement for the SAC in female meiosis is also seen in Xenopus female meiosis [55].

Cyclin B-Cdk1 may play a role in inhibiting APC/C activity in *Drosophila* female meiosis, as *Cyclin B* knockdown and conditional mutants of *Cyclin B* result in precocious homolog segregation [10]. In Xenopus oocytes, Cyclin B-Cdk1 has been shown to bind to and inhibit the activity of Separase [56], and the above phenotype may indicate a similar role in *Drosophila* oocytes. In such a model, Cyclin B loss results in APC/C-independent activation of Separase. This could be tested genetically, since it predicts that precocious anaphase observed in the Cyclin B knockdown would still occur in an APC/C knockdown background. While this experiment has not been performed, it was found that precocious anaphase in the *Cyclin B* knockdown oocyte is suppressed by loss of *Cyclin B3* [10]. As we discuss below, Cyclin B3 appears to activate the APC/C in meiosis. Therefore, the precocious anaphase resulting from Cyclin B loss appears to depend on the APC/C. This would better fit with a model in which Cyclin B-Cdk1 inhibits APC/C activity rather than Separase activity (**Figure 2**).

Loss of a single copy of the *Mtrm* gene results in precocious anaphase [6]. As for its role in NEB, Mtrm appears to function in metaphase I primarily as an inhibitor of Polo. Therefore Cyclin B-Cdk1 activity and Polo inhibition are both necessary for maintaining a metaphase I arrest.

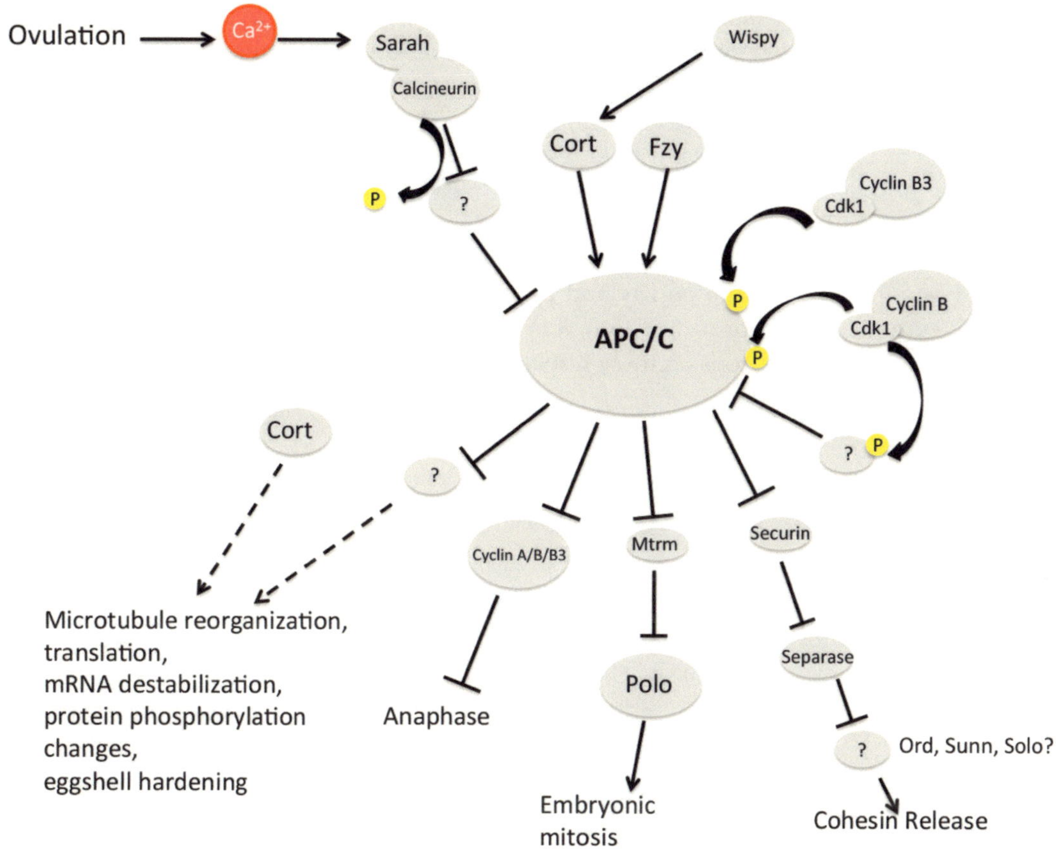


**Figure 2.** Model for regulation of and roles of the anaphase-promoting complex in *Drosophila* female meiosis. See text for details and references.

## 7. Egg activation and the resumption of meiosis

In *Drosophila*, ovulation is the trigger for egg activation, one of the major transition points in development. At activation, the egg undergoes multiple changes that set the stage for embryonic development. These include the completion of meiosis, global changes in mRNA stability, translation, protein phosphorylation, as well as changes in cytoskeletal organization and the completion of eggshell formation. All of these changes appear to require the *Cort* gene [40] (**Figure 2**).

*Cort* was originally discovered in a screen for maternal effect lethal mutations [57]. *Cort* mutants undergo normal metaphase I arrest at stage 14. Anaphase I is generally normal, though some eggs arrest at this point. The vast majority of *Cort* mutant eggs arrest in metaphase II [58, 59]. The cloning of *Cort* revealed it to encode a member of the Cdc20/Cdh1 family of APC/C activators. Cort interacts with the APC/C core and is required for the destruction of mitotic cyclins and Securin in the female germline [44, 60]. APC/$C^{Cort}$ appears to function in a partially redundant manner with the canonical APC/C$^{Fzy}$. *Fzy* hypomorphs arrest in anaphase II, with elevated cyclin levels, while *cort, fzy* double mutants arrest in meiosis I, with further elevated cyclins levels [44]. Though both APC/Cs target Cyclin B for destruction, they appear to have distinct sites of activity: Cort is responsible for spindle midzone degradation of Cyclin B, whereas Fzy is responsible for degradation of Cyclin B all along the spindle [44]. In addition to cyclins and Securin, APC/C$^{Cort}$ targets Mtrm for destruction, thereby promoting Polo activity that appears to be important for proper mitosis in the early embryo [61] (**Figure 2**). Cort recognizes a degron sequence on Mtrm that is related to but distinct from the D-box, found on Fzy and Fzr targeted proteins [61]. Given that APC/C$^{Cort}$ targets the mitotic cyclins for destruction, it is likely that it also recognizes canonical D-box and possibly KEN box degrons on these targets, but this has not been directly tested as yet.

As mentioned, Cort functions in multiple processes that depend on egg activation. Wild type oocytes contain arrays of microtubules around the cortex of the egg. These are broken down into shorter filaments at egg activation. *Cort* mutants fail to undergo this change in microtubule organization [58, 59]. How Cort affects this change in microtubule behavior is not known.

Cort is also implicated in the translation of specific mRNAs at egg activation. These include mRNA for patterning genes *Bicoid*, and possibly *Toll* and *Torso*. Reduced translation of *bicoid* mRNA correlates with reduced polyA tail length, suggesting that Cort promotes translation by promoting polyadenylation of specific mRNAs [59].

Cort is required for the destabilization of many mRNAs at the mid-blastula transition in cycle 14 of embryogenesis. At this transition, many maternal mRNAs are degraded and zygotic transcription is upregulated. Egg activation leads to a pathway in which the Pan gu (Png) kinase is activated and promotes the translation of the RNA-binding protein Smaug. Smaug is responsible for the destabilization of mRNAs at the transition to zygotic development [62]. It is currently unknown at what level in this pathway Cort functions.

Egg activation also involves global changes in protein phosphorylation, and Cort is implicated in a subset of these [63]. One protein that is dephosphorylated at egg activation dependent on Cort is Gnu, a component of the Png complex that is implicated in translational control at egg activation [63]. While the significance of this dephosphorylation is not clear, one possibility is that APC/C$^{Cort}$ activates the Png kinase complex by promoting dephosphorylation of Gnu, possibly by targeting for destruction a Gnu kinase. Png then promotes the translation of Smaug, leading to transcript destabilization. The identification of specific Cort targets at egg activation will be necessary to sort out the relationship amongst these Cort-dependent functions. It will also be important to determine if these functions of Cort depend on its role as an APC/C activator or if they represent novel functions of Cort.

## 8. Activation of the APC/C at ovulation

In many species, fertilization is the signal for egg activation, but in *Drosophila*, ovulation triggers this process (reviewed in [64]). Even though egg activation is not directly coupled to fertilization, the events are linked, as both depend on passage of the egg through the oviduct. When the egg enters the uterus it is positioned to allow sperm that is stored by the female after mating to enter the egg through an opening at the egg anterior called the micropyle. Meanwhile, ovulation triggers egg activation. The experimental application of physical pressure on a mature oocyte can induce egg activation, suggesting that physical pressure encountered during passage through the oviduct triggers egg activation [65]. Supporting this idea, inhibiter studies implicate mechanosensitive calcium channels, presumably activated during squeezing of the egg in the oviduct, in mediating an initial increase in calcium levels upon ovulation. This is followed by the IP3-mediated release of intracellular calcium stores in the egg. The result is a wave of calcium that passes through the oocyte cytoplasm [66]. The increase in calcium appears to then lead to activation of calcium-dependent enzymes such as Calcineurin (**Figure 2**).

The central role of Calcineurin in *Drosophila* egg activation was first revealed through the study of its interacting partner, Sarah [67–69]. Sarah is a positive regulator of Calcineurin in *Drosophila* meiosis, though it is a negative regulator in other contexts. *Sarah* mutants display a metaphase I arrest, and elevated Cyclin B, suggesting lack of APC/C activity. In addition, other events of egg activation are impaired, such as the translation of *bicoid* mRNA, and reorganization of cortical microtubules. *Calcineurin (CanB2)* mutants were subsequently found to have similar defects, indicating that *sarah* phenotypes are a result of loss of Calcineurin activity [67]. Phosphorylation of Sarah by Glycogen synthase kinase-3 (GSK-3/Shaggy) and by Polo, are necessary for Calcineurin activity, implicating these kinases as potential upstream regulators of Calcineurin, in addition to calcium [70].

In Xenopus, Calcineurin appears to promote anaphase by relieving APC/$C^{Fzy}$ of Emi2 inhibition [71, 72]. As discussed earlier, it is not yet clear if the *Drosophila* Emi, Rca1, is important for meiosis and therefore it is not clear if Calcineurin targets this protein or another to mediate APC/C activation (**Figure 2**).

The similarity between *CanB2* and *sarah* mutants on one hand and *cort* mutants on the other, support the idea that Calcineurin acts in a pathway with APC/$C^{Cort}$ (presumably upstream, as in Xenopus) to promote egg activation. The main difference between *calcineurin* and *cort* mutants is the timing of the meiotic arrest: meiosis II for *cort* mutants and meiosis I for Calcineurin mutants. As mentioned *cort, fzy* double mutants arrest in meiosis I [44], suggesting that Calcineurin is required upstream of both APC/C complexes.

While Cort regulates translation, possibly by affecting polyadenylation, Cort itself is subject to this form of regulation. Wispy is a female specific PolyA Polymerase required for translation, and transcript destabilization in *Drosophila* embryos [73, 74]. *Wispy* mutants arrest in meiosis I, at least in part due to failure to translate Cort [74]. A role in Fzy translation has not yet been reported (**Figure 2**).

In mitotic cells, APC/$C^{Fzy}$ activity is dependent on phosphorylation by Cdk1. Based on the mutant phenotype for *Cyclin B3*, Cyclin B3-Cdk1 may play this APC/C activating role in meiosis. *Cyclin B3* mutant females undergo meiotic arrest in metaphase or early anaphase of either meiosis I or II [10, 75]. Similarly, Cyclin B3 RNAi injection into early embryos results in an anaphase arrest, suggesting a role in APC/C activation in the mitotic divisions of the early embryo [75]. If Cyclin B3-Cdk1 activates the APC/C in meiosis, it remains to be determined if this involves direct phosphorylation of APC/C subunits, and whether APC/$C^{Cort}$ or APC/$C^{Fzy}$ or both are activated (**Figure 2**).

## 9. Chromosome cohesion and its release in meiosis

The key event of meiotic anaphase is the separation of attached homologs in meiosis I, and then sister chromatids in meiosis II. In mitotic cells in *Drosophila* and possibly all eukaryotes, sister chromatids are held together by the ring-like cohesin complex. At anaphase, the APC/C dependent cleavage the kleisin component of the cohesin complex (called Rad21), leads to chromosome segregation. In diverse organisms, from yeast to mammals, the cohesin complex is modified for meiosis. Most notably, a meiotic kleisin, Rec8, takes the place of Rad21. APC/$C^{Cdc20}$ activation at anaphase triggers the degradation of Securin, leading to Separase activation and consequent Separase-mediated cleavage of Rec8 (reviewed in [76]). Rec8 cleavage and cohesin disassembly in meiosis occurs in two steps. Prior to anaphase I, homologs are kept together by the combined effect of crossing over between homologs and sister chromatid cohesion distal to crossovers. At anaphase I, cohesion along chromatid arms is released to allow homolog segregation. At anaphase II, centromere-proximal cohesion is released to allow sister chromatids to separate [77].

As described above, the APC/C activator, Cort is required for anaphase II of female meiosis. FISH using an X-chromosome centromere-proximal probe revealed that these meiosis II figures each contain a single dot, indicating that centromere cohesion is maintained in *cort* mutants, implying that APC/$C^{Cort}$ is necessary for the release of centromere-proximal cohesion in anaphase II [44]. *Cort, fzy* double mutants as well as *Calcineurin* mutants produce an arrest in meiosis I [44], but it is not known if this arrest occurs prior to the release of arm cohesion. This could be easily tested using arm-specific FISH probes described below.

The roles of Securin and Separase in *Drosophila* meiosis were investigated, employing centromere-proximal and arm-specific FISH probes to monitor cohesion release in meiosis I and II [78]. The expression of a D-box, KEN-box mutant version of Securin in *Drosophila* oocytes that were depleted of endogenous Securin by RNAi, produced a delay or failure of homolog segregation in meiosis I and sister segregation in meiosis II. In the same study it was found that RNAi knockdown of the *Drosophila Separase* gene leads to an identical phenotype [78] (**Figure 2**). Interestingly, neither *Separase* knockdown nor stabilized Securin lead to a complete failure of cohesion release in meiosis, though both produced a complete and stable failure of cohesion release in the post-meiotic polar body chromosomes [78]. While it is possible that incomplete effects are due to a failure to completely inactivate Separase in these experiments, it could also be that a 2nd pathway for cohesion release operates in parallel with and partially redundant with the Securin/Separase pathway.

Knockdown of cohesin component *SMC3* leads to precocious homolog segregation (though not sister segregation) [78], implying a role for the cohesin complex in arm cohesion in *Drosophila*. Complicating this interpretation is the finding that many cohesin components, including SMC1 and SMC3 are important for synaptonemal complex (SC) assembly or maintenance [79]. The SC brings homologs into register in meiotic prophase and thus it is necessary for homolog pairing in meiosis I. Recent evidence has shown that core cohesin components are indeed required for cohesion, independent of their roles in SC assembly or maintenance. First, it was found that knockdown of either *SMC1* or the cohesin loading protein, *Eco*, leads to non-disjunction even in cases where a cross-over occurred [80]. Therefore, the non-disjunction is not due to failed synapsis and is presumably due to a failure of cohesion. Direct cytological evidence also supports an essential role for the core cohesin complex in meiotic cohesion. Using an arm-specific FISH probe, it was found that *SMC3* knockdown, but not knockdown of *Rad21*, results in absence of chromosome arm cohesion in meiotic prophase [78]. Interestingly, centromere-proximal cohesion appears to persist in these oocytes, either because of incomplete loss of cohesin at these sites or perhaps because a 2nd mechanism contributes to centromeric cohesion [78].

While the core components of the cohesin complex are required for meiotic cohesion, the identity of the Separase-cleavable component remains unknown. Fruit flies lack an obvious Rec8 homolog. A distant relative of Rec8, C(2)M, was found to function in the SC, but it appears to be released from chromosomes well before anaphase I. Furthermore, a form of C(2) M lacking putative Separase cleavage sites does not prevent anaphase, suggesting that C(2) M is not the cleavable kleisin [81]. Meanwhile, several lines of evidence demonstrate that the mitotic kleisin, Rad21, is also not the cleavable cohesin complex component in meiosis. First, a Separase cleavage site mutation in Rad21 fails to prevent anaphase in meiosis. Second, when the *Rad21* gene was replaced with a TEV-cleavable *Rad21* transgene, TEV cleavage prior to anaphase did not result in precocious anaphase [82]. Finally, knockdown of *Rad21* in meiosis did not lead to precocious release of arm cohesion [78].

While the Separase target in *Drosophila* meiosis is not known, 3 likely candidates have been identified, Ord, Sunn and Solo (**Figure 2**). All three genes were found to be required for proper homolog and sister chromatid disjunction in meiosis, and they each encode novel proteins that have been found to localize to meiotic chromosomes and to core cohesin complexes [83–87]. It will be important to assess these proteins for Separase-mediated cleavage, either by identifying and mutating putative Separase cleavage sites or by direct identification of cleavage products in post-meiotic eggs.

## 10. Conclusion

The ability to study mutant and knockdown phenotypes in female meiosis in recent years has led to a great advancement in our understanding of how cell cycle regulators work together to regulate meiosis. One of the big challenges for the future will be in discovering specific substrates for Cdk1 and other kinases, and for the APC/C. There is also a need to better understand how these central meiotic regulators are themselves regulated in meiosis.

## Author details

Mohammed Bourouh and Andrew Swan*

*Address all correspondence to: aswan@uwindsor.ca

Department of Biological Sciences, University of Windsor, Windsor, Ontario, Canada

## References

[1] Spradling AC. Developmental genetics of oogenessis. In: Bate M, Martinez-Arias A, editors. The Development of *Drosophila melanogaste*r. Cold Spring Harbor: Cold Spring Harbor laboratory Press; 1993. p. 1-70

[2] Bastock R, St Johnston D. Drosophila oogenesis. Current Biology. 2008;**18**(23):R1082-R1087

[3] Lee LA, Orr-Weaver TL. Regulation of cell cycles in Drosophila development: Intrinsic and extrinsic cues. Annual Review of Genetics. 2003;**37**:545-578

[4] Foe VE, Odell GM, Edgar BA. Mitosis and morphogenesis in the Drosophila embryo. In: Bate M, Martinez-Arias A, editors. The Development of *Drosophila melanogaster*. Cold Spring harbor: Cold Spring Harbor laboratory Press; 1993. p. 149-300

[5] Von Stetina JR, Orr-Weaver TL. Developmental control of oocyte maturation and egg activation in metazoan models. Cold Spring Harbor Perspectives in Biology. 2011;**3**(10): a005553

[6] Xiang Y, Takeo S, Florens L, Hughes SE, Huo LJ, Gilliland WD, et al. The inhibition of polo kinase by matrimony maintains G2 arrest in the meiotic cell cycle. PLoS Biology. 2007;**5**(12):e323

[7] Von Stetina JR, Tranguch S, Dey SK, Lee LA, Cha B, Drummond-Barbosa D. Alpha-Endosulfine is a conserved protein required for oocyte meiotic maturation in Drosophila. Development. 2008;**135**(22):3697-3706

[8] Ni JQ, Zhou R, Czech B, Liu LP, Holderbaum L, Yang-Zhou D, et al. A genome-scale shRNA resource for transgenic RNAi in Drosophila. Nature Methods. 2011;**8**(5):405-407

[9] Staller MV, Yan D, Randklev S, Bragdon MD, Wunderlich ZB, Tao R, et al. Depleting gene activities in early Drosophila embryos with the 'maternal-Gal4-shRNA' system. Genetics. 2013;**193**(1):51-61

[10] Bourouh M, Dhaliwal R, Rana K, Sinha S, Guo Z, Swan A. Distinct and overlapping requirements for Cyclins A, B and B3 in Drosophila female meiosis. G3 (Bethesda). 2016;**6**(11):3711-3724

[11] Jacobs HW, Knoblich JA, Lehner CF. Drosophila Cyclin B3 is required for female fertility and is dispensable for mitosis like Cyclin B. Genes & Development. 1998;**12**(23):3741-3751

[12] Sugimura I, Lilly MA. Bruno inhibits the expression of mitotic cyclins during the prophase I meiotic arrest of Drosophila oocytes. Developmental Cell. 2006;**10**(1):127-135

[13] Vardy L, Pesin JA, Orr-Weaver TL. Regulation of Cyclin A protein in meiosis and early embryogenesis. Proceedings of the National Academy of Sciences of the United States of America. 2009;**106**(6):1838-1843

[14] Gharbi-Ayachi A, Labbe JC, Burgess A, Vigneron S, Strub JM, Brioudes E, et al. The substrate of Greatwall kinase, Arpp19, controls mitosis by inhibiting protein phosphatase 2A. Science. 2010;**330**(6011):1673-1677

[15] Mochida S, Maslen SL, Skehel M, Hunt T. Greatwall phosphorylates an inhibitor of protein phosphatase 2A that is essential for mitosis. Science. 2010;**330**(6011):1670-1673

[16] Rangone H, Wegel E, Gatt MK, Yeung E, Flowers A, Debski J, et al. Suppression of scant identifies Endos as a substrate of greatwall kinase and a negative regulator of protein phosphatase 2A in mitosis. PLoS Genetics. 2011;**7**(8):e1002225

[17] Kim MY, Bucciarelli E, Morton DG, Williams BC, Blake-Hodek K, Pellacani C, et al. Bypassing the Greatwall-Endosulfine pathway: Plasticity of a pivotal cell-cycle regulatory module in *Drosophila melanogaster* and *Caenorhabditis elegans*. Genetics. 2012; **191**(4):1181-1197

[18] Archambault V, Zhao X, White-Cooper H, Carpenter AT, Glover DM. Mutations in Drosophila Greatwall/Scant reveal its roles in mitosis and meiosis and interdependence with Polo kinase. PLoS Genetics. 2007;**3**(11):e200

[19] Wang P, Pinson X, Archambault V. PP2A-twins is antagonized by greatwall and collaborates with polo for cell cycle progression and centrosome attachment to nuclei in drosophila embryos. PLoS Genetics. 2011;**7**(8):e1002227

[20] Von Stetina JR, Lafever KS, Rubin M, Drummond-Barbosa DA. Genetic screen for dominant enhancers of the cell-cycle regulator alpha-Endosulfine identifies matrimony as a strong functional Interactor in drosophila. G3 (Bethesda). 2011;**1**(7):607-613

[21] Wang P, Galan JA, Normandin K, Bonneil E, Hickson GR, Roux PP, et al. Cell cycle regulation of Greatwall kinase nuclear localization facilitates mitotic progression. The Journal of Cell Biology. 2013;**202**(2):277-293

[22] Wang P, Larouche M, Normandin K, Kachaner D, Mehsen H, Emery G, et al. Spatial regulation of greatwall by Cdk1 and PP2A-Tws in the cell cycle. Cell Cycle. 2016;**15**(4):528-539

[23] Petronczki M, Lenart P, Peters JM. Polo on the rise-from mitotic entry to cytokinesis with Plk1. Developmental Cell. 2008;**14**(5):646-659

[24] Das A, Shah SJ, Fan B, Paik D, DiSanto DJ, Hinman AM, et al. Spindle assembly and chromosome segregation requires central spindle proteins in drosophila oocytes. Genetics. 2016;**202**(1):61-75

[25] McKim KS, Jang JK, Manheim EA. Meiotic recombination and chromosome segregation in Drosophila females. Annual Review of Genetics. 2002;**36**:205-232

[26] Halpin D, Kalab P, Wang J, Weis K, Heald R. Mitotic spindle assembly around RCC1-coated beads in Xenopus egg extracts. PLoS Biology. 2011;**9**(12):e1001225

[27] Cesario J, McKim KS. RanGTP is required for meiotic spindle organization and the initiation of embryonic development in Drosophila. Journal of Cell Science. 2011;**124**(Pt 22): 3797-3810

[28] Carmena M, Wheelock M, Funabiki H, Earnshaw WC. The chromosomal passenger complex (CPC): From easy rider to the godfather of mitosis. Nature Reviews. Molecular Cell Biology. 2012;**13**(12):789-803

[29] Radford SJ, Jang JK, McKim KS. The chromosomal passenger complex is required for meiotic acentrosomal spindle assembly and chromosome biorientation. Genetics. 2012; **192**(2):417-429

[30] Colombie N, Cullen CF, Brittle AL, Jang JK, Earnshaw WC, Carmena M, et al. Dual roles of Incenp crucial to the assembly of the acentrosomal metaphase spindle in female meiosis. Development. 2008;**135**(19):3239-3246

[31] Radford SJ, Hoang TL, Gluszek AA, Ohkura H, McKim KS. Lateral and end-on kinetochore attachments are coordinated to achieve bi-orientation in Drosophila oocytes. PLoS Genetics. 2015;**11**(10):e1005605

[32] Jang JK, Rahman T, McKim KS. The kinesinlike protein Subito contributes to central spindle assembly and organization of the meiotic spindle in Drosophila oocytes. Molecular Biology of the Cell. 2005;**16**(10):4684-4694

[33] Piekny A, Werner M, Glotzer M. Cytokinesis: Welcome to the Rho zone. Trends in Cell Biology. 2005;**15**(12):651-658

[34] Resnick TD, Satinover DL, Macisaac F, Stukenberg PT, Earnshaw WC, Orr-Weaver TL, et al. INCENP and Aurora B promote meiotic sister chromatid cohesion through localization of the Shugoshin MEI-S332 in Drosophila. Developmental Cell. 2006;**11**(1):57-68

[35] Carmena M, Lombardia MO, Ogawa H, Earnshaw WC. Polo kinase regulates the localization and activity of the chromosomal passenger complex in meiosis and mitosis in *Drosophila melanogaster*. Open Biology. 2014;**4**(11):140162

[36] Nogueira C, Kashevsky H, Pinto B, Clarke A, Orr-Weaver TL. Regulation of centromere localization of the Drosophila Shugoshin MEI-S332 and sister-chromatid cohesion in meiosis. G3 (Bethesda). 2014;**4**(10):1849-1858

[37] Goto H, Kiyono T, Tomono Y, Kawajiri A, Urano T, Furukawa K, et al. Complex formation of Plk1 and INCENP required for metaphase-anaphase transition. Nature Cell Biology. 2006;**8**(2):180-187

[38] Kabeche L, Compton DA. Cyclin A regulates kinetochore microtubules to promote faithful chromosome segregation. Nature. 2013;**502**:110-113

[39] Peters JM. The anaphase promoting complex/cyclosome: A machine designed to destroy. Nature Reviews. Molecular Cell Biology. 2006;**7**(9):644-656

[40] Pesin JA, Orr-Weaver TL. Regulation of APC/C activators in mitosis and meiosis. Annual Review of Cell and Developmental Biology. 2008;**24**:475-499

[41] Reis A, Chang HY, Levasseur M, Jones KT. APCcdh1 activity in mouse oocytes prevents entry into the first meiotic division. Nature Cell Biology. 2006;**8**(5):539-540

[42] Malhotra S, Vinod PK, Mansfeld J, Stemmann O, Mayer TU. The Anaphase-Promoting Complex/Cyclosome is essential for entry into meiotic M-phase. Developmental Cell. 2016;**36**(1):94-102

[43] Raff JW, Jeffers K, Huang JY. The roles of Fzy/Cdc20 and Fzr/Cdh1 in regulating the destruction of cyclin B in space and time. The Journal of Cell Biology. 2002;**157**(7):1139-1149

[44] Swan A, Schupbach T. The Cdc20 (Fzy)/Cdh1-related protein, Cort, cooperates with Fzy in cyclin destruction and anaphase progression in meiosis I and II in Drosophila. Development. 2007;**134**(5):891-899

[45] Sako K, Suzuki K, Isoda M, Yoshikai S, Senoo C, Nakajo N, et al. Emi2 mediates meiotic MII arrest by competitively inhibiting the binding of Ube2S to the APC/C. Nature Communications. 2014;**5**:3667

[46] Grosskortenhaus R, Sprenger F. Rca1 inhibits APC-Cdh1(Fzr) and is required to prevent cyclin degradation in G2. Developmental Cell. 2002;**2**(1):29-40

[47] JQ W, Kornbluth S. Across the meiotic divide—CSF activity in the post-Emi2/XErp1 era. Journal of Cell Science. 2008;**121**(Pt 21):3509-3514

[48] Ivanovska I, Lee E, Kwan KM, Fenger DD, Orr-Weaver TL. The Drosophila MOS ortholog is not essential for meiosis. Current Biology. 2004;**14**(1):75-80

[49] Lara-Gonzalez P, Westhorpe FG, Taylor SS. The spindle assembly checkpoint. Current Biology. 2012;**22**(22):R966-R980

[50] Touati SA, Wassmann K. How oocytes try to get it right: Spindle checkpoint control in meiosis. Chromosoma. 2016;**125**(2):321-335

[51] Gilliland WD, Hughes SE, Cotitta JL, Takeo S, Xiang Y, Hawley RS. The multiple roles of mps1 in Drosophila female meiosis. PLoS Genetics. 2007;**3**(7):e113

[52] Malmanche N, Owen S, Gegick S, Steffensen S, Tomkiel JE, Sunkel CE. Drosophila BubR1 is essential for meiotic sister-chromatid cohesion and maintenance of synaptonemal complex. Current Biology. 2007;**17**:1489-1497

[53] Buffin E, Emre D, Karess RE. Flies without a spindle checkpoint. Nature Cell Biology. 2007;**9**:565-572

[54] Batiha O, Swan A. Evidence that the spindle assembly checkpoint does not regulate APC(Fzy) activity in Drosophila female meiosis. Genome/National Research Council Canada = Genome/Conseil national de recherches Canada. 2012;**55**(1):63-67

[55] Shao H, Li R, Ma C, Chen E, Liu XJ. Xenopus oocyte meiosis lacks spindle assembly checkpoint control. The Journal of Cell Biology. 2013;**201**(2):191-200

[56] Gorr IH, Boos D, Stemmann O. Mutual inhibition of separase and Cdk1 by two-step complex formation. Molecular Cell. 2005;**19**(1):135-141

[57] Schupbach T, Wieschaus E. Female sterile mutations on the second chromosome of *Drosophila melanogaster*. I. Maternal effect mutations. Genetics. 1989;**121**(1):101-117

[58] Page AW, Orr-Weaver TL. The Drosophila genes grauzone and cortex are necessary for proper female meiosis. Journal of Cell Science. 1996;**109**:1707-1715

[59] Lieberfarb ME, Chu T, Wreden C, Theurkauf W, Gergen JP, Strickland S. Mutations that perturb poly(A)-dependent maternal mRNA activation block the initiation of development. Development. 1996;**122**(2):579-588

[60] Pesin JA, Orr-Weaver TL. Developmental role and regulation of cortex, a meiosis-specific anaphase-promoting complex/cyclosome activator. PLoS Genetics. 2007;**3**(11):e202

[61] Whitfield ZJ, Chisholm J, Hawley RS, Orr-Weaver TL. A meiosis-specific form of the APC/C promotes the oocyte-to-embryo transition by decreasing levels of the polo kinase inhibitor matrimony. PLoS Biology. 2013;**11**(9):e1001648

[62] Tadros W, Goldman AL, Babak T, Menzies F, Vardy L, Orr-Weaver T, et al. SMAUG is a major regulator of maternal mRNA destabilization in Drosophila and its translation is activated by the PAN GU kinase. Developmental Cell. 2007;**12**(1):143-155

[63] Krauchunas AR, Sackton KL, Wolfner MF. Phospho-regulation pathways during egg activation in *Drosophila melanogaster*. Genetics. 2013;**195**(1):171-180

[64] Horner VL, Wolfner MF. Transitioning from egg to embryo: Triggers and mechanisms of egg activation. Developmental Dynamics: An Official Publication of the American Association of Anatomists. 2008;**237**(3):527-544

[65] Horner VL, Wolfner MF. Mechanical stimulation by osmotic and hydrostatic pressure activates Drosophila oocytes in vitro in a calcium-dependent manner. Developmental Biology. 2008;**316**(1):100-109

[66] Kaneuchi T, Sartain CV, Takeo S, Horner VL, Buehner NA, Aigaki T, et al. Calcium waves occur as Drosophila oocytes activate. Proceedings of the National Academy of Sciences of the United States of America. 2015;**112**(3):791-796

[67] Takeo S, Hawley RS, Aigaki T. Calcineurin and its regulation by Sra/RCAN is required for completion of meiosis in Drosophila. Developmental Biology. 2010;**344**(2):957-967

[68] Horner VL, Czank A, Jang JK, Singh N, Williams BC, Puro J, et al. The Drosophila calcipressin sarah is required for several aspects of egg activation. Current Biology. 2006;**16**(14):1441-1446

[69] Takeo S, Tsuda M, Akahori S, Matsuo T, Aigaki T. The calcineurin regulator sra plays an essential role in female meiosis in Drosophila. Current Biology. 2006;**16**(14):1435-1440

[70] Takeo S, Swanson SK, Nandanan K, Nakai Y, Aigaki T, Washburn MP, et al. Shaggy/glycogen synthase kinase 3beta and phosphorylation of Sarah/regulator of calcineurin are essential for completion of Drosophila female meiosis. Proceedings of the National Academy of Sciences of the United States of America. 2012;**109**(17):6382-6389

[71] Nishiyama T, Yoshizaki N, Kishimoto T, Ohsumi K. Transient activation of calcineurin is essential to initiate embryonic development in *Xenopus laevis*. Nature. 2007;**449**(7160):341-345

[72] Machida YJ, Dutta A. The APC/C inhibitor, Emi1, is essential for prevention of rereplication. Genes & Development. 2007;**21**(2):184-194

[73] Cui J, Sackton KL, Horner VL, Kumar KE, Wolfner MF. Wispy, the Drosophila homolog of GLD-2, is required during oogenesis and egg activation. Genetics. 2008;**178**(4):2017-2029

[74] Benoit P, Papin C, Kwak JE, Wickens M, Simonelig M. PAP- and GLD-2-type poly(A) polymerases are required sequentially in cytoplasmic polyadenylation and oogenesis in Drosophila. Development. 2008;**135**(11):1969-1979

[75] Yuan K, O'Farrell PH. Cyclin B3 is a mitotic Cyclin that promotes the metaphase-anaphase transition. Current Biology. 2015;**25**(6):811-816

[76] Peters JM, Nishiyama T. Sister chromatid cohesion. Cold Spring Harbor Perspectives in Biology. 2012;**4**(11)

[77] Revenkova E, Jessberger R. Keeping sister chromatids together: Cohesins in meiosis. Reproduction. 2005;**130**(6):783-790

[78] Guo Z, Batiha O, Bourouh M, Fifield E, Swan A. Role of Securin, Separase and Cohesins in female meiosis and polar body formation in drosophila. Journal of Cell Science. 2016;**129**(3):531-542

[79] Tanneti NS, Landy K, Joyce EF, McKim KSA. Pathway for synapsis initiation during zygotene in Drosophila oocytes. Current Biology: CB. 2011;**21**(21):1852-1857

[80] Weng KA, Jeffreys CA, Bickel SE. Rejuvenation of meiotic cohesion in oocytes during prophase I is required for chiasma maintenance and accurate chromosome segregation. PLoS Genetics. 2014;**10**(9):e1004607

[81] Heidmann D, Horn S, Heidmann S, Schleiffer A, Nasmyth K, Lehner CF. The Drosophila meiotic kleisin C(2)M functions before the meiotic divisions. Chromosoma. 2004;**113**(4):177-187

[82] Urban E, Nagarkar-Jaiswal S, Lehner CF, Heidmann SK. The Cohesin subunit Rad21 is required for Synaptonemal complex maintenance, but not sister chromatid cohesion, during drosophila female meiosis. PLoS Genetics. 2014;**10**(8):e1004540

[83] Bickel SE, Wyman DW, OrrWeaver TL. Mutational analysis of the Drosophila sister-chromatid cohesion protein ORD and its role in the maintenance of centromeric cohesion. Genetics. 1997;**146**(4):1319-1331

[84] Yan R, McKee BD. The cohesion protein SOLO associates with SMC1 and is required for synapsis, recombination, homolog bias and cohesion and pairing of centromeres in Drosophila meiosis. PLoS Genetics. 2013;**9**(7):e1003637

[85] Yan RH, Thomas SE, Tsai JH, Yamada Y, McKee BD. SOLO: A meiotic protein required for centromere cohesion, coorientation, and SMC1 localization in *Drosophila melanogaster*. The Journal of Cell Biology. 2010;**188**(3):335-349

[86] Khetani RS, Bickel SE. Regulation of meiotic cohesion and chromosome core morphogenesis during pachytene in Drosophila oocytes. Journal of Cell Science. 2007;**120**(Pt 17):3123-3137

[87] Krishnan B, Thomas SE, Yan R, Yamada H, Zhulin IB, McKee BD. Sisters unbound is required for meiotic centromeric cohesion in *Drosophila melanogaster*. Genetics. 2014;**198**(3):947-965

Chapter 4

# Non-genetic Transgenerational Inheritance of Acquired Traits in *Drosophila*

Brian Xia and J. Steven de Belle

Additional information is available at the end of the chapter

http://dx.doi.org/10.5772/intechopen.71643

**Abstract**

It is increasingly recognized that acquired traits may be transgenerationally transmitted through non-DNA sequence-based elements, with epigenetics as perhaps the most important mechanism. Here we review examples of non-genetic transgenerational inheritance in *Drosophila*, highlighting transgenerational programming of metabolic status and longevity, one particular histone modification as an evolutionarily conserved underlying mechanism, and important implications of such studies in understanding health and diseases.

**Keywords:** aging, *Drosophila*, H3K27me3, metabolic state, PRC2, transgenerational epigenetic inheritance

## 1. Introduction

Epigenetics is the science of non-DNA sequence-based modifications of gene expression and, subsequently, phenotypic variability at both the genomic and organismal levels [1]. Studies over the past several decades have distinguished DNA methylation, histone modification, and non-coding RNA-based processes as the key mechanisms underlying epigenetic regulation. Epigenetic inheritance has been observed across species, including prokaryotes, plants, and animals [2–8], with an epigenetic trait defined as "a stably heritable phenotype resulting from changes in a chromosome without alterations in the DNA sequence" [9]. Interestingly, certain epigenetically-regulated phenotypes can propagate across multiple generations, leading to the concept of transgenerational epigenetic inheritance (TEI) [4–8]. This emerging concept has triggered numerous debates and revived old controversies in the scientific community as to whether acquired traits may be transmitted across generations. Nonetheless, it has profoundly reshaped our understanding of biology, particularly human diseases, as

| Year | Intervention/treatment (F0 only) | Phenotypic/genomic response | Generation with effect | Authors |
|---|---|---|---|---|
| 2007 | Tumor suppressor gene mutation | Tumor risk | F2 but not F3 | Xing et al. |
| 2009 | Chronic pentylenetetrazole treatment of adult males | Transcriptomic profile in CNS | F2 | Sharma and Singh |
| 2010 | Old age | Memory loss | F2 | Burns and Mery |
| 2012 | Low male availability during mating | Number of offspring (to quantify fitness) | F2 & F3 | Brommer et al. |
| 2013 | Post-eclosion feeding of virgin females with a high-sugar diet | Body composition in larvae | F2 | Buescher et al. |
| 2015 | Gamma radiation in young adult males | Longevity & rate of development | F2 but not F3 | Shameer et al. |
| 2015 | Yeast concentration in diets used to raise larvae through development | Somatic rDNA instability & copy number variation | F2 & up to F60 | Aldrich and Maggert |
| 2016 | Post-eclosion feeding of both virgin males and females with various diets | Longevity & reproduction | F2 & F3 | Xia and de Belle |
| 2016 | Extended olfactory training with young adults | Approach bias to the same trained odors | F2 | Williams |
| 2016 | High fat diet to raise larvae through development | Pupal body weight | F2 | Dew-Budd et al. |
| 2016 | Different food conditions used to raise male larvae and adults | Longevity | F2 | Roussou et al. |
| 2016 | Post-eclosion dietary, genetic and pharmacological treatments of both virgin males and females | Longevity & H3K27me3 levels | F2 | Xia et al. |
| 2017 | Epialleles, as defined by differential levels of H3K27me3 | Eye color | F5 & up to F10 | Ciabrelli et al. |
| 2017 | Grandmaternal age | Embryonic & embryonic to adult viability | F2* | Bloch Qazi et al. |
| 2017 | Genetic manipulation of parental metabolism | Triglyceride levels & transcriptional profile | F2 | Palu et al. |

*Potential transgenerational effects were not clearly-defined and quantified.

**Table 1.** Primary research papers describing TEI in ***Drosophila*** where phenotypic and/or genomic responses were investigated in the F2 or later generations.

stable epigenetic marks may record environmental challenges through modified gene expression patterns and ensure long-lasting, while reversible responses in the absence of the initial triggering events [10–17]. Importantly, the adaptive and reversible nature of epigenetic regulation may offer exciting therapeutic targets to help prevent or treat most, if not all, chronic diseases, including cardiovascular disease (CVD), diabetes, neurodegenerative diseases, and cancers [10–13, 16, 18–20].

The fruit fly (*Drosophila melanogaster*) offers multiple advantages for assaying TEI, in particular to characterize the underlying epigenetic mechanisms, and to identify gene targets for drug discovery. First, the short rearing period and lifespan of fruit flies facilitate transgenerational experiments over multiple generations within a reasonable time scale. Second, various examples of transgenerational inheritance have been established in *Drosophila* (**Table 1**) that enable rapid identification and characterization of underlying epigenetic mechanisms. Third, all major epigenetic mechanisms are present in this model system [1], although DNA methylation in flies appears to be different from many other eukaryotic organisms and is present only at very low levels in adults [21, 22]. Importantly, $N^6$-methyladenine may complement the function of DNA methylation in flies [23]. Finally, *Drosophila* has been increasingly used for modeling human diseases and drug discovery [24–28]. The *Drosophila* heart has been used to model several different aspects of human CVDs, including congenital heart disease and cardiomyopathy [29–31]. *Drosophila* is a recently-established model system for obesity and diabetes [26, 32, 33]. It has also been widely used to model cognitive diseases [34, 35], and various cancers [36].

TEI has been thoroughly reviewed, focusing mostly on data obtained from mammals [5, 6, 8, 37–39]. Here, such studies from *Drosophila* are discussed, in particular, to highlight transgenerational programming of metabolic status and longevity, and tri-methylation of histone H3 at lysine 27 (H3K27me3) as an evolutionarily conserved epigenetic mechanism underlying TEI.

## 2. Transgenerational inheritance at the organismal level

### 2.1. Metabolism

The current Western diet has been defined by increased consumption of meat products, dairy items, grains, and sugar-infused drinks [40]. Having profound effects on glycemic load, fatty acid composition, macronutrient composition, micronutrient density, acid-base balance, sodium-potassium ratio, and fiber content, this diet may underlie the growing prevalence of chronic diseases in Western society, especially CVD, obesity, diabetes, and dementia [41–44]. Often, multiple conditions manifest themselves simultaneously in afflicted individuals, suggesting shared elements in disease pathology. Obesity and other metabolic disorders, for example, are associated with various secondary disease indications as the underlying cellular and organismal metabolism is fundamental to nearly all necessary biological processes [45]. The prominent role of nutrition and other environmental factors in the development of metabolic disorders offers a promising model to identify and characterize the underlying epigenetic mechanisms, leading to diet optimization and nutrition-responsive therapies to combat chronic diseases (cf. [42]). Thus, nutrition has been studied extensively regarding TEI of diabetes and other metabolic disorders across various animal models [14, 46–48]. Metabolic dysfunctions are often measured through development and glucose/insulin homeostasis after nutritional or dietary interventions including overnutrition, high-fat, low-protein (LP), and high-sugar (HS) diets. Typically, the well-controlled application of dietary manipulations and well-established hallmarks of various metabolic disorders offer a tractable yet indispensable approach to studying TEI in many animal models.

*Drosophila* shares key metabolic pathways and characteristics with vertebrates (cf. [29, 32]). Glucose has been well-studied in the context of metabolic status given its pivotal role in insulin signaling since the 1930s [49]. *Drosophila* utilizes trehalose, the disaccharide of glucose, as its primary form of hemolymph (insect equivalent of blood) sugar [50, 51]. Regulating glycometabolism and maintaining viability in response to shifting external factors [51], trehalose is broken down through the catalyzing activity of trehalase into accessible glucose molecules. Thus, hemolymph trehalose and glucose levels may be quantified to assay glycometabolism in *Drosophila* [29, 32]. As the primary source of body fat from *Drosophila* to humans [52], triglycerides (TAGs) may be quantified for monitoring gluconeogenesis, the metabolic pathway responsible for glucose generation from non-carbohydrate substrates [29, 32]. Both AKT and 4EBP proteins are phosphorylated in response to insulin signaling [53, 54]. AKT is particularly well-characterized as a core component of the PI3K/AKT/mTOR pathway, which is linked to cell cycle regulation, cancer, and longevity [55]. Quantification of phosphorylated-AKT and phosphorylated-4EBP levels has been used to measure insulin sensitivity or resistance [29]. The availability of these assays to characterize both metabolic homeostasis and underlying pathways has supported the use of *Drosophila* to examine TEI of metabolic status after nutritional or genetic manipulations in the founding (F0) generation [32, 56, 57].

Buescher et al. recorded elevated trehalose, glycogen, and TAG levels as well as reduced body weight in adult female F0 flies after feeding on an HS diet for 7 days post-eclosion [32]. Glucose levels were found to be affected by the HS treatment, suggestive of gluconeogenesis dysregulation. Interestingly, trehalose and glucose levels were elevated in the first generation (F1) male larvae, along with a decrease in glycogen levels. Consistently, gene expression analyses demonstrated decreased expression of the genes involved in fat body lipolysis and gluconeogenesis, and increased expression of the ones involved in gut lipolysis, fatty acid synthesis, sugar transport and glycolysis. These results have confirmed the traditional models of insulin signaling, in which impaired insulin sensitivity leads to global increases in circulating blood sugars and decreases in sugar storage. Both glucose and trehalose levels were elevated, with TAG unaffected in the F2 male larvae; trehalose was elevated while TAG was decreased, with glucose unaffected in the F2 female larvae, supporting the existence of gender-dependent differences in transgenerational inheritance of metabolic programming. These results have demonstrated the long-lasting and transgenerational effects of early-life (post-eclosion) nutrition on metabolic status, establishing *Drosophila* as a useful model system to study TEI of nutritional programming of metabolic homeostasis and disorders.

Then, Dew-Budd et al. assayed the effects of gender and genetic lineage on transgenerational inheritance of certain metabolic phenotypes after rearing male (F0; paternal ancestry) and female (maternal ancestry) larvae of 10 (to measure pupal body weight) or 3 (metabolic composition and egg size) independent genetic lines on a high-fat diet [57]. Substantial differences in body weight, metabolic composition, or egg size were observed in both F1 and F2 generations between paternal and maternal ancestries or among different F0 genotypes. Interestingly, phenotypic changes in the F0 flies appeared not to be a consistent predictor of these hallmarks in their untreated F1 and F2 descendants. Therefore, "personalized" consideration of ancestral contributions may be needed to understand and prevent metabolic diseases such as obesity and diabetes.

Palu et al. have employed loss-of-function mutants to induce obesity, assayed with elevated TAGs, in F0 parents and then check TAG levels in heterozygous F1 and wild-type F2 offspring [56]. Loss of *AKHR* (encoding adipokinetic hormone receptor) leads to reduced fat body lipid mobilization and elevated TAG accumulation, as adipokinetic hormone functions analogously to the fasting hormone glucagon in mammals [52]. Mutant *AKHR* F0 and wild-type flies in reciprocal crosses produced heterozygous F1 offspring. These F1 heterozygotes were then crossed to wild-type females or males to generate four types of genetically distinct wild-type F2 (+/+) progeny, corresponding to mutant *AKHR* F0 grandpaternal or grandmaternal and heterozygous F1 paternal or maternal ancestors. Both male and female F0 mutants displayed elevated TAG levels, which were then normalized in the F1 heterozygotes, possessing a functional copy of *AKHR*. Interestingly, this Mendelian model of inheritance was not always followed in the F2 generation with low TAG levels observed in the grandpaternal/maternal group, while normal in the other three groups. Consistently, *ACC*, encoding a conserved Acetyl-CoA carboxylase that acts as the rate-limiting step in fatty acid synthesis [58], was found to be dysregulated in this particular F2 group. These results suggest that genetic manipulation of parental metabolism can provide an effective approach for studying TEI of metabolic state.

### 2.2. Aging

Aging has been increasingly recognized as a malleable process and the largest risk factor for most aging-related diseases (ARDs). It is no accident that the rapid increase in life expectancy worldwide is concomitant with the epidemic progression of many of these life-threatening and costly diseases [59, 60]. Recent work has demonstrated that many factors, including environmental conditions (e.g., diet) and genetic mutations, can impact the aging process across species [61–63]. In particular, anti-aging interventions often delay or prevent multiple ARDs in animal models [62–64], stimulating the emerging interdisciplinary field of geroscience to study the connection between aging and diseases, and to develop novel multi-disease preventative and therapeutic interventions by targeting the aging process itself [59, 65]. There are clear practical and ethical complications associated with studying aging and its transgenerational inheritance directly in human populations. The timescale of conducting such longitudinal studies would be unreasonable, at best. The shortage of isogenic replicates (e.g., twins) and imprecise environmental manipulation in human models also pose a significant problem in terms of reproducibility and subsequent mechanistic studies.

*Drosophila* presents itself as an excellent model to study aging, especially its transgenerational inheritance and the underlying mechanisms, owing to its relatively short lifespan, genetic homology with other models and humans, and suite of enriched investigative tools. *Drosophila* has an average lifespan of 2–3 months yet undergoes key parallel developmental stages similar to those of humans [25]. Studies on its life cycle have revealed a number of highly conserved pathways involved in organismal development. The *Hox* genes, for example, which control segment identity during embryonic development, were first identified in *Drosophila* after observation of mutant flies growing legs in the place of antennae [66]. *Hox* genes were later found to be conserved in humans and also linked to congenital disorders, including synpolydactyly and hand-foot-genital syndrome [67–69]. In addition, the key aging pathways, including mechanistic target of rapamycin, sirtuin, and insulin/insulin growth factor 1 signaling, are well conserved in fruit flies [70–72]. Finally, tissue-specific and time-dependent genetic manipulations may be readily achievable for most genes, for instance, using the 22,270 transgenic lines (currently, covering ~88% of all predicted protein-coding genes) from the Vienna *Drosophila* Resource Center [73]. Therefore, *Drosophila* is well-suited for both correlational and mechanistic studies focusing on transgenerational programming of longevity after nutritional or environmental manipulations in the F0 parents [74–77].

Gamma radiation causes DNA damage and mutations, leading to various health dysfunctions and subsequent lifespan reduction [78, 79]. High doses of gamma irradiation were found to decrease longevity in the F0 flies and further propagate to the F1 and F2, but not to the F3 generation [76]. In contrast, low doses extended longevity across the F0–F2 generations, consistent with the concept of hormesis, by which low exposure to harmful agents (irradiation, caloric restriction, heat stress, and free radicals) improves general health and longevity [79–81]. Related studies have revealed several underlying mechanisms including insulin and glucose metabolism, proteasome activity and histone deacetylation [81, 82]. Histone deacetylation may be particularly relevant in this context as an epigenetic modification involved with many biological processes and human diseases, including CVD, metabolic disorders, and cancers [83–85].

Our recent work has established the first animal model of early-life nutrition-mediated programming of longevity and its transgenerational inheritance [74]. Newly-eclosed F0 virgin flies were reared on one of three different diets (low-protein or LP, intermediate-protein, and high-protein) for the first 7 days post-eclosion. Longevity was assayed for males and females, both virgin and mated, across the F0–F2 generations, allowing us to determine the potential impact of gender and mating on transgenerational inheritance of longevity. Our results suggest that early-life nutrition-induced programming effects on longevity may be transmitted to the F1 generation through intergenerational effects and further to the F2 generation through transgenerational effects, independently of gender and mating. The programming effects, although diminishing, were still present in the F3 generation for the low- and intermediate-protein diets. These observations suggest that early-life nutrition may produce long-lasting and transgenerationally heritable effects on the aging process across multiple generations. Notably, these long-lasting programming effects may be derived from both maternal and paternal contributions, as we treated both newly-eclosed F0 males and females to induce potentially maximal alterations. In contrast, a similar treatment was applied only to the females to examine transgenerational programming of metabolic status [32]. Most rodent studies also used either males or females, instead of both [48]. This design would not distinguish potentially different contributions from males and females, something that requires further investigation. Interestingly, transgenerational glucose intolerance in mice (*Mus musculus*) may be transmitted via the maternal or paternal line through different mechanisms [86, 87], suggesting that transgenerational nutritional programming effects may potentially be additive when induced in both males and females.

A more recent study has demonstrated that distinct dietary manipulations in the larval stage or throughout adulthood may also induce transgenerational programming of longevity [75]. The F2 male offspring were found to be long-lived if F0 male adults were subjected to dietary restriction, but not to starvation, whereas the same outcome was observed if F0 male larvae were exposed to starvation, but not to dietary restriction. The authors also generated two separate groups of F2 males, from the F1 male (paternal) or female (maternal) offspring of the F0 male larvae exposed to various food media. Extended longevity was observed in both groups of F2 males, but greater extension was seen in the F2 maternal males with one laboratory strain. By contrast, the starvation-induced transgenerational effects were observed only in the F2 paternal males with a different strain. Therefore, cross-generational inheritance of nutrition-mediated longevity changes may be passed through either the male or female line or both, depending on genetic background. Unfortunately, it is unclear whether the observed gender-dependent differences resulted from intergenerational or transgenerational inheritance, as longevity was not assessed in the F0 and F1 generations.

### 2.3. Fitness

Fitness refers to the reproductive success of an organism over the duration of its lifetime, and has often been linked to genetic regulation. Recent studies, though sparse, have prompted the idea that non-genetic or epigenetic mechanisms may modulate fitness across generations [74, 88–90]. Studying the interplay between genetics and epigenetics through fitness may help us understand various complex traits and disorders [89]. *Drosophila* is particularly suitable for studying TEI of fitness for its rapid maturation following eclosion and high fecundity among model organisms [91].

Brommer et al. have reported that sexual conflict (male availability) may impact the fitness of future progeny up to the F3 generation [90]. Female fitness was quantified by lifetime production of offspring, and male fitness by total offspring produced in a six-day period. For the F0 generation, female flies underwent either a low (one male for 1 day followed by no male for 3 days) or high male (one male for 1 day followed by a different male for 3 days) exposure treatment. This four-day cycle was repeated for the duration of the females' lifespan to measure lifetime fecundity. The same process was repeated for the F1 and F2 generations, thus producing eight groups of F3 flies with distinct ancestral history. All F3 generation daughters experienced the treatment of high male exposure. All comparisons, when made relative to the low versus high male treatments experienced by the F0 females, provided a measure of transgenerational inheritance of fitness. The results indicated that low male exposure treatment in the F0 females did not affect female fecundity across the F1–F3 generations, but increased male fitness in the F1 generation and decreased male fitness in the F2 and F3 generations.

In the same study where we assayed transgenerational nutrition-mediated programming of longevity (see above), we also explored the transgenerational effects of the same early-life diets on lifetime fecundity (egg production) as a measure of fitness and the potential trade-off between longevity and fecundity [74]. Lifetime fecundity was found to be decreased across the F0–F2 generations after raising the F0 virgin male and female flies on the LP diet for 7 days before their mating, while increased transgenerationally after the same treatment with the intermediate-protein diet. Fecundity was also increased in the F0 and F1 generations after the same treatment with the high-protein diet, but the increasing effect was not seen in the F2 generation. These results demonstrate that early-life dietary changes affect fitness of the same generation and the reproductive success of future generations with certain dietary changes. Interestingly, correlation analyses on longevity and fecundity data revealed no evidence for trade-off between them across the F0–F2 generations. This finding argues that lab-raised flies, with abundant food supplies at all times, may have evolved to abandon such trade-off constraints through hundreds of generations. Therefore, transgenerational nutritional programming of fitness may be achieved independently of longevity, raising the interesting possibility of elevating both longevity and fitness with proper nutrition across generations.

Bloch Qazi et al. recently reported the cross-generational effects of grandmaternal and maternal age on offspring viability and development up to the F2 generation [88]. The study, however, appeared not to distinguish between intergenerational (grandmaternal to maternal and maternal to F2 offspring) and transgenerational (grandmaternal to F2) effects. The complicated design with three interacting factors (i.e., grandmaternal age, maternal age, and stress) and subsequent analyses with mixed-model ANOVAs made it challenging to make conclusions about a straight forward transgenerational effect, although the P value was smaller than 0.05 in three of analyses for the "grandmaternal" factor (in the presence of the intergenerational effect or "maternal" factor).

### 2.4. Memory

Many behavioral traits, including cognitive functions, may be transgenerationally affected by experiences and environmental factors in mammals, most likely through epigenetic mechanisms [92]. Memory is an essential cognitive function which declines during aging and is impaired in most neurodegenerative diseases such as Alzheimer's disease; it is subjected to various epigenetic regulations, providing novel therapeutic avenues to combat cognitive disorders [12, 93]. Therefore, studying TEI of memory is of immense importance to our understanding of mental health and diseases. A *Drosophila* memory TEI model is established by two recent studies [94, 95] and further corroborated by a similar report in which increased startle responses to the conditioned odor after paternal F0 olfactory fear conditioning was observed in the subsequent adult F1 and F2 mice [96].

A widely-used dual-odor discriminative Pavlovian conditioning assay involves training groups of flies to associate one odor ($CS^+$; conditioned stimulus) with aversive electric or mechanical shocks (US; unconditioned stimulus), and the other odor ($CS^-$) as a non-associative control [97–99]. Aged (25-day-old) flies produced F1 offspring with memory impairment detectable in young adults (3–5 days old), and this impairment was transmitted to the F2 generation [95]. The transgenerational effect was specific to short-term memory (STM; as tested 15 min after training), and appeared to be caused by oxidative stress in both F0 maternal and paternal flies. Although the same authors did not evaluate memory in aged F0 parents, an earlier study [100] demonstrated that aging specially impaired middle-term memory (MTM), which starts to form within 15 min after training and is considered to be an aging-sensitive component of STM [101, 102]. In addition to concluding that offspring cognitive ability may be influenced by parental age [95], these studies collectively argue that aged F0 parents may acquire a loss of oxidative stress-sensitive STM, and this acquired memory loss can be transgenerationally inherited at least to the F2 generation. This new explanation also provides a possible mechanistic direction for future investigation, as MTM formation requires normal function of the *amnesiac* (*amn*) gene that encodes a precursor neuropeptide encompassing fly homologs of mammalian pituitary adenylate cyclase activating peptide (PACAP) and growth hormone-releasing hormone (GHRH; see below for further discussion) [103].

In a more recent study, F1 and F2 flies, without any training and prior exposure, displayed selective preference toward the same CS odors which were used during 5 days of discriminative training of F0 parents [94]. This preference was selective for the salient CS odors experienced by the F0 parents but not the specific CS-US association, as the F1 and F2 flies did not differentiate between odors that were originally used to train their F0 parents under an aversive (with electric shocks as US) *vs.* appetitive (with corn meal and sucrose as US) conditions. Consistently, discriminative conditioning appeared to increase the perceived salience of the $CS^+$ odors [104]. Importantly, the observed odor-selective preference in the F1 flies required normal function of *amn* and preserved function of dorsal paired medial neurons in which *amn* is predominantly expressed [105]. Thus, the *amn* gene may be involved with transgenerational inheritance of acquired loss of STM in aged F0 parents [95] and odor-selective preference from discriminative training in the F0 flies [94]. In agreement with this idea, PACAP and/or GHRH stimulate growth hormone release [106], while down-regulation of growth hormone may be involved with cross-generational toxicity [107]. The *amn* gene also plays an important role in the behavioral response to intoxicating levels of alcohol [108], while alcohol abuse has been known to be transgenerationally heritable [109]. Collectively, these studies support *Drosophila* as a useful model to study transgenerational inheritance of memory impairment triggered by environmental factors (e.g., aging) and behavioral traits acquired from experiences (e.g., training), and epigenetic regulation of *amn*-encoded peptides as one potential underlying mechanism.

## 3. Transgenerational inheritance at the molecular and genomic level

Despite advancement of high-throughput sequencing and the recent surge of research on TEI, there are currently few studies focusing on the transgenerational effects at the molecular and genomic level, and thus the underlying mechanisms remain largely obscure [6, 8, 37–39, 92, 110]. Several recent studies in flies, however, may shed some light on this situation [77, 111–115].

Chronic treatment (7-day feeding and 7-day withdraw) of the F0 males with pentylenetetrazole (PTZ), an FDA-revoked convulsant drug, caused locomotor deficits and long-term alterations in the CNS (central nervous system) transcriptome [116]. A follow-up study from the same group [113] demonstrated that the F0 males (with PTZ treatment) displayed a CNS transcriptomic profile closest to the F2 males; and differentially expressed genes in the F1 males, F1 females, and F2 males showed significant overlap with the PTZ-impacted genes in the F0 males. Interestingly, further clustering analysis of CNS and testis transcriptome profiles and concordant analysis of differentially expressed genes between them implied gametic involvement in the observed transgenerational effect in gene expression. These results suggest that the acquired somatic transcriptomic alteration in F0 PTZ-treated males may be passed via sperm at least to the F2 generation. This is the first report to study transgenerational inheritance of genome-wide transcriptomic profile as a "phenotype," acquired through drug treatment in the F0 generation.

In another study, a high-protein diet led to somatic rDNA instability and copy number reduction in F0 parental flies [111]. As the insulin/insulin-like growth factor and TOR signaling pathways regulate ribosome biogenesis and rDNA expression for nutrient availability [117], genetic and pharmacological manipulation of insulin/TOR signaling produced similar effects, corroborating the results from dietary treatment. Importantly, rDNA copy number reduction remained in the F2 generation and was still present in flies maintained on standard food for 6 years. These results suggest that the genome rearrangement in F0 flies acquired through feeding on the high-protein diet occurred in both somatic and germ cells, and was transgenerationally heritable for over 150 generations. This outcome revealed a robust and long-lasting transgenerational consequence of adult diets. In a remarkable recent study, early-life protein restriction in mice induced a linear correlation between growth restriction and DNA methylation at certain rDNA copies that lasted into adulthood [118]. These findings, establishing rDNA as a genomic target of nutritional availability across species, are of obvious importance for human health and diseases, as copy number variations have been linked to many chronic diseases such as schizophrenia and Alzheimer's disease [119–121].

Another curious study has shown that a dominant and hyperactive mutation in the *hopscotch* gene (*Hop*$^{Tum-l}$), encoding the *Drosophila* JAK kinase, caused epigenetic alterations in F0 parental flies that were transgenerationally heritable and thus influenced tumorigenesis in their F1 and F2 offspring [114]. Interestingly, the transcriptional repressor *Krueppel*, known to repress transcription of the *fushi-tarazu* gene which encodes a homeodomain protein required for embryonic segment number and cell fate [122], is a *Hop*$^{Tum-l}$ enhancer [123]. *Krueppel* mutations caused increased DNA methylation in the *fushi-tarazu* promoter region. This effect was transmitted across generations in the presence of *Hop*$^{Tum-l}$ [114]. Therefore, DNA methylation may be altered by *Krueppel* mutations, functioning as heritable epigenetic markings in *Drosophila*. JAK hyper-activation may then interfere with epigenetic reprogramming, allowing the changed DNA methylation (epimutation) to propagate across generations and influence tumor susceptibility.

## 4. Polycomb repressive complex 2 (PRC2) mediates H3K27me3 as a conserved epigenetic mechanism underlying transgenerational inheritance

Despite decades of intense studies linking all key types of epigenetic regulation (i.e., DNA methylation, histone modifications and non-coding RNAs) to TEI, direct and convincing experimental evidence in support of underlying mechanisms and governing principles is rare [2–8, 37–39]. The difficulties lie in the time-consuming nature of such studies, and lack of well-established epimutations, clearly-defined phenotypic contributions and stably-inherited epigenetic markings across multiple generations. Here we highlight two recent persuasive studies in *Drosophila* that have characterized one particular histone modification (H3K27me3) as part of an evolutionarily conserved epigenetic mechanism underling transgenerational inheritance [77, 112].

H3K27me3 is a repressive methylation mark on histone H3 established by PRC2 through its core catalytic subunit, the H3K27-specific methyltransferase encoded by the E(z) gene in flies [124] and EZH2 in mammals [125]. PRC2 is evolutionarily conserved across species, including unicellular alga (*Chlamydomonas reinhardtii*) and budding yeast (*Cryptococcus neoformans*) [124–127]. Genes marked with higher-than-normal levels of H3K27me3 in human and mouse spermatozoa continue to show repression during gametogenesis, embryogenesis, and development, suggestive of a role of this histone modification during TEI [128–130]. Furthermore, paternal diet affects H3K27me3 marks at specific loci in their offspring, implying that such nutrition-induced epigenetic modifications may be selectively retained across generations in mice [131]. Finally, TEI of longevity has been reported for H3K4me3 in worm (*Caenorhabditis elegans*) [132], and the bivalent chromatin domains covered by H3K27me3 and H3K4me3 marks have been implicated in aging and ARDs in humans [133, 134]. These results collectively suggest that H3K27me3 may function as an evolutionarily conserved epigenetic mechanism underlying transgenerational inheritance. Our recent work and that of Ciabrelli et al. have directly validated the concept in the context of nutrition-mediated longevity programming, transgene expression, and endogenous genetic variation [77, 112]. Further strengthening the idea, H3K27me3 markings have been found to propagate across generations from the maternal (and likely paternal) germline and survive reprogramming events during early embryogenesis in flies [115].

Our most recent study examined E(z)-mediated H3K27me3 as one potential epigenetic mechanism underlying transgenerational inheritance of longevity [77]. It was prompted by our earlier work to establish nutritional programming of longevity and its transgenerational inheritance [74], and by recent studies supporting the notion that PRC2-mediated H3K27me3 may regulate aging across species. H3K27me3 repressive markings and an epigenomic PRC2 signature marked by EZH2 and SUZ12 (another core component of PRC2) binding have been found to be associated with age-associated differentially methylated regions and aging-associated genes in human embryonic stem cells and various other cell lines, implicating this repressive epigenetic marker as a common mechanism of aging in humans [135]. Consistently, Polycomb repression is associated with healthy aging in humans [136], and replicative senescence of stem cells, an *in vitro* aging model [137, 138]. H3K27me3 and H3K4me3 are also the frequent antagonistic partners found on the bivalent chromatin domains which may be implicated in aging and ARDs in humans [133, 134]. In addition, heterozygous mutations of E(z) increase longevity while also reducing H3K27me3 levels in adult flies, suggesting that PRC2-dependent H3K27me3 may regulate aging in *Drosophila* [139]. Interestingly, E(z)-mediated H3K27me3 is required for paternal transmission of obesity through reprogramming of metabolic genes in flies [140], supporting its potential role in transgenerational reprogramming. Finally, UTX-1 (an H3K27-specific histone demethylase) has been shown to regulate aging, and H3K4me3-mediated TEI of longevity has been reported in *C. elegans* [132, 141].

E(z) protein level was significantly upregulated in F0 flies, and back to normal in F2 flies, after post-eclosion treatment of F0 flies with the LP diet [77]. In contrast, the resulting increase of E(z)-dependent H3K27me3 was seen in the F0 parents and their F2 offspring. Correspondingly, longevity was reduced in both F0 and F2 flies. These results suggest that early-life dietary insults may trigger E(z)-mediated H3K27me3 changes via misregulation of E(z), and consequently nutrition-induced H3K27me3 dysfunction may be transmitted across generations and underlie TEI of nutritional programming of longevity. First, E(z)-mediated H3K27me3 was found to be necessary for TEI of longevity programming, as early-life RNAi-mediated specific knockdown of E(z) only in the F0 parents extended longevity while reducing H3K27me3 activity, and early-life specific inhibition of E(z) enzymatic function with EPZ-6438 (a highly EZH2-selective inhibitor) also extended lifespan while rendering the H3K27me3 level low across generations. Importantly, the effects of RNAi-mediated knockdown on H3K27me3 and longevity were specific, as (I) similar effects were observed with two independent RNAi transgenes, (II) the E(z) protein level was normal in the F2 generation after its knockdown in the F0 parents, and (III) longevity, E(z), and H3K27me3 levels were not affected without heat shock to induce RNAi transgenes. Similarly, the EPZ-6438-induced effects were specific, as (I) EPZ-6438, as a phase II clinical drug, is highly EZH2 selective and considered safe [142], and (II) E(z) protein was unaffected by EPZ-6438 even in the F0 parents. In addition, H3K27me3 was found to be sufficient for TEI of longevity programming, as EPZ-6438 greatly alleviated the longevity-reducing effect of the LP diet, while counterbalancing its upregulation of H3K27me3 across the F0 to F2 generations. Our data have convincingly demonstrated that E(z)-mediated H3K27me3 activity may play a critical role in the general health of an organism and function as one epigenetic mechanism underlying TEI of early-life nutrition-mediated longevity programming. Our findings have also provided the first proof-of-concept for an epigenetic therapy to confer transgenerational health benefits in a model system, manifested through improved longevity.

Another important aspect of our study was early-life rather than adult-oriented interventions. The critical period refers to a time frame in which an organism's nervous system is especially susceptible to environmental modification. This phenomenon is common to nearly all multicellular model organisms as it primes the organism to environmental stimuli and programs physiological pathways responsible for maintaining general health. Studies have linked abnormalities in the critical period to the development of autism spectrum disorder [143], attention deficit hyperactivity disorder [144], schizophrenia [145, 146], obesity [147], and other ARDs [148]. Indeed, the Developmental Origins of Health and Disease hypothesis (DOHaD) postulates that the current mainstream adult-oriented therapies may be less efficacious than those delivered during the developmental phases of life [149, 150]. Our study has provided direct validation of this concept through the delivery of EPZ-6438 at various time points throughout adult life to alleviate LP-induced longevity reduction. The alleviation effect was found to be greatest, intermediate, or very mild when the drug was delivered within the first 7 days, from day 3–10, or from day 10–17 after eclosion, respectively. The effect was even seen in the F2 generation when the inhibitor was delivered within the first 7 days post-eclosion. These data support the DOHaD approach for studying ARDs in *Drosophila* and the use of a developmentally appropriate time period for intervention. Our follow-up experiments indicated that early-life administration of EPZ-6438 can also prevent multiple LP-induced ARDs (i.e., cardiomyopathy, type 2 diabetes, and aging-related memory loss) throughout adult life.

This represents a novel proof-of-concept of an early-life multi-disease therapy, leveraging epigenetic reprogramming to provide life-long protection against multiple – possibly all – ARDs (Xia et al., unpublished results).

To study epigenetic phenomena in flies, Ciabrelli et al. employed a transgene inserted in chromosome arm *2 L* (*Fab2L*) to establish stable and isogenic epilines that carried distinct epialleles as defined by differential levels of PRC2-dependent H3K27me3 [112]. The *Fab2L* transgene contains the reporter gene *mini-white*, whose expression determines red pigmentation in the eye, under the control of *Fab-7*, a 3.6-kb genomic region that includes a PRE (Polycomb response element). Despite being located on a different chromosome (*3R*), the endogenous *Fab-7* region can affect PRE-responsive repression of the *Fab2L* transgene through long-range 3D chromatin interactions [151, 152], producing variable *mini-white* expression-dependent eye colors among individual flies. These epigenetic differences were somatic and not transgenerationally heritable, but enhancing long-range interactions between *Fab2L* and the endogenous *Fab-7* through removal of one copy of *Fab-7* induced a plastic epigenetic state, allowing the authors to establish the stable and isogenic epilines with the most repressed (white) or the most derepressed (red) eye phenotypes through 15 generations of selection for eye color.

Their subsequent characterization indicated that (I) these epilines carried either silent or active epialleles of *Fab2L*, as determined by high or low levels of PRC2-responsive H2K27me3; (II) these epialleles could be stably and dominantly transmitted to naïve flies, with acquired epigenetic states stably maintained at least until the F10 generations through self-crossing; (III) epiallele maintenance required 3D chromatin interactions, with both epialleles fully and specifically reversed to a non-selective state after complete removal of the endogenous *Fab-7*; (IV) epiallele inheritance also followed the rules of paramutation under natural environment conditions, with environmental factors (e.g., temperature and humidity) affecting the phenotypes of the epialleles; and (V) the paradigm could apply to a naturally occurring phenotype (i.e., antenna-to-leg homeotic transformation [153]) of a spontaneous neomorphic mutation of the homeotic *Antennapedia* gene. This important work, with well-established stable and isogenic epialleles as defined by distinct levels of H3K27me3 markings, has overcome many shortcomings of earlier studies of transgenerational inheritance, such as weak effects fading away within a few generations, ill-defined contributions to the observed phenotypes, and unclear epigenetic markings (cf. [114, 154]). The results have convincingly demonstrated stable transgenerational H3K27me3-mediated inheritance of transgene expression and endogenous genetic variation in fruit flies [112], corroborating our study of establishing the same epigenetic mechanism underlying transgenerational inheritance of nutrition-programmed longevity [77].

In this mode of TEI, PRC2 functions through H3K27me3 repressive markers to acquire specific epigenetic states in response to environmental stimuli or triggers. Alternative states are defined by different levels of H3K27me3 to affect gene expression and epigenetic phenotypes [77, 112, 131]. Polycomb-mediated repression at specific loci and/or long-range chromatin interactions act together to maintain acquired states *in cis* [112], and distinct levels of H3K27me3, as deposited in the maternal oocytes [155], resist epigenetic reprogramming during early embryogenesis and are transmitted across generations, enabling transgenerational inheritance of acquired states and phenotypes [115]. The acquisition and establishment of epigenetic states may occur rapidly during developmentally appropriate time periods [77, 131] or gradually through phenotypic selection [112]. Deposit of H3K27me3 appears to be locus-specific in response to environmental factors (cf. [131]). The extent and robustness of its inheritance may be environmental factor- and trait-dependent, with the transgenerational effects upon acquired complex traits (e.g., aging) quickly adapting to further environmental changes and decaying away in a few generations (cf. [77]), or upon simple traits (e.g., transgene expression) being relatively resistant to further environmental modifications and transmitting across many generations (cf. [112]).

## 5. Conclusion

*Drosophila* as a versatile model organism is profoundly advancing our understanding of TEI and its underlying mechanisms. Short lifespan, well-conserved epigenetic mechanisms, and powerful genetic tools have facilitated TEI studies at molecular, genomic, and organismal levels after various environmental and genetic manipulations (**Table 1**). Many studies have employed dietary interventions at the larval or early-adult life stages, or throughout adulthood, similar to those in mammals [48, 156]. Early-life nutrition in particular has been linked to adult health and diseases, prompting the increasingly-recognized DOHaD approach for studying various ARDs including CVD, obesity, diabetes, dementia, and certain cancers [4, 150, 156]. Importantly, these existing TEI models have enabled exciting investigations of the underlying molecular and epigenetic mechanisms. Here, we have highlighted PRC2-mediated H3K27me3 markings as an evolutionarily conserved epigenetic mechanism underlying transgenerational inheritance [77, 112, 115].

## 6. Recommendations

TEI research is a relatively new science. H3K27me3-mediated inheritance is providing a platform to address many important questions about TEI in future studies. What are the signals and underlying molecular mechanisms responding to the initial environmental stimuli? How do these signals trigger an epigenetic process and establish corresponding epigenetic states? How can such specific epigenetic states, likely originating in somatic cells, be transmitted to germ cells to enable transgenerational inheritance? What are the molecular mechanisms that maintain transgenerational inheritance? Is H3K27me3 unique in that it may resist epigenetic reprogramming [115]? Is H3K27me3 a common epigenetic mechanism responsible for non-genetic transgenerational inheritance across species? We anticipate that the *Drosophila* model will continue to broaden our understanding of TEI biology and related human diseases in particular.

## Acknowledgements

We thank Drs. Dustin Schones and Wendong Huang (City of Hope Cancer Research Center) for valuable discussion and comments, and Dr. Shouzhen Xia (Dart Neuroscience LLC) for critical reading and editing of our manuscript. This work was supported by internal funding to SD.

## Author details

Brian Xia[1]* and J. Steven de Belle[2]

*Address all correspondence to: bxia@mit.edu

1 Massachusetts Institute of Technology, Cambridge, MA, USA

2 Dart Neuroscience LLC, San Diego, CA, USA

## References

[1] Allis DC, Jenuwein T, Reinberg D, editors. Epigenetics. New York: CSHL Press; 2007. 500 p

[2] Eaton SA, Jayasooriah N, Buckland ME, Martin DI, Cropley JE, Suter CM. Roll over Weismann: Extracellular vesicles in the transgenerational transmission of environmental effects. Epigenomics. 2015;**7**:1165-1171. DOI: 10.2217/epi.15.58

[3] Chen Q, Yan W, Duan E. Epigenetic inheritance of acquired traits through sperm RNAs and sperm RNA modifications. Nature Reviews. Genetics. 2016;**17**:733-743. DOI: 10.1038/nrg.2016.106

[4] Hanson MA, Skinner MK. Developmental origins of epigenetic transgenerational inheritance. Environ Epigenet. 2016;**2**:dvw002. DOI: 10.1093/eep/dvw002

[5] Miska EA, Ferguson-Smith AC. Transgenerational inheritance: Models and mechanisms of non-DNA sequence-based inheritance. Science. 2016;**354**:59-63. DOI: 10.1126/science.aaf4945

[6] Sharma A. Transgenerational epigenetics: Integrating soma to germline communication with gametic inheritance. Mechanisms of Ageing and Development. 2017;**163**:15-22. DOI: 10.1016/j.mad.2016.12.015

[7] Jablonka E, Raz G. Transgenerational epigenetic inheritance: Prevalence, mechanisms, and implications for the study of heredity and evolution. The Quarterly Review of Biology. 2009;**84**:131-176

[8] Gapp K, Bohacek J. Epigenetic germline inheritance in mammals: Looking to the past to understand the future. Genes, Brain, and Behavior. 2017. DOI: 10.1111/gbb.12407

[9] Berger SL, Kouzarides T, Shiekhattar R, Shilatifard A. An operational definition of epigenetics. Genes & Development. 2009;**23**:781-783. DOI: 10.1101/gad.1787609

[10] Bay JL, Morton SM, Vickers MH. Realizing the potential of adolescence to prevent transgenerational conditioning of noncommunicable disease risk: Multi-sectoral design frameworks. Healthcare (Basel). 2016;**4**:E39. DOI: 10.3390/healthcare4030039

[11] Benedetti R, Conte M, Iside C, Altucci L. Epigenetic-based therapy: From single- to multi-target approaches. The International Journal of Biochemistry & Cell Biology. 2015;**69**:121-131. DOI: 10.1016/j.biocel.2015.10.016

[12] Delgado-Morales R, Agís-Balboa RC, Esteller M, Berdasco M. Epigenetic mechanisms during ageing and neurogenesis as novel therapeutic avenues in human brain disorders. Clinical Epigenetics. 2017;**9**:67. DOI: 10.1186/s13148-017-0365-z

[13] Li C, Casanueva O. Epigenetic inheritance of proteostasis and ageing. Essays in Biochemistry. 2016;**60**:191-202. DOI: 10.1042/EBC20160025

[14] Stegemann R, Buchner DA. Transgenerational inheritance of metabolic disease. Seminars in Cell & Developmental Biology. 2015;**43**:131-140. DOI: 10.1016/j.semcdb.2015.04.007

[15] Sun C, Burgner DP, Ponsonby A-L, Saffery R, Huang R-C, Vuillermin PJ, Cheung M, Craig JM. Effects of early-life environment and epigenetics on cardiovascular disease risk in children: Highlighting the role of twin studies. Pediatric Research. 2013;**73**:523-530. DOI: 10.1038/pr.2013.6

[16] Van der Harst P, de Windt LJ, Chambers JC. Translational perspective on epigenetics in cardiovascular disease. Journal of the American College of Cardiology 2017;**70**:590-606. DOI: 10.1016/j.jacc.2017.05.067

[17] Vickers MH. Developmental programming and transgenerational transmission of obesity. Annals of Nutrition & Metabolism. 2014;**64**(Suppl 1):26-34. DOI: 10.1159/000360506

[18] Chen Z, Li S, Subramaniam S, Shyy JY-J, Chien S. Epigenetic regulation: A new frontier for biomedical engineers. Annual Review of Biomedical Engineering. 2017;**19**:195-219. DOI: 10.1146/annurev-bioeng-071516-044720

[19] Rasool M, Malik A, Naseer MI, Manan A, Ansari S, Begum I, Qazi MH, Pushparaj PN, Abuzenadah AM, Al-Qahtani MH, Kamal MA, Pushparaj PN, Gan SH. The role of epigenetics in personalized medicine: Challenges and opportunities. BMC Medical Genomics. 2015;**8**(Suppl 1):S5. DOI: 10.1186/1755-8794-8-S1-S5

[20] Yan W, Herman JG, Guo M. Epigenome-based personalized medicine in human cancer. Epigenomics. 2016;**8**:119-133. DOI: 10.2217/epi.15.84

[21] Capuano F, Mülleder M, Kok R, Blom HJ, Ralser M. Cytosine DNA methylation is found in *Drosophila melanogaster* but absent in *Saccharomyces cerevisiae, Schizosaccharomyces pombe,* and other yeast species. Analytical Chemistry. 2014;**86**:3697-3702. DOI: 10.1021/ac500447w

[22] Panikar CS, Rajpathak SN, Abhyankar V, Deshmukh S, Deobagkar DD. Presence of DNA methyltransferase activity and CpC methylation in *Drosophila melanogaster*. Molecular Biology Reports. 2015;**42**:1615-1621. DOI: 10.1007/s11033-015-3931-5

[23] Zhang G, Huang H, Liu D, Cheng Y, Liu X, Zhang W, Yin R, Zhang D, Zhang P, Liu J, Li C, Liu B, Luo Y, Zhu Y, Zhang N, He S, He C, Wang H, Chen D. N6-methyladenine DNA modification in *Drosophila*. Cell. 2015;**161**:893-906. DOI: 10.1016/j.cell.2015.04.018

[24] Konsolaki M.Fruitful research: Drug target discovery for neurodegenerative diseases in *Drosophila*. Expert Opinion on Drug Discovery. 2013;**8**:1503-1513. DOI: 10.1517/17460441.2013.849691

[25] Pandey UB, Nichols CD. Human disease models in *Drosophila melanogaster* and the role of the fly in therapeutic drug discovery. Pharmacological Reviews. 2011;**63**:411-436. DOI: 10.1124/pr.110.003293

[26] Graham P, Pick L. *Drosophila* as a model for diabetes and diseases of insulin resistance. Current Topics in Developmental Biology. 2017;**121**:397-419. DOI: 10.1016/bs.ctdb.2016.07.011

[27] Millburn GH, Crosby MA, Gramates LS, Tweedie S. FlyBase consortium. FlyBase portals to human disease research using *Drosophila* models. Disease Models & Mechanisms. 2016;**9**:245-252. DOI: 10.1242/dmm.023317

[28] Ugur B, Chen K, Bellen HJ. *Drosophila* tools and assays for the study of human diseases. Disease Models & Mechanisms. 2016;**9**:235-244. DOI: 10.1242/dmm.023762

[29] Na J, Musselman LP, Pendse J, Baranski TJ, Bodmer R, Ocorr K, Cagan R. A *Drosophila* model of high sugar diet-induced cardiomyopathy. PLoS Genetics. 2013;**9**:e1003175. DOI: 10.1371/journal.pgen.1003175

[30] Ocorr K, Vogler G, Bodmer R. Methods to assess *Drosophila* heart development, function and aging. Methods. 2014;**68**:265-272. DOI: 10.1016/j.ymeth.2014.03.031

[31] Zhu J-Y, Fu Y, Nettleton M, Richman A, Han Z. High throughput in vivo functional validation of candidate congenital heart disease genes in *Drosophila*. eLife. 2017;**6**:e22617. DOI: 10.7554/eLife.22617

[32] Buescher JL, Musselman LP, Wilson CA, Lang T, Keleher M, Baranski TJ, Duncan JG. Evidence for transgenerational metabolic programming in *Drosophila*. Disease Models & Mechanisms. 2013;**6**:1123-1132. DOI: 10.1242/dmm.011924

[33] Brookheart RT, Duncan JG. *Drosophila melanogaster*: An emerging model of transgenerational effects of maternal obesity. Molecular and Cellular Endocrinology. 2016;**435**:20-28. DOI: 10.1016/j.mce.2015.12.003

[34] Akinola OB. Sweet old memories: A review of the experimental models of the association between diabetes, senility and dementia. Metabolic Brain Disease. 2016;**31**:1003-1010. DOI: 10.1007/s11011-016-9876-0

[35] Fernandez-Funez P, de Mena L, Rincon-Limas DE. Modeling the complex pathology of Alzheimer's disease in *Drosophila*. Experimental Neurology 2015;**274**:58-71. DOI: 10.1016/j.expneurol.2015.05.013

[36] Sonoshita M, Cagan RL. Modeling human cancers in *Drosophila*. Current Topics in Developmental Biology. 2017;**121**:287-309. DOI: 10.1016/bs.ctdb.2016.07.008

[37] Blake GE, Watson ED. Unravelling the complex mechanisms of transgenerational epigenetic inheritance. Current Opinion in Chemical Biology. 2016;**33**:101-107. DOI: 10.1016/j.cbpa.2016.06.008

[38] Van Otterdijk SD, Michels KB. Transgenerational epigenetic inheritance in mammals: How good is the evidence? The FASEB Journal. 2016;**30**:2457-2465. DOI: 10.1096/fj.201500083

[39] Prokopuk L, Western PS, Stringer JM. Transgenerational epigenetic inheritance: Adaptation through the germline epigenome? Epigenomics. 2015;**7**:829-846. DOI: 10.2217/epi.15.36

[40] Halton TL, Willett WC, Liu S, Manson JE, Stampfer MJ, Potato HFB. French fry consumption and risk of type 2 diabetes in women. The American Journal of Clinical Nutrition. 2006;**83**:284-290

[41] Hsu TM, Kanoski SE. Blood-brain barrier disruption: Mechanistic links between western diet consumption and dementia. Frontiers in Aging Neuroscience. 2014;**6**:88. DOI: 10.3389/fnagi.2014.00088

[42] Medina-Remón A, Kirwan R, Lamuela-Raventós RM, Estruch R. Dietary patterns and the risk of obesity, type 2 diabetes mellitus, cardiovascular diseases, asthma, and neurodegenerative diseases. Critical Reviews in Food Science and Nutrition. 2016:1-35. DOI: 10.1080/10408398.2016.1158690

[43] Martinez KB, Leone V, Chang EB. Western diets, gut dysbiosis, and metabolic diseases: Are they linked? Gut Microbes. 2017;**8**:130-142. DOI: 10.1080/19490976.2016.1270811

[44] Cordain L, Eaton SB, Sebastian A, Mann N, Lindeberg S, Watkins BA, O'Keefe JH, Brand-Miller J. Origins and evolution of the western diet: Health implications for the 21st century. The American Journal of Clinical Nutrition. 2005;**81**:341-354

[45] San Martin R, Brito J, Siques P, León-Velarde F. Obesity as a conditioning factor for high-altitude diseases. Obesity Facts. 2017;**10**:363-372. DOI: 10.1159/000477461

[46] Jimenez-Chillaron JC, Ramon-Krauel M, Ribo S, Diaz R. Transgenerational epigenetic inheritance of diabetes risk as a consequence of early nutritional imbalances. The Proceedings of the Nutrition Society. 2016;**75**:78-89. DOI: 10.1017/S0029665115004231

[47] Frias AE, Grove KL. Obesity: A transgenerational problem linked to nutrition during pregnancy. Seminars in Reproductive Medicine. 2012;**30**:472-478. DOI: 10.1055/s-0032-1328875

[48] Aiken CE, Ozanne SE. Transgenerational developmental programming. Human Reproduction Update. 2014;**20**:63-75. DOI: 10.1093/humupd/dmt043

[49] Falta W, Boller R. Insulärer und insulinresistenter diabetes. Klinische Wochenschrift. 1931;**10**:438-443. DOI: 10.1007/BF01736348

[50] Becker A, Schlöder P, Steele JE, Wegener G. The regulation of trehalose metabolism in insects. Experientia. 1996;**52**:433-439

[51] Matsuda H, Yamada T, Yoshida M, Nishimura T. Flies without trehalose. The Journal of Biological Chemistry. 2015;**290**:1244-1255. DOI: 10.1074/jbc.M114.619411

[52] Grönke S, Müller G, Hirsch J, Fellert S, Andreou A, Haase T, Jäckle H, Kühnlein RP. Dual lipolytic control of body fat storage and mobilization in *Drosophila*. PLoS Biology. 2007;**5**: e137. DOI: 10.1371/journal.pbio.0050137

[53] Teleman AA, Hietakangas V, Sayadian AC, Cohen SM. Nutritional control of protein biosynthetic capacity by insulin via Myc in *Drosophila*. Cell Metabolism. 2008;**7**:21-32. DOI: 10.1016/j.cmet.2007.11.010

[54] Boucher J, Kleinridders A, Kahn CR. Insulin receptor signaling in normal and insulin-resistant states. Cold Spring Harbor Perspectives in Biology. 2014;**6**:a009191. DOI: 10.1101/cshperspect.a009191

[55] Risso G, Blaustein M, Pozzi B, Mammi P, Srebrow A. Akt/PKB: One kinase, many modifications. The Biochemical Journal. 2015;**468**:203-214. DOI: 10.1042/BJ20150041

[56] Palu RAS, Praggastis SA, Thummel CS. Parental obesity leads to metabolic changes in the F2 generation in *Drosophila*. Molecular Metabolism. 2017;**6**:631-639. DOI: 10.1016/j.molmet.2017.03.012

[57] Dew-Budd K, Jarnigan J, Reed LK. Genetic and sex-specific transgenerational effects of a high fat diet in *Drosophila melanogaster*. PLoS One. 2016;**11**:e0160857. DOI: 10.1371/journal.pone.0160857

[58] Parvy J-P, Napal L, Rubin T, Poidevin M, Perrin L, Wicker-Thomas C, Montagne J. *Drosophila melanogaster* acetyl-CoA-carboxylase sustains a fatty acid-dependent remote signal to waterproof the respiratory system. PLoS Genetics. 2012;**8**:e1002925. DOI: 10.1371/journal.pgen.1002925

[59] Kennedy BK, Berger SL, Brunet A, Campisi J, Cuervo AM, Epel ES, Franceschi C, Lithgow GJ, Morimoto RI, Pessin JE, Rando TA, Richardson A, Schadt EE, Wyss-Coray T, Sierra F. Geroscience: Linking aging to chronic disease. Cell. 2014;**159**:709-713. DOI: 10.1016/j.cell.2014.10.039

[60] Partridge L. Intervening in ageing to prevent the diseases of ageing. Trends in Endocrinology and Metabolism. 2014;**25**:555-557. DOI: 10.1016/j.tem.2014.08.003

[61] Gale CR, Hagenaars SP, Davies G, Hill WD, Liewald DCM, Cullen B, Penninx BW, International Consortium for Blood Pressure GWAS, CHARGE Consortium Aging and Longevity Group, Boomsma DI, Pell J, AM MI, Smith DJ, Deary IJ, Harris SE. Pleiotropy between neuroticism and physical and mental health: Findings from 108 038 men and women in UK Biobank. Translational Psychiatry. 2016;**6**:e791. DOI: 10.1038/tp.2016.56

[62] Arriola Apelo SI, Lamming DW. Rapamycin: An inhibiTOR of aging emerges from the soil of Easter Island. The Journals of Gerontology. Series A, Biological Sciences and Medical Sciences. 2016;**71**:841-849. DOI: 10.1093/gerona/glw090

[63] Fontana L, Partridge L. Promoting health and longevity through diet: From model organisms to humans. Cell. 2015;**161**:106-118. DOI: 10.1016/j.cell.2015.02.020

[64] Gillespie ZE, Pickering J, Eskiw CH. Better living through chemistry: Caloric restriction (CR) and CR mimetics alter genome function to promote increased health and lifespan. Frontiers in Genetics. 2016;**7**:142. DOI: 10.3389/fgene.2016.00142

[65] Sierra F. The emergence of geroscience as an interdisciplinary approach to the enhancement of health span and life span. Cold Spring Harbor Perspectives in Medicine. 2016;**6**: a025163. DOI: 10.1101/cshperspect.a025163

[66] Lappin TRJ, Grier DG, Thompson A, Halliday HL. HOX genes: Seductive science, mysterious mechanisms. The Ulster Medical Journal. 2006;**75**:23-31

[67] Quinonez SC, Innis JW. Human HOX gene disorders. Molecular Genetics and Metabolism. 2014;**111**:4-15. DOI: 10.1016/j.ymgme.2013.10.012

[68] Innis JW. Role of HOX genes in human development. Current Opinion in Pediatrics. 1997;**9**:617-622

[69] Goodman FR, Scambler PJ. Human HOX gene mutations. Clinical Genetics. 2001;**59**:1-11

[70] Kapahi P, Zid BM, Harper T, Koslover D, Sapin V, Benzer S. Regulation of lifespan in *Drosophila* by modulation of genes in the TOR signaling pathway. Current Biology. 2004;**14**:885-890. DOI: 10.1016/j.cub.2004.03.059

[71] Frankel S, Ziafazeli T, Rogina B. dSir2 and longevity in *Drosophila*. Experimental Gerontology. 2011;**46**:391-396. DOI: 10.1016/j.exger.2010.08.007

[72] Partridge L, Alic N, Bjedov I, Piper MD. Aging in *Drosophila*: The role of the insulin/Igf and TOR signalling network. Experimental Gerontology. 2011;**46**:376-381. DOI: 10.1016/j.exger.2010.09.003

[73] Dietzl G, Chen D, Schnorrer F, Su K-C, Barinova Y, Fellner M, Gasser B, Kinsey K, Oppel S, Scheiblauer S, Couto A, Marra V, Keleman K, Dickson BJ. A genome-wide transgenic RNAi library for conditional gene inactivation in *Drosophila*. Nature. 2007;**448**:151-156. DOI: 10.1038/nature05954

[74] Xia B, de Belle JS. Transgenerational programming of longevity and reproduction by post-eclosion dietary manipulation in *Drosophila*. Aging 2016;**8**:1115-1134. DOI :10.18632/aging.100932

[75] Roussou IG, Savakis C, Tavernarakis N, Metaxakis A. Stage dependent nutritional regulation of transgenerational longevity. Nutr Healthy Aging. 2016;**4**:47-54. DOI: 10.3233/NHA-160012

[76] Shameer PM, Sowmithra K, Harini BP, Chaubey RC, Jha SK, Shetty NJ. Does exposure of male *Drosophila melanogaster* to acute gamma radiation influence egg to adult development time and longevity of $F_1$-$F_3$ offspring? Entomological Science. 2015;**18**:368-376. DOI: 10.1111/ens.12120

[77] Xia B, Gerstin E, Schones DE, Huang W, de Belle JS. Transgenerational programming of longevity through E(z)-mediated histone H3K27 trimethylation in *Drosophila*. Aging 2016;**8**:2988-3008. DOI: 10.18632/aging.101107

[78] Tindall KR, Stein J, Hutchinson F. Changes in DNA base sequence induced by gamma-ray mutagenesis of lambda phage and prophage. Genetics. 1988;**118**:551-560

[79] Moskalev AA, Plyusnina EN, Shaposhnikov MV. Radiation hormesis and radioadaptive response in *Drosophila melanogaster* flies with different genetic backgrounds: The role of cellular stress-resistance mechanisms. Biogerontology. 2011;**12**:253-263. DOI: 10.1007/s10522-011-9320-0

[80] Calabrese EJ. Hormesis: A fundamental concept in biology. Microbial Cell. 2014;**1**:145-149. DOI: 10.15698/mic2014.05.145

[81] Rattan SIS. Hormesis in aging. Ageing Research Reviews. 2008;**7**:63-78. DOI: 10.1016/j.arr.2007.03.002

[82] Calabrese EJ. Hormetic mechanisms. Critical Reviews in Toxicology. 2013;**43**:580-606. DOI: 10.3109/10408444.2013.808172

[83] Tang J, Yan H, Zhuang S. Histone deacetylases as targets for treatment of multiple diseases. Clinical Science. 2013;**124**:651-662. DOI: 10.1042/CS20120504

[84] Haberland M, Montgomery RL, Olson EN. The many roles of histone deacetylases in development and physiology: Implications for disease and therapy. Nature Reviews. Genetics. 2009;**10**:32-42. DOI: 10.1038/nrg2485

[85] Ceccacci E, Minucci S. Inhibition of histone deacetylases in cancer therapy: Lessons from leukaemia. British Journal of Cancer. 2016;**114**:605-611. DOI: 10.1038/bjc.2016.36

[86] Ding G-L, Wang F-F, Shu J, Tian S, Jiang Y, Zhang D, Wang N, Luo Q, Zhang Y, Jin F, Leung PCK, Sheng J-Z, Huang H-F. Transgenerational glucose intolerance with Igf2/H19 epigenetic alterations in mouse islet induced by intrauterine hyperglycemia. Diabetes. 2012;**61**:1133-1142. DOI: 10.2337/db11-1314

[87] Jimenez-Chillaron JC, Isganaitis E, Charalambous M, Gesta S, Pentinat-Pelegrin T, Faucette RR, Otis JP, Chow A, Diaz R, Ferguson-Smith A, Patti ME. Intergenerational transmission of glucose intolerance and obesity by in utero undernutrition in mice. Diabetes. 2009;**58**:460-468. DOI: 10.2337/db08-0490

[88] Bloch Qazi MC, Miller PB, Poeschel PM, Phan MH, Thayer JL, Medrano CL. Transgenerational effects of maternal and grandmaternal age on offspring viability and performance in *Drosophila melanogaster*. Journal of Insect Physiology. 2017;**100**:43-52. DOI: 10.1016/j.jinsphys.2017.05.007

[89] Bonduriansky R, Crean AJ, Day T. The implications of nongenetic inheritance for evolution in changing environments. Evolutionary Applications. 2012;**5**:192-201. DOI: 10.1111/j.1752-4571.2011.00213.x

[90] Brommer JE, Fricke C, Edward DA, Chapman T. Interactions between genotype and sexual conflict environment influence transgenerational fitness in *Drosophila melanogaster*. Evolution. 2012;**66**:517-531. DOI: 10.1111/j.1558-5646.2011.01449.x

[91] Strømnaes Ø. Sexual maturity in *drosophila*. Nature. 1959;**183**:409-410. DOI: 10.1038/183409

[92] Bohacek J, Mansuy IM. Molecular insights into transgenerational non-genetic inheritance of acquired behaviours. Nature Reviews. Genetics. 2015;**16**:641-652. DOI: 10.1038/nrg3964

[93] Singh P, Srivas S, Thakur MK. Epigenetic regulation of memory – therapeutic potential for disorders. Current Neuropharmacology. 2017. DOI: 10.2174/1570159X15666170404144522

[94] Williams ZM. Transgenerational influence of sensorimotor training on offspring behavior and its neural basis in *Drosophila*. Neurobiology of Learning and Memory. 2016;**131**:166-175. DOI: 10.1016/j.nlm.2016.03.023

[95] Burns JG, Mery F. Transgenerational memory effect of ageing in *Drosophila*. Journal of Evolutionary Biology. 2010;**23**:678-686. DOI: 10.1111/j.1420-9101.2010.01932.x

[96] Dias BG, Ressler KJ. Parental olfactory experience influences behavior and neural structure in subsequent generations. Nature Neuroscience. 2014;**17**:89-96. DOI: 10.1038/nn.3594

[97] Mery F, Kawecki TJ. A cost of long-term memory in *Drosophila*. Science. 2005;**308**:1148. DOI: 10.1126/science.1111331

[98] Tully T, Preat T, Boynton SC, Del Vecchio M. Genetic dissection of consolidated memory in *Drosophila*. Cell. 1994;**79**:35-47. DOI: 10.1016/0092-8674(94)90398-0

[99] Tully T, Quinn WG. Classical conditioning and retention in normal and mutant *Drosophila melanogaster*. Journal of Comparative Physiology A. 1985;**157**:263-277

[100] Tamura T, Chiang A-S, Ito N, Liu H-P, Horiuchi J, Tully T, Saitoe M. Aging specifically impairs amnesiac-dependent memory in *Drosophila*. Neuron. 2003;**40**:1003-1011. DOI: 10.1016/S0896-6273(03)00732-3

[101] Ramaswami M. Specifying the age-sensitive component of a short-term memory. Neuron. 2003;**40**:877-879. DOI: 10.1016/S0896-6273(03)00759-1

[102] Margulies C, Tully T, Dubnau J. Deconstructing memory in *Drosophila*. Current Biology. 2005;**15**:R700-R713. DOI: 10.1016/j.cub.2005.08.024

[103] Feany MB, Quinn WG. A neuropeptide gene defined by the *Drosophila* memory mutant amnesiac. Science. 1995;**268**:869-873

[104] Xia S, Tully T. Segregation of odor identity and intensity during odor discrimination in *Drosophila* mushroom body. PLoS Biology. 2007;**5**:e264. DOI: 10.1371/journal.pbio.0050264

[105] Waddell S, Armstrong JD, Kitamoto T, Kaiser K, Quinn WG. The amnesiac gene product is expressed in two neurons in the *Drosophila* brain that are critical for memory. Cell. 2000;**103**:805-813. DOI: 10.1016/S0092-8674(00)00183-5

[106] Montero M, Yon L, Kikuyama S, Dufour S, Vaudry H. Molecular evolution of the growth hormone-releasing hormone/pituitary adenylate cyclase-activating polypeptide gene family: Functional implication in the regulation of growth hormone secretion. Journal of Molecular Endocrinology. 2000;**25**:157-168

[107] Yu L, Jia Y, Su G, Sun Y, Letcher RJ, Giesy JP, Yu H, Han Z, Liu C. Parental transfer of tris(1,3-dichloro-2-propyl) phosphate and transgenerational inhibition of growth of zebrafish exposed to environmentally relevant concentrations. Environmental Pollution. 2017;**220**:196-203. DOI: 10.1016/j.envpol.2016.09.039

[108] Moore MS, DeZazzo J, Luk AY, Tully T, Singh CM, Heberlein U. Ethanol intoxication in *Drosophila*: Genetic and pharmacological evidence for regulation by the cAMP signaling pathway. Cell. 1998;**93**:997-1007. DOI: 10.1016/S0092-8674(00)81205-2

[109] Yohn NL, Bartolomei MS, Blendy JA. Multigenerational and transgenerational inheritance of drug exposure: The effects of alcohol, opiates, cocaine, marijuana, and nicotine. Progress in Biophysics and Molecular Biology. 2015;**118**:21-33. DOI: 10.1016/j.pbiomolbio.2015.03.002

[110] Pembrey M, Saffery R, Bygren LO. Network in epigenetic epidemiology. Human transgenerational responses to early-life experience: Potential impact on development, health and biomedical research. Journal of Medical Genetics. 2014;**51**:563-572. DOI: 10.1136/jmedgenet-2014-102577

[111] Aldrich JC, Maggert KA. Transgenerational inheritance of diet-induced genome rearrangements in *Drosophila*. PLoS Genetics. 2015;**11**:e1005148. DOI: 10.1371/journal.pgen.1005148

[112] Ciabrelli F, Comoglio F, Fellous S, Bonev B, Ninova M, Szabo Q, Xuéreb A, Klopp C, Aravin A, Paro R, Bantignies F, Cavalli G. Stable Polycomb-dependent transgenerational inheritance of chromatin states in *Drosophila*. Nature Genetics. 2017;**49**:876-886. DOI: 10.1038/ng.3848

[113] Sharma A, Singh P. Detection of transgenerational spermatogenic inheritance of adult male acquired CNS gene expression characteristics using a *Drosophila* systems model. PLoS One. 2009;**4**:e5763. DOI: 10.1371/journal.pone.0005763

[114] Xing Y, Shi S, Le L, Lee CA, Silver-Morse L, Li WX. Evidence for transgenerational transmission of epigenetic tumor susceptibility in *Drosophila*. PLoS Genetics. 2007;**3**:1598-1606. DOI: 10.1371/journal.pgen.0030151

[115] Zenk F, Loeser E, Schiavo R, Kilpert F, Bogdanović O, Iovino N. Germ line-inherited H3K27me3 restricts enhancer function during maternal-to-zygotic transition. Science. 2017;**357**:212-216. DOI: 10.1126/science.aam5339

[116] Mohammad F, Singh P, Sharma A. A *Drosophila* systems model of pentylenetetrazole induced locomotor plasticity responsive to antiepileptic drugs. BMC Systems Biology. 2009;**3**:11. DOI: 10.1186/1752-0509-3-11

[117] Grewal SS. Insulin/TOR signaling in growth and homeostasis: A view from the fly world. The International Journal of Biochemistry & Cell Biology. 2009;**41**:1006-1010. DOI: 10.1016/j.biocel.2008.10.010

[118] Holland ML, Lowe R, Caton PW, Gemma C, Carbajosa G, Danson AF, Carpenter AAM, Loche E, Ozanne SE, Rakyan VK. Early-life nutrition modulates the epigenetic state of specific rDNA genetic variants in mice. Science. 2016;**353**:495-498. DOI: 10.1126/science.aaf7040

[119] Usher CL, McCarroll SA. Complex and multi-allelic copy number variation in human disease. Briefings in Functional Genomics. 2015;**14**:329-338. DOI: 10.1093/bfgp/elv028

[120] Cuccaro D, De Marco EV, Cittadella R, Cavallaro S. Copy number variants in Alzheimer's disease. Journal of Alzheimer's Disease. 2017;**55**:37-52. DOI: 10.3233/JAD-160469

[121] Rutkowski TP, Schroeder JP, Gafford GM, Warren ST, Weinshenker D, Caspary T, Mulle JG. Unraveling the genetic architecture of copy number variants associated with schizophrenia and other neuropsychiatric disorders. Journal of Neuroscience Research. 2017;**95**:1144-1160. DOI: 10.1002/jnr.23970

[122] Weiner AJ, Scott MP, Kaufman TC. A molecular analysis of fushi tarazu, a gene in *Drosophila* Melanogaster that encodes a product affecting embryonic segment number and cell fate. Cell. 1984;**37**:843-851

[123] Shi S, Calhoun HC, Xia F, Li J, Le L, Li WX. JAK signaling globally counteracts heterochromatic gene silencing. Nature Genetics. 2006;**38**:1071-1076. DOI: 10.1038/ng1860

[124] Jones RS, Gelbart WM. Genetic analysis of the enhancer of zeste locus and its role in gene regulation in *Drosophila melanogaster*. Genetics. 1990;**126**:185-199

[125] Laible G, Wolf A, Dorn R, Reuter G, Nislow C, Lebersorger A, Popkin D, Pillus L, Jenuwein T. Mammalian homologues of the Polycomb-group gene enhancer of zeste mediate gene silencing in *Drosophila* heterochromatin and at *S. cerevisiae* telomeres. The EMBO Journal. 1997;**16**:3219-3232. DOI: 10.1093/emboj/16.11.3219

[126] Dumesic PA, Homer CM, Moresco JJ, Pack LR, Shanle EK, Coyle SM, Strahl BD, Fujimori DG, Yates JR 3rd, Madhani HD. Product binding enforces the genomic specificity of a yeast Polycomb repressive complex. Cell 2015;**160**:204-218. DOI: 10.1016/j.cell.2014.11.039

[127] Shaver S, Casas-Mollano JA, Cerny RL, Cerutti H. Origin of the Polycomb repressive complex 2 and gene silencing by an E(z) homolog in the unicellular alga *Chlamydomonas*. Epigenetics. 2010;**5**:301-312. DOI: 10.4161/epi.5.4.11608

[128] Ihara M, Meyer-Ficca ML, Leu NA, Rao S, Li F, Gregory BD, Zalenskaya IA, Schultz RM, Meyer RG. Paternal poly (ADP-ribose) metabolism modulates retention of inheritable sperm histones and early embryonic gene expression. PLoS Genetics. 2014;**10**:e1004317. DOI: 10.1371/journal.pgen.1004317

[129] Erkek S, Hisano M, Liang C-Y, Gill M, Murr R, Dieker J, Schübeler D, van der Vlag J, Stadler MB, Peters AHFM. Molecular determinants of nucleosome retention at CpG-rich sequences in mouse spermatozoa. Nature Structural & Molecular Biology 2013;**20**:868-875. DOI: 10.1038/nsmb.2599

[130] Brykczynska U, Hisano M, Erkek S, Ramos L, Oakeley EJ, Roloff TC, Beisel C, Schübeler D, Stadler MB, Peters AH. Repressive and active histone methylation mark distinct promoters in human and mouse spermatozoa. Nature Structural & Molecular Biology. 2010;**17**:679-687. DOI: 10.1038/nsmb.1821

[131] Carone BR, Fauquier L, Habib N, Shea JM, Hart CE, Li R, Bock C, Li C, Gu H, Zamore PD, Meissner A, Weng Z, Hofmann HA, Friedman N, Rando OJ. Paternally induced transgenerational environmental reprogramming of metabolic gene expression in mammals. Cell. 2010;**143**:1084-1096. DOI: 10.1016/j.cell.2010.12.008

[132] Greer EL, Maures TJ, Ucar D, Hauswirth AG, Mancini E, Lim JP, Benayoun BA, Shi Y, Brunet A. Transgenerational epigenetic inheritance of longevity in *Caenorhabditis elegans*. Nature. 2011;**479**:365-371. DOI: 10.1038/nature10572

[133] Watson CT, Disanto G, Sandve GK, Breden F, Giovannoni G, Ramagopalan SV. Age-associated hyper-methylated regions in the human brain overlap with bivalent chromatin domains. PLoS One. 2012;**7**:e43840. DOI: 10.1371/journal.pone.0043840

[134] Rakyan VK, Down TA, Maslau S, Andrew T, Yang T-P, Beyan H, Whittaker P, McCann OT, Finer S, Valdes AM, Leslie RD, Deloukas P, Spector TD. Human aging-associated DNA hypermethylation occurs preferentially at bivalent chromatin domains. Genome Research. 2010;**20**:434-439. DOI: 10.1101/gr.103101.109

[135] Dozmorov MG. Polycomb repressive complex 2 epigenomic signature defines age-associated hypermethylation and gene expression changes. Epigenetics. 2015;**10**:484-495. DOI: 10.1080/15592294.2015.1040619

[136] Kim S, Welsh DA, Myers L, Cherry KE, Wyckoff J, Jazwinski SM. Non-coding genomic regions possessing enhancer and silencer potential are associated with healthy aging and exceptional survival. Oncotarget. 2015;**6**:3600-3612. DOI: 10.18632/oncotarget.2877

[137] Wagner W, Ho AD, Zenke M. Different facets of aging in human mesenchymal stem cells. Tissue Engineering. Part B, Reviews. 2010;**16**:445-453. DOI: 10.1089/ten.TEB.2009.0825

[138] Schellenberg A, Lin Q, Schüler H, Koch CM, Joussen S, Denecke B, Walenda G, Pallua N, Suschek CV, Zenke M, Wagner W. Replicative senescence of mesenchymal stem cells causes DNA-methylation changes which correlate with repressive histone marks. Aging. 2011;**3**: 873-888. DOI: 10.18632/aging.100391

[139] Siebold AP, Banerjee R, Tie F, Kiss DL, Moskowitz J, Harte PJ. Polycomb repressive complex 2 and Trithorax modulate *Drosophila* longevity and stress resistance. Proceedings of the National Academy of Sciences of the United States of America. 2010;**107**:169-174. DOI: 10.1073/pnas.0907739107

[140] Öst A, Lempradl A, Casas E, Weigert M, Tiko T, Deniz M, Pantano L, Boenisch U, Itskov PM, Stoeckius M, Ruf M, Rajewsky N, Reuter G, Iovino N, Ribeiro C, Alenius M, Heyne S, Vavouri T, Pospisilik JA. Paternal diet defines offspring chromatin state and intergenerational obesity. Cell. 2014;**159**:1352-1364. DOI: 10.1016/j.cell.2014.11.005

[141] Maures TJ, Greer EL, Hauswirth AG, Brunet A. The H3K27 demethylase UTX-1 regulates *C. elegans* lifespan in a germline-independent, insulin-dependent manner. Aging Cell. 2011;**10**: 980-990. DOI: 10.1111/j.1474-9726.2011.00738.x

[142] Knutson SK, Warholic NM, Wigle TJ, Klaus CR, Allain CJ, Raimondi A, Porter Scott M, Chesworth R, Moyer MP, Copeland RA, Richon VM, Pollock RM, Kuntz KW, Keilhack H. Durable tumor regression in genetically altered malignant rhabdoid tumors by inhibition of methyltransferase EZH2. Proceedings of the National Academy of Sciences of the United States of America. 2013;**110**:7922-7927. DOI: 10.1073/pnas.1303800110

[143] Takahashi T, Yoshimura Y, Hiraishi H, Hasegawa C, Munesue T, Higashida H, Minabe Y, Kikuchi M. Enhanced brain signal variability in children with autism spectrum disorder during early childhood. Human Brain Mapping. 2016;**37**:1038-1050. DOI: 10.1002/hbm.23089

[144] Halperin JM, Bédard A-CV, Curchack-Lichtin JT. Preventive interventions for ADHD: A neurodevelopmental perspective. Neurotherapeutics. 2012;**9**:531-541. DOI: 10.1007/s13311-012-0123-z

[145] Iannitelli A, Quartini A, Tirassa P, Bersani G. Schizophrenia and neurogenesis: A stem cell approach. Neuroscience and Biobehavioral Reviews. 2017;**80**:414-442. DOI: 10.1016/j.neubiorev.2017.06.010

[146] Selemon LD, Zecevic N. Schizophrenia: A tale of two critical periods for prefrontal cortical development. Translational Psychiatry. 2015;**5**:e623. DOI: 10.1038/tp.2015.115

[147] Gillman MW. Early infancy as a critical period for development of obesity and related conditions. Nestlé Nutrition Workshop Series. Paediatric Programme. 2010;**65**:13-20; discussion 20. DOI: 10.1159/000281141

[148] Cameron N, Demerath EW. Critical periods in human growth and their relationship to diseases of aging. American Journal of Physical Anthropology. 2002;(Suppl 35):159-184. DOI: 10.1002/ajpa.10183

[149] Hanson MA, Low FM, Gluckman PD. Epigenetic epidemiology: The rebirth of soft inheritance. Annals of Nutrition & Metabolism. 2011;**58**(Suppl 2):8-15. DOI: 10.1159/000328033

[150] Barker DJP. Developmental origins of adult health and disease. Journal of Epidemiology and Community Health. 2004;**58**:114-115

[151] Tolhuis B, Blom M, Kerkhoven RM, Pagie L, Teunissen H, Nieuwland M, Simonis M, de Laat W, van Lohuizen M, van Steensel B. Interactions among Polycomb domains are guided by chromosome architecture. PLoS Genetics 2011;**7**:e1001343. DOI: 10.1371/journal.pgen.1001343

[152] Bantignies F, Roure V, Comet I, Leblanc B, Schuettengruber B, Bonnet J, Tixier V, Mas A, Cavalli G. Polycomb-dependent regulatory contacts between distant Hox loci in *Drosophila*. Cell. 2011;**144**:214-226. DOI: 10.1016/j.cell.2010.12.026

[153] Talbert PB, Garber RL. The *Drosophila* homeotic mutation Nasobemia (AntpNs) and its revertants: An analysis of mutational reversion. Genetics. 1994;**138**:709-720

[154] Heard E, Martienssen RA. Transgenerational epigenetic inheritance: Myths and mechanisms. Cell. 2014;**157**:95-109. DOI: 10.1016/j.cell.2014.02.045

[155] Iovino N, Ciabrelli F, Cavalli G. PRC2 controls *Drosophila* oocyte cell fate by repressing cell cycle genes. Developmental Cell. 2013;**26**:431-439. DOI: 10.1016/j.devcel.2013.06.021

[156] Langley-Evans SC. Nutrition in early life and the programming of adult disease: A review. Journal of Human Nutrition and Dietetics. 2015;**28**(Suppl 1):1-14. DOI: 10.1111/jhn.12212

Chapter 5

# *Drosophila* Imaginal Discs as a Playground for Genetic Analysis: Concepts, Techniques and Expectations for Biomedical Research

Cristina M. Ostalé, Ana Ruiz-Gómez, Patricia Vega,
Mireya Ruiz-Losada, Carlos Estella and
Jose F. de Celis

Additional information is available at the end of the chapter

http://dx.doi.org/10.5772/intechopen.72758

**Abstract**

*Drosophila* imaginal discs are epithelial tissues perfectly suited to use them as a playground to define the functional contribution of genes to epithelial development and organ morphogenesis. The more we know about the discs and the mechanisms directing their development, the best prepared we are to assign specific "functions" to individual genes based on phenotypic observations. Conversely, and thinking from the perspective of the gene, the more we know about its function, the best inferences we could make about the mechanisms underlying imaginal disc *development*. This reciprocal relationship, coupled to the arsenal of possible experimental approaches available in Drosophila genetics, genomics and cellular biology, makes these tissues excellent systems to address biological problems with biomedical relevance. In this review, an overview of three interconnected aspects related to the use of *Drosophila* imaginal discs as an experimental system to analyze gene function is given: (i) imaginal discs biology, with a focus in the genetic mechanisms involved in pattern formation; (ii) concepts and available experimental tools for the analyses of gene function and (iii) uses of *Drosophila* and the imaginal discs for addressing biomedical problems.

**Keywords:** *Drosophila*, genetic analysis, growth control, pattern formation, imaginal disc

## 1. Introduction

Imaginal discs are epithelial tissues that grow within the larva of holometabolous insects and differentiate most of the external parts of the adult during metamorphosis [1]. They are named after the adult structures they form, for example, the wing imaginal disc makes the wing and the thorax, while the leg discs develop the leg appendages and the pleura. Each disc has a characteristic size, shape, histology and fate map, and they are connected to the larval epidermis and to the tracheal system of the larvae [1]. Imaginal discs are a favorite subject of study for developmental and cell biologists, and the analysis of their characteristics has shaped key concepts in developmental biology, including the notions of cell determination, cell autonomy and positional information [2]. The study of imaginal discs is also contributing to identify and characterize the cellular and biochemical mechanisms underlying these concepts.

A key peculiarity that in part account for the success of the imaginal discs as experimental model systems is that they retain a considerable developmental plasticity during larval development. Thus, when let unperturbed, each imaginal disc will undergo with a clock-like precision cell divisions, growth and territorial specification, forming a fixed inventory of cuticular structures during differentiation. Simultaneously, the discs remain extremely plastic and reactive to experimental manipulations during most of their development. This developmental plasticity is particularly manifested when the discs are cut and transplanted into adult host, where disc fragments can reconstitute the missing parts ("regeneration") and even alter their identity ("transdetermination") [2, 3]. The ability to regenerate has been more recently observed *in situ* in experiments in which disc fragments are mechanically removed or where particular regions are eliminated through the induction of cell death [3–5].

Imaginal discs are also extremely reactive to genetic manipulations, and altering the expression of genes encoding a wide variety of proteins related to epithelial development in imaginal cells results in precise adult phenotypes [6]. The responsive nature of imaginal discs to genetic and other experimental manipulations is one of the reasons explaining why imaginal discs have been repetitively used in developmental biology. In fact, they have traditionally been either drivers or early adopters of novel experimental approaches directed to unravel the genetic and cellular basis of epithelial biology and organ morphogenesis [2]. In this manner, imaginal discs not only played a key role in the transition from experimental embryology to developmental genetics but also in the posterior move from formal mechanistic interpretations to increasingly detailed molecular and cellular descriptions.

Another aspect that explains the success of imaginal discs as experimental tools at different historical periods is the richness of biological processes participating in their development and differentiation. Thus, most common developmental operations, including cell proliferation and death, cell growth and differentiation, pattern formation and tissue mechanics and morphogenesis, as well as their underlying molecular mechanisms, can be analyzed in the discs. Because these processes are common to the development of all multicellular organisms and also regulated by conserved genes and molecular interactions, the discs are a most convenient system to dissect genetically complex developmental mechanisms. In this review, some key aspects of imaginal disc biology are summarized, and the experimental tools available to analyze the contribution of genes to the development of imaginal disc development are described. We will also summarize how to capitalize on the knowledge we have about the discs to address problems with biomedical impact.

http://dx.doi.org/10.5772/intechopen.72758

## 2. Genetic regulation of imaginal disc development

Imaginal discs are versatile, responsive and modular tissues for genetic experimentation. Internal regulatory processes not only determine their development but they also communicate and interact with other larval organs to influence growth and developmental timing [7–10]. In the discs, patterning and growth are interconnected aspects driven by conserved signaling pathways and complex transcriptional regulatory networks that coordinate gene expression along a field of epithelial cells. Although each imaginal disc has its own peculiarities, their modes of development share multiple common aspects, including regulated cell proliferation and the progressive generation of gene expression domains. Thus, a common feature of all discs is the existence of a continuous deployment and refinement of gene expression patterns that culminate in the allocation of cellular fates to individual cells or fields of cells. This process is linked to the position that each cell occupies in the epithelium, but it is mostly a consequence of each cell developmental history, as defined by the gene regulatory networks that were operating in its progenitors. The mechanistic links between gene activity and the patterned distribution of differentiated cells make possible to use phenotypic approaches to identify and characterize the function of individual genes through genetic analysis.

The origin of imaginal cells is the embryonic ectoderm. It is in this epithelial layer where a set of gene regulatory events defines the position of groups of cells as imaginal precursors [1, 11] (**Figure 1**). The specification of each imaginal primordium follows the same logic of gene regulatory events that will direct their subsequent development. Thus, the segmented embryonic ectoderm contains a Cartesian system of positional information along the antero-posterior (A/P) and dorso-ventral (D/V) axes defined by the expression of ligands

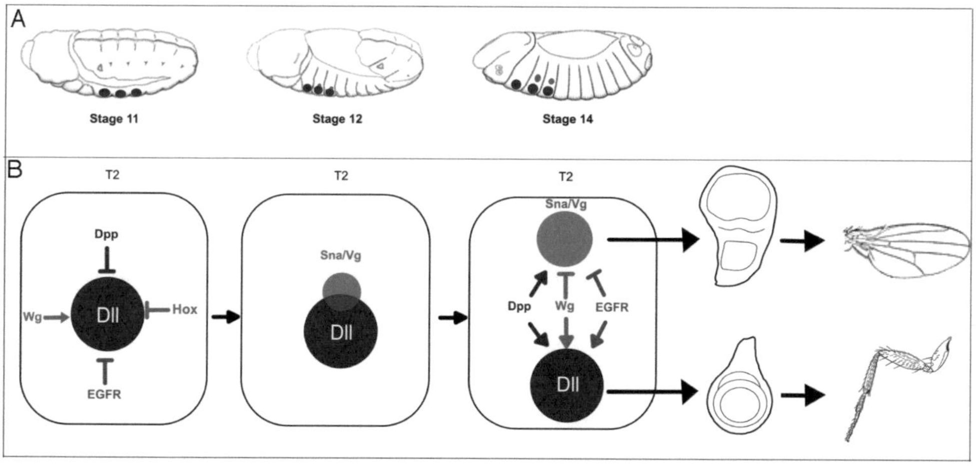


**Figure 1.** Specification of the thoracic appendage primordia. (A) Schematic representation of Drosophila embryogenesis indicating the position of the primordia for the thoracic discs in the T1, T2 and T3 segments. (B) Development of the disc primordium in the T2 segment. At stage 11, *Dll* expression (dark grey circle) is activated by Wg and repressed dorsally and ventrally by the Dpp and EGFR pathways, respectively. Some hours later, the wing primordium marked by the expression of *vg* and *sna* originates (light grey circle) and includes both *Dll*-expressing cells and *Dll* non-expressing cells. At stage 14, the wing and leg primordia are fully separated. The genetic inputs into *Dll* and *sna*/vg are schematized. Note that these genes integrate differentially the activity of the Wg, Dpp and EGFR pathways.

belonging to the Hedgehog (Hh), wingless (Wg), decapentaplegic (Dpp) and epidermal growth factor receptor (EGFR) pathways [1, 12–15] (**Figure 1**). The restricted expression of these ligands results in the generation of overlapping signaling domains in which pathway-specific transcription factors are expressed or active (**Figure 1**). These transcription factors act through *cis*-regulatory modules (CRM) present in a set of genes involved in the early determination of embryonic cells as imaginal precursors. Some of the best-characterized genes belonging to this class are *eyeless* (*eye*), *vestigial* (*vg*), *Distal-less* (*Dll*), *escargot* (*esg*), *buttonhead* (*btd*) and *Sp1* [11, 16–19]. The expression of these genes becomes restricted to groups of cells positioned with respect to parasegmental boundaries and located in precise A/P and D/V positions [20].

Complementary to the A/P and D/V coordinate systems, which are common to all embryonic segments, the ectoderm also bears a segment-specific code of homeotic genes resulting from the differential expression of the *Bithorax* (Bx) and *Antennapedia* (Antp) gene complexes [21]. The genes of the *Bx* complex prevent the development of imaginal primordia from the abdominal segments and confer different identities to the thoracic segments 2 and 3. Similarly, the genes of the Antp complex confer segmental identity to the imaginal precursors present in the cephalic segments [21]. Once the position, size and identity of each primordium are determined, its development implies cell proliferation and invagination from the embryonic surface [1]. For all subsequent stages, the discs will grow as hollow sacs within the interior of the larvae that remain connected through a stalk to the larval epidermis. From the primordia that were specified in the embryo to the mature third instar discs, there happens a considerable increase in size due to cell proliferation, and this increase is always accompanied by the generation of gene expression domains that progressively become coincident with the adult structures; each cell will give rise during metamorphosis (**Figure 1**). The same signaling pathways that participate in the specification of the primordia drive the generation of gene expression domains. The continuous interplay between signaling and transcription is a common aspect to the development of all discs. In this manner, at each time-point in the development of the discs, localized domains of signaling are converted into territories of gene expression, which in turn drive the generation of novel signaling domains directing further domains of gene expression (**Figure 2**).

The examples of the leg and wing discs illustrate how signaling-transcription networks coupled to the increase in the size of the epithelium have been adapted to generate diverse expression patterns associated to cell fate allocations. The leg and wing discs originate from the same primordium located in the ventral region of each hemisegment of the mesothorax (parasegment 5). These early primordia can first be recognized as a group of ~30 cells in each thoracic hemisegment that express the homeobox gene *Dll* [22]. *Dll* expression is activated by Wg and repressed dorsally and ventrally by the Dpp and EGFR pathways, respectively [23]. These *Dll*-expressing cells give rise to all regions of the adult thorax, including both the ventral (i.e., legs) and dorsal (i.e., wing and haltere) appendages [22, 24]. As the embryo develops, a group of cells of these early primordia moves dorsally to form the dorsal primordia (wing and haltere), lose *Dll* expression and activate the wing-promoting genes, *vestigial* (*vg*) and *snail* (*sna*) [16, 22, 25] (**Figure 1**). The remaining *Dll*-expressing cells in each thoracic hemisegment give rise to the leg discs [24, 26] (**Figure 1**). At the molecular level, the signals that govern the separation between the dorsal and ventral primordia are integrated by the CRMs of the *Dll* and *sna* genes. Dpp is required for the expression of *Dll* and *sna*, whereas Wnt and EGFR signalling repress wing fate and promote leg primordia formation [14, 15].

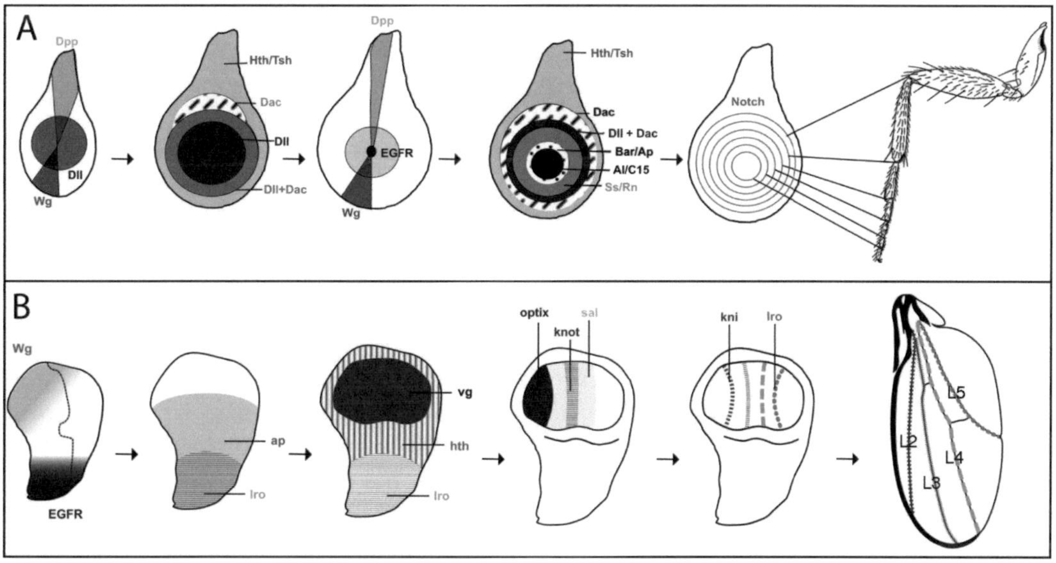


**Figure 2.** Overview of leg and wing imaginal disc patterning. (A) Schematic representation of the leg disc at different developmental stages, from early L2 (left) to late L3 (right). Initially, Wg (dark grey) and Dpp (light grey) activate *Dll* expression in the distal domain of the leg (grey circle) and repress *dac*. As the disc grows, *dac* expression (striped and grey circle) escapes the repression of Wg and Dpp, and the initial PD axis is established. Later on, the distal part of the leg is further subdivided in nested expression patterns by the activity of the EGFR pathway, which ligand is expressed in the distal-most tip of the leg disc. The combinatorial code of the PD axis patterning genes localized the activation of the Notch pathway (circles) in concentric rings that prefigure the joints of the adult leg. (B) Schematic representation of the wing disc at different developmental stages, from L2 (left) to late L3 (right). The L2 disc is subdivided into a proximal domain where the EGFR pathway is active (dark grey) and an anterior distal domain where the Wg pathway is active (light grey)—EGFR signaling activates the expression of the Iroquois genes (*Iro*) and apterous (*ap*). The confrontation of ap expressing and not expressing cells triggers Notch activation and vestigial (*vg*) expression in the primordium of the wing blade. Proximal cells express the homothorax gene (hth; stripped) and will contribute to the hinge region of the wing. The wing blade is subdivided into expression domains due to the activity of the Hedgehog and Decapentaplegic signaling pathways. The expression of optix (black), spalt (*sal*; light grey) and knot (dark grey) is shown. The expression of Knirps (*kni*) and Iroquois (*Iro*) is then activated in provein territories (lines) from where later each longitudinal vein (L2-L5) differentiates.

From this point onwards, both groups of cells follow their development independently, using the information already present in the primordium to drive subsequent developmental steps. In the case of the wing disc, the first subdivision imposed over the antero-posterior compartment initial subdivision is between proximal and distal cells (**Figure 2**). Proximal cells express the EGFR ligand *vein* (*Vn*), whereas distal cells express the gene *wingless* (*wg*) [15, 27, 28]. Decreasing EGFR signalling levels from distal to proximal regions of the disc regulates the formation of nested expression domains of genes encoding transcription factors. In more proximal cells, the genes of the Iroquois (*Iro*) and *Spalt* (*sal*) complexes are expressed and contributed to the determination of the future thorax [27, 29–31]. In a more extended domain, the expression of *apterous* (*ap*) is also activated in response to the EGFR pathway [15, 32, 33], and the Notch signaling pathway is activated at the boundary between cells expressing and not expressing *ap* and drives the expression of *vestigial* (*vg*), a cofactor necessary for the specification of wing cells [25, 34–38]. From now onwards, the territories fated to become the thorax, hinge and blade will expand and further divide into smaller subdomains of gene expression related to pattern elements such as veins and sensory organs (**Figure 2**). In the expanding wing blade, for example, most of the expression territories are established with respect to the A/P compartment boundary due to the activity of the Hh and Dpp signaling pathways [39]. In this manner, the central region of the wing is patterned by the differential response to Hh signaling of a battery of genes required for the positioning of the central wing veins (L3 and L4) and the central intervein (L3/4 intervein) [40]. Simultaneously, the lateral regions of the wing blade are patterned through a set of transcription factors which expression is regulated by the Dpp signaling pathway (**Figure 2**).

In the case of the leg, the restricted expression of Dpp and Wg in dorsal and ventral domains of the leg imaginal disc directs the formation of the proximo-distal (P/D) axis. Initially, high levels of Wg and Dpp activate *Dll* in the center of the leg disc and repress the expression of *dachshund* (*dac*). As the discs grow, *dac* escapes the repression of Wg and Dpp in the medial domain of the leg disc and is activated by *Dll* [23]. Once these genes are activated, their expression is locked by autoregulatory mechanisms. In the periphery of the disc, where combined low levels of Wg and Dpp are found, the expression of *homothorax* (*Hth*) and *teashirt* (*Tsh*) is activated. Later on, the activity of the EGFR pathway in the distal domain of the leg disc is required to activate a series of secondary P/D targets, including the tarsal restricted genes *aristalless* (*al*), *Bar* (*B*) and *rotound* (*rn*) [11, 23] (**Figure 2**). An important consequence of the P/D subdivision is the activation of the Notch pathway in concentric rings that subsequently correspond to the joints, movable structures separating adjacent leg segments [41]. Joint development is controlled by the action of subsidiary Notch target genes, such as the transcription factors encoded by *oddskipped* family genes (*odd*) and *dysfusion* (*dys*) in proximal and distal joints, respectively. In particular, Dys regulates the expression of several Rho-GTPase regulators and pro-apoptotic genes that together sculpt the tarsal joints [42, 43].

In summary, as the discs grow in size, its pattern is progressively established and prefigurates as expression domains the position where different structures, such as veins, sensory organs and tarsal joints, that will differentiate during metamorphosis. This process relies upon regulatory mechanisms that link signaling with transcriptional regulation along the epithelium. Subsequently, each disc will initiate its differentiation and morphogenesis during pupal development, in a course that includes extensive morphogenetic movements and fusion between imaginal discs. The culmination of imaginal disc development is the generation of precise patterns of differentiation. As these patterns are under strict regulation of genes encoding transcription factors and signaling components, changes in the expression pattern or activity of these genes result in precise alterations of the pattern of cell differentiation and organ morphogenesis. These alterations, the mutant phenotypes, can be used as diagnostic criteria to define the requirements of the gene and to annotate its function in relation to the processes affected. Most of the impact of *Drosophila* in biological research is due to the availability of methods to analyze gene function *in vivo*. These methods rely on the power to generate and analyze mutations and to detect the timing and patterning of gene expression. It is through the analysis of the consequences of manipulating gene activity that specific functions could be assigned to particular genes.

## 3. Genetic analysis in *Drosophila*

When genetic analysis was first used to characterize the contribution of a gene to a particular process, the definition of "gene function" was abstract, referring more to the requirement of the gene than to the actual biochemical function of the protein it encoded, which was for the most part unknown. Thus, genes encoding transcription factors or signaling molecules could be classified as "segmentation genes," because they displayed mutant phenotypes affecting the segmentation of the embryo [44]. By looking at the particularities of the mutant phenotype, these genes could be further classified into discrete classes that later were shown to correspond to the different levels in the hierarchy of regulatory interactions driving segmentation [44]. During most of the 20th century, methods to analyze a gene were blind to a large extent. In this manner, mutations generated randomly were selected because they failed to complement a particular allele or gene deficiency [45, 46] or because they displayed or modified a phenotype in the tissue of interest [47–50]. Favorite methods to generate mutations were chemical (EMS), physical (ionizing radiations) and later through the mobilization of transposable elements [51, 52].

The availability of the *Drosophila* DNA sequence in 2000 [53], combined with the development of novel techniques to generate mutations in the following years, marked a shift in the way genes were studied. When the entire genetic map of the fly was open to scrutinize, it was necessary to develop new methods that allowed the generation of gene-specific mutations from the sequence ("reverse genetics"). In this manner, homologous recombination was adopted to fly genetics [54, 55], although the technical complexity and low frequency of these events precluded a generalized adoption by the fly community. More impact had the implementation of RNA interference (RNAi), which combined with the Gal4/UAS system, already extensively used to generate gain-of-function conditions [56] allowed a targeted reduction of mRNA levels of the gene of interest in specific tissues [57]. The generation of genome-wide collections of UAS-RNAi lines further allowed screening systematically the genome, either by searching for mutant phenotypes in the tissue of interest or for modifiers of a particular genetic condition [58, 59]. RNAi is being massively used as a first approach to identify the functional requirements of a gene or gene family of interest, but still has the problems of generating only hypomorphic conditions and the existence of off-target effects caused by sequence similarities.

In more recent years, the adaptation of the CRISPR/Cas9 technology to the fly is allowing an unprecedented level of precision and easiness with which a gene can be targeted [60–63]. This method is based on the use of the nuclease Cas9 guided by small RNAs (gRNAs) to generate double-strand breaks (DSB) at a target genomic locus, allowing its use as a highly efficient and specific system for gene edition. Using CRISPR/Cas9 allows targeted manipulation of a given gene in different manners. Thus, the CRISPR/Cas 9 system can be employed to generate sequence-specific DSB to disrupt the target locus when a single gRNA is used, resulting in the generation of small insertion or deletions (In/Dels) through the error-prone process of non-homologous end-joining (NHEJ) repair in the coding sequence. This approach can be used to disrupt coding genes, leading to an array of mutations ranging from hypomorphs to amorph alleles (**Figure 3**), caused by frameshifts in the reading frame, premature stop codons or triplet insertions or deletions [63–65]. The NHEJ repair system can also be directed by two gRNAs to delete a specific fragment flanking the targeted sequences [63, 66]. The resulting DNA change consists in a deletion of a longer sequence, which could include an entire open reading frame, an exon or also non-coding sequences such as candidate regulatory regions, being also an useful approach to analyze transcriptional regulation *in situ* (**Figure 3**).

A second application of the CRISPR-Cas9 system is to stimulate HDR (*h*omology-*d*irect *r*epair) by using homologous DNA sequences as template for the DNA repair and two gRNAs, allowing precise

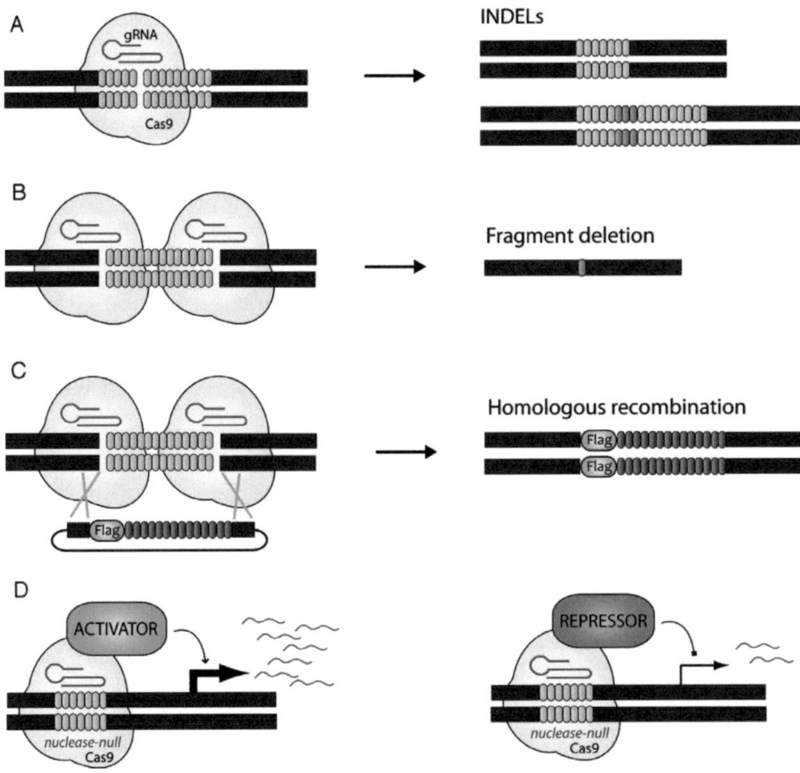


**Figure 3.** Applications of CRISPR/Cas9 for gene editing in Drosophila. Schematic representation of Cas9-DNA interactions leading to changes in the genomic sequence (A–C) or to changes in gene expression (D). The genomic DNA is represented as thick lines, Cas9 is represented by a grey shape, guide RNA (gRNA) are shown as grey lines and nucleotides targeted by gRNA as grey ovals. (A) Single gRNA leads to DNA double strand breaks (left) that are corrected by non-homologous end joining (right) leading to INDELs. (B) Double gRNAs lead to nearby double strand breaks (left) that when corrected result in a deletion of the DNA located between adjacent breaks (right). (C) The use of two gRNAs in the presence of sequences with homology to the target DNA (left) leads to the substitution of this DNA (knock-out) by the selected sequence (knock-in), in this case a tagged version of a coding region (Flag). (D) The targeting of inactive forms of Cas9 (nuclease-null Cas9) fused with activating (left) or repressor (right) domains leads to the corresponding changes in gene expression from adjacent promoters.

genome editing by generating sequence substitutions following a deletion (knock-out/knock-in). This method relies on supplying a homologous repair DNA (ssDNA or dsDNA) engineered with the desired modifications [62, 67–69]. This approach allows not only the induction of particular mutations in the genome but also the integration of ssDNA oligonucleotides for epitope TAGs into protein coding genes, allowing the tagging of endogenous proteins (**Figure 3**).

More interestingly, when stimulating HDR using longer homologous repair dsDNA (homology arms), the removed genomic region can be replaced with site-specific recombinase sequences such as attP sites [67], allowing subsequent integrations at this genomic position of modified versions of the gene or orthologous genes and sequences encoding for tagged proteins (**Figure 3**). Creation of a DSB dramatically increases the frequency of homologous recombination [70], allowing the expression *in situ* of protein variants expressed under the same regulatory DNA as the endogenous gene (**Figure 3**). In addition, *F*lippase *r*ecognition *t*arget (FRT) sites could be introduced flanking the region on interest to allow tissue-specific or clonal deletion of specific sequences [67] (**Figure 3**). Engineered alleles can be subsequently interrogated by their impact on any particular part of the fly or used to follow the expression of the protein of interest due to the incorporation of different TAGs (**Figure 3**).

Beyond genome engineering, CRISPR/Cas9 has also been used to regulate endogenous gene expression in both cells and organisms without causing any mutation. For this approach, a nuclease-dead or inactive Cas9 is fused to a transcriptional activator or repressor domain and can be recruited to specific target DNA by its gRNAs, allowing activation through CRISPR or repression by CRISPR interference (**Figure 3**) [71–73]. In addition, an inactive Cas9 co-expressed with a gRNA can also be used for immunoprecipitation of specific DNA regions as a variant form of engineered DNA binding-mediated chromatin immunoprecipitation (enChIP), and associated proteins can be subsequently identified by mass spectrometry (enChIP-MS) [74]. In summary, CRISPR/Cas9 technology is providing a precise and efficient method for sequence-specific targeting of Cas9, resulting in genomic alterations, changes in gene expression and even the isolation of protein complexes bound to specific DNA regions. The application of this technology is triggering an unprecedented level of precision in genetic and genomic analysis in *Drosophila* research.

## 4. Uses of *Drosophila* and the imaginal discs to address biomedical problems

The precision by which the genome can be modified, combined with the knowledge we have about the cellular and molecular fundaments of imaginal disc development, justify the use of these epithelial tissues as experimental models to address a variety of biological questions, including biomedical ones. According to the Homophila database, around 75% of human disease genes cataloged in OMIM (On-line Mendelian Inheritance in Man) database have close homologs in the fruit fly [75]. This high degree of evolutionary conservation makes even more compelling the use of *Drosophila melanogaster* to develop experimental models of human diseases [76, 77].

The generation of *Drosophila* models for human diseases includes a variety of approaches, including blind and unbiased large-scale genetic screens, functional analysis of *Drosophila* genes orthologous for a known human disease gene, the expression of human genes in *Drosophila* tissues and the construction of genetic models that reproduce some characteristics of complex human syndromes such as cancer, kidney and metabolic diseases among many others [78–86]. These approaches are focused on either the gene or genes causing the disease or, complementary, the tissue where the disease is manifested. A successful example of the first approach, unbiased genetic screens, involved a mosaic screen of newly generated lethal mutations in the X chromosome that allowed the identification of 21 novel genes associated with human diseases for which no mutations were previously known [87]. The analysis of *Drosophila* genes with human counterparts contributing to human diseases has been mostly applied in the context of several neurodegenerative diseases, and some examples are the analysis of *parkin* and *Pink1*, genes related to Parkinson disease, and *sphingosine-1-P-lyase* for Charcôt-Marie-Tooth neuropathy [88]. In these cases, the *Drosophila* model allows a first approximation to study the functional relevance of the gene, including the consequences of its loss, its expression, the subcellular localization of the protein and its biochemical characteristics. Similarly, the expression of human proteins in flies has mostly targeted neurodegenerative diseases such as Alzheimer disease, Tau-induced neurodegeneration (Tauopathy), polyglutamine diseases including Huntington disease and spinocerebellar ataxia, among others [89]. In this manner, it was shown that the overexpression of the human Aβ42 protein in the central nervous system of *Drosophila* causes amyloid deposition, progressive learning defects, extensive neurodegeneration and shortened lifespan, all of the hallmarks of Alzheimer's disease [90].

Apart from offering a convenient experimental system to identify the function of a gene or the consequences of the expression of relevant protein variants, *Drosophila* has been instrumental in expanding the catalog of relevant genes contributing to the function and outcome of a particular genetic condition through the use of "genetic modifier screens". These experiments consist in the search for mutations that can either increase or suppress the phenotype caused by a genetic condition of interest, under the assumption that these modifying alleles might identify additional components relevant for the function of the protein of interest. Such approach has been applied to a variety of genetic backgrounds, either in blind genetic screens, i.e., through the generation of random mutations in a particular genetic background, or by introducing in these backgrounds the expression of RNAi directed against all genes or particular candidates. Using these approaches, it was found that mutations in Neprilysin 2 modify the progressive retinal degeneration concomitant with plaque formation caused by the overexpression of human Aβ42 in the *Drosophila* eye [91].

*Drosophila* can also be effectively used for drug screens as well as in target discovery [77, 92]. Screening for novel drugs in flies enables for the selection of candidates with physiological characteristics that are difficult to analyze by cell culture or biochemical assays. In addition to high-throughput screening of potential therapeutics, *Drosophila* is a powerful tool for studying the molecular mechanism of a specific drug *in vivo*. The diminished effort and costs of analyzing the targets and possible "off-targets" in a *Drosophila* model make it useful as a first validation organism for establishing the efficacy and toxicity of a drug *in vivo*. In fact, some approved treatments, such as the kinase inhibitor vandetanib (ZD6474) for treating medullary thyroid carcinoma patients, were validated in *Drosophila* previous to clinical trials [93]. When using *Drosophila* for developing a treatment, it is important to consider the possible differences in pharmacokinetics of a drug in this organism and the differences in tissular distribution, which may affect the optimal doses of the compound. Also toxicity might be different, although a strong correlation has been shown [94]. Nowadays, there are several companies that have been using *Drosophila melanogaster* as human disease model for screening for therapeutic drugs such as Aktogen, En Vivo Pharmaceuticals, Genescient Corp and Medros Pharmaceuticals. Indeed, the ease of screening big samples of individuals has also been employed for drug discovery in *Drosophila* disease models. For example, a library of 2000 compounds was checked in a Fragile-X syndrome *Drosophila* model for pharmacological rescue [95], and 9 molecules were found to rescue lethality, among them three belonging to GABAergic inhibitory pathway. This and further studies led to the performance of human trials of GABAergic treatment (reviewed in [96]).

## 5. Conclusions

Several characteristics converge to sustain and reinforce the use of *Drosophila* in the post-genomic era as a motor for biomedical research. The availability of novel techniques to manipulate and scrutinize the genome with unprecedented sophistication and precision can be used in experimental models, such as the imaginal discs, that are particularly well suited for genetic and molecular analyses. The convergence of technical improvements used in a favorable experimental system is sustained by the functional conservation of the relevant genes and the cellular processes they regulate and by the multiple adaptations that this system allows, including genetic, biochemical, cellular and pharmacological experimental approaches.

## Acknowledgements

We are very grateful to A. López-Varea and N. Esteban for excellent technical support. This work was supported by MINECO grant BFU2015-64220-P to JFdC. Institutional support from Fundación Ramón Areces and Banco de Santander is also acknowledged.

## Author details

Cristina M. Ostalé†, Ana Ruiz-Gómez†, Patricia Vega, Mireya Ruiz-Losada, Carlos Estella and Jose F. de Celis*

*Address all correspondence to: jfdecelis@cbm.csic.es

Centro de Biología Molecular Severo Ochoa, CSIC and Universidad Autónoma de Madrid, Madrid, Spain

†These authors contributed equally

## References

[1] Cohen SM. Imaginal disc development. In: Bate M, Martinez Arias A, editors. The Development of *Drosophila melanogaster*. New York: Cold Spring Harbor Laboratory Press. 1993. pp 747-841

[2] Beira JV, Paro R. The legacy of Drosophila imaginal discs. Chromosome. 2016;**125**:573-592

[3] Worley MI, Setiawan L, Hariharan IK. Regeneration and transdetermination in Drosophila imaginal discs. Annual Review of Genetics. 2012;**46**:289-310

[4] Repiso A, Bergantinos C, Corominas M, Serras F. Tissue repair and regeneration in Drosophila imaginal discs. Development, Growth & Differentiation. 2011;**53**:177-185

[5] Diaz-Garcia S, Baonza A. Pattern reorganization occurs independently of cell division during Drosophila wing disc regeneration in situ. Proceedings of the National Academy of Sciences of the United States of America. 2013;**110**:13032-13037

[6] Molnar C, Resnik-Docampo M, Organista MF, Martín M, Hevia CF, de Celis JF. Signalling pathways in development and human disease: A Drosophila wing perspective.

In: Plaseska-Karanfilska D, editor. Human Genetic Diseases. InTech; 2011. ISBN: 978-953-307-936-3

[7] Neto-Silva R, Wells B, Johnston L. Mechanisms of growth and homeostasis in the Drosophila wing. Annual Review of Cell and Developmental Biology. 2009;**25**:197-220

[8] Garelli A, Gontijo AM, Miguela V, Caparros E, Dominguez M. Imaginal discs secrete insulin-like peptide 8 to mediate plasticity of growth and maturation. Science. 2012; **336**:579-582

[9] Colombani J, Andersen DS, Leopold P. Secreted peptide Dilp8 coordinates Drosophila tissue growth with developmental timing. Science. 2012;**336**:582-585

[10] Gokhale RH, Hayashi T, Mirque CD, Shingleton AW. Intra-organ growth coordination in Drosophila is mediated by systemic ecdysone signaling. Developmental Biology. 2016; **418**:135-145

[11] Estella C, Voutev R, Mann RS. A dynamic network of morphogens and transcription factors patterns the fly leg. Current Topics in Developmental Biology. 2012;**98**:173-198

[12] Kubota K, Goto S, Hayashi S. The role of Wg signaling in the patterning of embryonic leg primordium in Drosophila. Developmental Biology. 2003;**257**:117-126

[13] Goto S, Hayashi S. Cell migration within the embryonic limb primordium of Drosophila as revealed by a novel fluorescence method to visualize mRNA and protein. Development Genes and Evolution. 1997;**207**:194-198

[14] Goto S, Hayashi S. Specification of the embryonic limb primordium by graded activity of Decapentaplegic. Development. 1997;**124**:125-132

[15] Wang SH, Simcox A, Campbell G. Dual role for Drosophila epidermal growth factor receptor signaling in early wing disc development. Genes & Development. 2000;**14**:2271-2276

[16] Fuse N, Hirose S, Hayashi S. Determination of wing cell fate by the escargot and snail genes in Drosophila. Development. 1996;**122**:1059-1067

[17] Estella C, Rieckhof G, Calleja M, Morata G. The role of *buttonhead* and *Sp1* in the development of the ventral imaginal discs of Drosophila. Development. 2003;**130**:5929-5941

[18] Potier D et al. Identification of cis-regulatory modules encoding temporal dynamics during development. BMC Genomics. 2014;**15**:534

[19] Cordoba S, Requena D, Jory A, Saiz A, Estella C. The evolutionarily conserved transcription factor Sp1 controls appendage growth through Notch signaling. Development. 2016;**143**:3623-3631

[20] Bate CM, Martinez-Arias A. The embryonic origin of imaginal discs in Drosophila. Development. 1991;**112**:755-761

[21] Morata G. How Drosophila appendages develop. Nature Reviews Molecular Cell Biology. 2001;**2**:89-97

[22] Cohen B, Simcox AA, Cohen SM. Allocation of the thoracic imaginal primordia in the Drosophila embryo. Development. 1993;**117**:397-608

[23] Kojima T. The mechanism of Drosophila leg development along the proximodistal axis. Development, Growth & Differentiation. 2004;**46**:115-129

[24] McKay DJ, Estella C, Mann RS. The origins of the Drosophila leg revealed by the cis-regulatory architecture of the Distalless gene. Development. 2009;**136**:61-71

[25] Williams JA, Bell JB, Carroll SB. Control of Drosophila wing and haltere development by the nuclear vestigial gene product. Genes & Development. 1991;**5**:2481-2495

[26] Campbell G, Tomlinson A. The roles of the homeobox genes *aristaless* and *Distal-less* in patterning the legs and wings of Drosophila. Development. 1998;**125**:4483-4493

[27] Zecca M, Struhl G. Subdivision of the Drosophila wing imaginal disc by EGFR-mediated signalling. Development. 2002;**129**:1357-1368

[28] Zecca M, Struhl G. Control of growth and patterning of the Drosophila wing imaginal disc by EGFR-mediated signaling. Development. 2002;**129**:1369-1376

[29] Diez del Corral R, Aroca P, Gómez-Skarmeta JL, Cavodeassi F, Modolell J. The Iroquois homeodomain proteins are required to specify body wall identity in Drosophila. Genes & Development. 1999;**13**:1754-1761

[30] Cavodeassi F, Rodriguez I, Modolell J. Dpp signalling is a key effector of the wing-body wall subdivision of the Drosophila mesothorax. Development. 2002;**129**:3815-3823

[31] Grieder NC, Morata G, Affolter M, Gehring WJ. Spalt major controls the development of the notum and of wing hinge primordia of the *Drosophila melanogaster* wing imaginal disc. Developmental Biology. 2009;**329**:315-326

[32] Milan M, Cohen SM. Temporal regulation of *apterous* activity during development of the Drosophila wing. Development. 2000;**127**:3069-3078

[33] Diaz-Benjumea F, Cohen SM. Interactions between dorsal and ventral cells in the imaginal disc directs wing development in Drosophila. Cell. 1993;**75**:741-752

[34] Williams JA, Paddock SW, Carroll SB. Pattern formation in a secondary field: A hierarchy of regulatory genes subdivides the developing Drosophila wing disc into discrete subregions. Development. 1993;**117**:571-584

[35] Diaz-Benjumea FJ, Cohen SM. Serrate signals through Notch to establish a wingless-dependent organized at the dorsal/ventral compartment boundary of the Drosophila wing. Development. 1995;**121**:4215-4225

[36] Kim J, Irvine KD, Carroll SB. Cell recognition, signal induction, and symmetrical gene activation at the dorsal-ventral boundary of the developing Drosophila wing. Cell. 1995;**82**:795-802

[37] Williams JA, Paddock SW, Vorwek K, Carroll SB. Organization of wing formation and induction of a wing-patterning gene at the dorso/ventral compartment boundary. Nature. 1994;**368**:299-305

[38] Klein T, Martinez-Arias A. Interactions among Delta, Serrate and Fringe modulate Notch activity during Drosophila wing development. Development. 1998;**125**:2951-2962

[39] de Celis JF. Positioning and differentiation of veins in the Drosophila wing. International Journal of Developmental Biology. 1998;**42**:335-343

[40] Mohler J, Seecoomar M, Agarwal S, Bier E, Hsai J. Activation of knot specifies the 3-4 intervein region in the Drosophila wing. Development. 2000;**127**:55-63

[41] Suzanne M. Molecular and cellular mechanisms involved in leg joint morphogenesis. Seminars in Cell & Developmental Biology. 2016;**55**:131-138

[42] Hao I, Green RB, Dunaevsky O, Lengyel JA, Rauskolb C. The odd-skipped family of zinc finger genes promotes Drosophila leg segmentation. Developmental Biology. 2003; **263**:282-295

[43] Cordoba S, Estella C. The bHLH-PAS transcription factor dysfusion regulates tarsal joint formation in response to Notch activity during Drosophila leg development. PLoS Genetics. 2014;**10**:e1004621

[44] Nüsslein-Volhard C, Roth S. Cellular Basis of Morphogenesis. New York: John Wiley and Sons. 1989. pp. 37-55

[45] Ashburner M, Aaron CS, Tsubota S. The genetics of a small autosomal region of *Drosophila melanogaster*, including the structural gene for alcohol dehydrogenase. V. Characterization of X-ray-induced *Adh* null mutations. Genetics. 1982;**102**:421-435

[46] Ashburner M et al. An exploration of the sequence of a 2.9-Mb region of the genome of *Drosophila melanogaster*: The Adh region. Genetics. 1999;**153**:179-219

[47] Therrien M, Morrison DK, Wong AM, Rubin GM. A genetic screen for modifiers of a kinase suppressor of Ras-dependent rough eye phenotype in Drosophila. Genetics. 2000; **156**:1231-1242

[48] Simon MA. Signal transduction during the development of the Drosophila R7 photoreceptor. Developmental Biology. 1994;**166**:431-442

[49] Raftery LA, Twombly V, Wharton K, Gelbart WM. Genetic screens to identify elements of the decapentaplegic signaling pathway in Drosophila. Genetics. 1995;**139**:241-254

[50] Jenny FH, Basler K. Powerful Drosophila screens that paved the wingless pathway. Fly (Austin). 2014;**8**:218-225

[51] Bellen HJ, Tong C, Tsuda H. 100 years of Drosophila research and its impact on vertebrate neuroscience: A history lesson for the future. Nature Reviews. Neuroscience. 2010;**11**:514-522

[52] St Johnston D. The art and design of genetic screens: *Drosophila melanogaster*. Nature Reviews. Genetics. 2002;**3**:176-188

[53] Adams MD et al. The genome sequence of *Drosophila melanogaster*. Science. 2000;**287**: 2185-2195

[54] Rong YS. Gene targeting by homologous recombination: A powerful addition to the genetic arsenal for Drosophila geneticists. Biochemical and Biophysical Research Communications. 2002;**297**:1-5

[55] Maggert KA, Gong WJ, Golic KG. Methods for homologous recombination in Drosophila. Methods in Molecular Biology. 2008;**420**:155-174

[56] Brand AH, Perrimon N. Targeted gene expression as a means of altering cell fates and generating dominant phenotypes. Development. 1993;**118**:401-415

[57] Boutros M, Ahringer J. The art and design of genetic screens: RNA interference. Nature Reviews. Genetics. 2008;**9**:554-566

[58] Dietzl G et al. A genome-wide transgenic RNAi library for conditional gene inactivation in Drosophila. Nature. 2007;**448**:151-156

[59] Mummery-Widmer JL et al. Genome-wide analysis of Notch signalling in Drosophila by transgenic RNAi. Nature. 2009;**458**:987-992

[60] Bassett AR, Tibbit C, Ponting CP, Liu JL. Mutagenesis and homologous recombination in Drosophila cell lines using CRISPR/Cas9. Biology Open. 2014;**3**:42-49

[61] Bassett AR, Tibbit C, Ponting CP, Liu JL. Highly efficient targeted mutagenesis of Drosophila with the CRISPR/Cas9 system. Cell Reports. 2013;**4**:220-228

[62] Gratz SJ et al. Genome engineering of Drosophila with the CRISPR RNA-guided Cas9 nuclease. Genetics. 2013;**194**:1029-1035

[63] Kondo S, Ueda R. Highly improved gene targeting by germline-specific Cas9 expression in Drosophila. Genetics. 2013;**195**:715-721

[64] Yu AM, McVey M. Synthesis-dependent microhomology-mediated end joining accounts for multiple types of repair junctions. Nucleic Acids Research. 2010;**38**:5706-5717

[65] Wang H et al. One-step generation of mice carrying mutations in multiple genes by CRISPR/Cas-mediated genome engineering. Cell. 2013;**153**:910-918

[66] Pathak T, Trivedi D, Hasan G. CRISPR-Cas-induced mutants identify a requirement for dSTIM in larval dopaminergic cells of *Drosophila melanogaster*. G3. 2017;**7**:923-933

[67] Baena-Lopez LA, Alexandre C, Mitchell A, Pasakarnis L, Vincent JP. Accelerated homologous recombination and subsequent genome modification in Drosophila. Development. 2013;**140**:4818-4825

[68] Gratz SJ et al. Highly specific and efficient CRISPR/Cas9-catalyzed homology-directed repair in Drosophila. Genetics. 2014;**196**:961-971

[69] Xue Z et al. Efficient gene knock-out and knock-in with transgenic Cas9 in Drosophila. G3. 2014;**4**:925-929

[70] Choulika A, Perrin A, Dujon B, Nicolas JF. Induction of homologous recombination in mammalian chromosomes by using the I-SceI system of *Saccharomyces cerevisiae*. Molecular Cell. Biology. 1995;**15**:1968-1973

[71] Ghosh S, Tibbit C, Liu JL. Effective knockdown of Drosophila long non-coding RNAs by CRISPR interference. Nucleic Acids Research. 2016;**44**:e84

[72] Lin S, Ewen-Campen B, Ni X, Housden BE, Perrimon N. In vivo transcriptional activation using CRISPR/Cas9 in Drosophila. Genetics. 2015;**201**:433-442

[73] Ewen-Campen B et al. Optimized strategy for in vivo Cas9-activation in Drosophila. Proceedings of the National Academy of Sciences of the United States of America. 2017;**114**:9409-9414

[74] Fujita T, Fujii H. Efficient isolation of specific genomic regions and identification of associated proteins by engineered DNA-binding molecule-mediated chromatin immunoprecipitation (enChIP) using CRISPR. Biochemical and Biophysical Research Communications. 2013;**439**:132-136

[75] Reiter LT, Potocki L, Chien S, Gribskov M, Bier E. A systematic analysis of human disease-associated gene sequences in *Drosophila melanogaster*. Genome Research. 2001; **11**:1114-1125

[76] Lehner B. Genotype to phenotype: Lessons from model organisms for human genetics. Nature Reviews. Genetics. 2013;**14**:168-178

[77] Yadav AK, Srikrishna S, Gupta SC. Cancer drug development using Drosophila as an in vivo tool: From bedside to bench and back. Trends in Pharmacological Sciences. 2016;**37**:789-806

[78] Bier E. Drosophila, the golden bug, emerges as a tool for human genetics. Nature Reviews. Genetics. 2005;**6**:9-23

[79] Lu B, Vogel H. Drosophila models of neurodegenerative diseases. Annual Review of Pathology. 2009;**4**:315-342

[80] McGary KL et al. Systematic discovery of nonobvious human disease models through orthologous phenotypes. Proceedings of the National Academy of Sciences of the United States of America. 2010;**107**:6544-6549

[81] Pastor-Pareja JC, Xu T. Dissecting social cell biology and tumors using Drosophila genetics. Annual Review of Genetics. 2013;**47**:51-74

[82] Graham P, Pick L. Drosophila as a model for diabetes and diseases of insulin resistance. Current Topics in Developmental Biology. 2017;**121**:397-419

[83] Jeibmann A, Paulus W. *Drosophila melanogaster* as a model organism of brain diseases. International Journal of Molecular Sciences. 2009;**10**:407-440

[84] Gilbert LI. Drosophila is an inclusive model for human diseases, growth and development. Molecular and Cellular Endocrinology. 2008;**293**:25-31

[85] Huang X, Warren JT, Buchanan J, Gilbert LI, Scott MP. Drosophila Niemann-Pick type C-2 genes control sterol homeostasis and steroid biosynthesis: A model of human neurodegenerative disease. Development. 2007;**134**:3733-3742

[86] Oishi K et al. Transgenic Drosophila models of Noonan syndrome causing PTPN11 gain-of-function mutations. Human Molecular Genetics. 2006;**15**:543-553

[87] Yamamoto S et al. A drosophila genetic resource of mutants to study mechanisms underlying human genetic diseases. Cell. 2014;**159**:200-214

[88] Atkinson D et al. Sphingosine 1-phosphate lyase deficiency causes Charcot-Marie-Tooth neuropathy. Neurology. 2017;**88**:533-542

[89] Oortveld MA et al. Human intellectual disability genes form conserved functional modules in Drosophila. PLoS Genetics. 2013;**9**:e1003911

[90] Iijima K et al. Dissecting the pathological effects of human Abeta40 and Abeta42 in Drosophila: A potential model for Alzheimer's disease. Proceedings of the National Academy of Sciences of the United States of America. 2004;**101**:6623-6628

[91] Finelli A, Kelkar A, Song HJ, Yang H, Konsolaki M. A model for studying Alzheimer's Abeta42-induced toxicity in *Drosophila melanogaster*. Molecular and Cellular Neurosciences. 2004;**26**:365-375

[92] Dar AC, Das TK, Shokat KM, Cagan RL. Chemical genetic discovery of targets and anti-targets for cancer polypharmacology. Nature. 2012;**486**:80-84

[93] Vidal M, Wells S, Ryan A, Cagan R. ZD6474 suppresses oncogenic RET isoforms in a Drosophila model for type 2 multiple endocrine neoplasia syndromes and papillary thyroid carcinoma. Cancer Research. 2005;**65**:3538-3541

[94] Rand MD. Drosophotoxicology: The growing potential for Drosophila in neurotoxicology. Neurotoxicology and Teratology. 2010;**1**:74-83

[95] Chang S et al. Identification of small molecules rescuing fragile X syndrome phenotypes in Drosophila. Nature Chemical Biology. 2008;**4**:256-263

[96] Lozano R, Hare EB, Hagerman RJ. Modulation of the GABAergic pathway for the treatment of fragile X syndrome. Neuropsychiatric Disease and Treatment. 2014;**10**:1769-1779

Section 2

# *Drosophila* - Model for Therapeutics

Chapter 6

# The Fruit Fly, *Drosophila melanogaster*: The Making of a Model (Part I)

Mariateresa Allocca, Sheri Zola and Paola Bellosta

Additional information is available at the end of the chapter

http://dx.doi.org/10.5772/intechopen.72832

**Abstract**

The fruit fly, *Drosophila melanogaster* (Meigen, 1830) has been established as a cornerstone for research into a wide array of subjects including diseases, development, physiology, and genetics. Thanks to an abundance of genetic tools, publicly available fly stocks, and databases, as well as their considerable biological similarity to mammalian systems, *Drosophila* has been solidified as a key model organism for elucidating many aspects of human disease. Herein is presented an overview of what makes *Drosophila* such an appealing model organism. In Part I of this chapter, basic *Drosophila* biology is reviewed and the most relevant genetic tools available to *Drosophila* researchers are covered. Then in part II, we outline the use of Drosophila as a model organism to study a wide array of pathologies in which *Drosophila* has been used, along with key advances made in the specific field using the fly as a model organism.

**Keywords:** animal model, cancer, diseases, *Drosophila*, genetic techniques, heart, immunology, kidney, metabolic disorders, neurodegeneration

## 1. Introduction

Searching PubMed with the key words "*Drosophila melanogaster* for model diseases", we find more than 2800 papers describing the use of this small and friendly invertebrate to study human diseases. This is quite remarkable considering that the genome of this animal is separated from ours by 795 million years. But what makes this organism so significant for the study of human diseases? First of all, the entire *Drosophila* genome has been sequenced [1] making it very simple to study and manipulate a particular gene. The *Drosophila* genome is 60% homologous to ours; in addition about 75% of the genes responsible for human diseases have a homolog in flies [2]. In addition, their small size (2–3 mm), short generation time, the easy and inexpensive way to culture them in the laboratory, and their powerful genetic tools have established *Drosophila* as one of the leading animal models for education and biomedical research [3]. Indeed *Drosophila* can be used anywhere from teaching basic genetics, to primary school, to understanding the more complicated metabolic pathways controlling fundamental physiological and pathological conditions.

It is little more than 100 years since Thomas Hunt Morgan and his colleagues, including his pupil Calvin Bridge and his wife Lillian Vaughan Morgan, redefined important concepts of *Drosophila* in the famous "fly-room" at Columbia University (**Figure 1**). He clarified the theory of inheritance previously defined by Mendel and the identification of the gene *white* earned him the Nobel Prize for Physiology and Medicine in 1933 for the role of chromosomes in heredity [4]. Since then research using fruit flies has contributed to numerous discoveries allowing for the identification of components of fundamental pathways that regulate the biology of animals as well as humans. Accomplishments that have been recognized over the years by subsequent other Nobel Laureates in Medicine and Physiology using *Drosophila* include:

- In 1946 to Hermann Joseph Muller for the use of X-ray irradiation to produce in vivo mutations.
- In 1995 to Edward B. Lewis, Christine Nuesslein-Volhard and Eric F. Wieschaus for their contribution to the discovery of the genetic control of early embryonic development.
- In 2011 to Bruce A. Beutler and Jules A. Hoffmann, for their success in defining innate immunity.
- In 2017 to Jeffrey C. Hall, Micheal Rosbash and Michael W. Young for their contributions to the molecular mechanisms that control the circadian rhythm.

Fly work has also benefitted from the strong commitment of *Drosophila* researchers to follow what is called the fly worker ethos, by adhering to the definition established by Bilder and Irvine [5]. This social behavior already set by Morgan and described by Kohler in 1994, describes a set of principles defined by the sharing of unpublished and published reagents,

**Figure 1.** Photograph of Morgan's Fly Room at Columbia University, around 1920. Courtesy of American Philosophical Society.

an open communication among members of the community, and the distribution of advanced genetic tools that have greatly helped the rapid advance of *Drosophila*'s powerful genetics. Since *Drosophila* strains cannot be maintained as frozen embryos, the obvious problem is that they must be kept as living cultures with routine work called "fly pushing". The ethos philosophy allowed the creation of several stock centers around the world that provide more than 80,000 *Drosophila* stock variants to customers. The Bloomington Drosophila Research Center (BDSC), hosted by the Indiana University in the USA, maintains the largest public collection of *Drosophila* lines and also supports fly research with basic protocols of fly work. In parallel, the Drosophila Genomics Resource Center (DGRC) collects vectors and cDNAs to further distribute them at a small fee to the community. There are also centers in Japan, China, and Europe that provide useful lines for screening, including an RNAi library, the large Vienna Drosophila RNAi Center (VDRC) and the TRIP-RNAi Harvard collections, all available through BDSC. FlyBase is the first database of integrated genetic and genomic data about *Drosophila melanogaster*, that also includes data from other species of *Drosophilidae*, created as an initiative to have all information about the methods for gene expression, development and physiology of *Drosophila*. FlyBase provided an example for the more specific databases like the Berkeley Drosophila Genome Project (BDGP) and modEncode. Nowadays the scope is to create a database for all Model Organisms that includes information not only from *Drosophila melanogaster* but also database information from *Rattus rattus* (Linnaeus, 1758), *Mus musculus* (Linnaeus, 1758), zebrafish *Danio renio* (Hamilton, 1882), the worm *Caenorhabditis elegans* (Maupas, 1900) and the yeast *Saccharomyces cerevisiae* (Meyen ex E.C. Hansen, 1883), reposted at the National Institute of Health (NIH).

Finally, *Drosophila* is also used as a model organism for educational purposes to illustrate in classrooms or to the public the relevance of genetics in biomedical research, or to explain concepts like inheritance. Sites like droso4school a promotion from the Manchester Fly Facility [6] or the Journal of Visual Experiments with the JoVe Science Education Data Base and Flymove are aimed at disseminating the relevance of *Drosophila* research to encourage and facilitate scientists to engage with primary and middle schools and with lay people.

In this two-part chapter, some of the many aspects that make *Drosophila* such a fundamental model organism are covered. Here, Part I outlines the basic biology and life cycle of the fly before summarizing some of the remarkable genetic tools available to *Drosophila* researchers. Part II will provide an overview of key disease states that *Drosophila* is used to model and some significant advances made in those fields.

## 2. Basic biology/life cycle

As a holometabolous insect, *Drosophila melanogaster* undergoes several drastic changes in body plan throughout its life. Progressing from an egg (embryo), to larva, then pupa, and finally adult, each stage provides a unique platform for studying a wide variety of diseases and conditions. Moreover, this animal's development is the result of a tight coordination between signals from hormones, (*in primis* ecdysone and prothoracicotropic hormone), nutrients (amino acids) and *Drosophila* insulin like peptides (Dilps) that act together to allow proper physiological growth of the animal.

### 2.1. Life cycle and regulation of development

The *Drosophila* life cycle lasts approximately 10 days at 25°C (**Figure 2**). A single fertile female can lay hundreds of eggs and *Drosophila* embryogenesis lasts approximately 24 h. During that time, the entire larval body plan is established though the expression of a number of critical genes, starting with several proteins transcribed from maternally derived mRNAs that were deposited at specific locations in the embryo such as *bicoid* and *dorsal* [7]. These proteins diffuse across it to establish both the anterior- posterior axis (e.g. bicoid) and the dorsal- ventral axis (e.g. dorsal). The diffusion of these proteins across the embryo forms gradients of each, and the varying levels of each protein will activate the transcription of specific cascades of genes including *gap genes*, *pair-wise genes*, *segment polarity genes*, and *hox genes* that will divide the embryo into segments, regions, and eventually structures [8]. Upon completing embryonic development, a first instar larva hatches from the egg and begins to eat. At this stage it is necessary for the larva to consume food not just for growth, but also to convert into storage as fats and sugars in the fat body, from where it will be used to sustain the larva though metamorphosis. As larvae grow, they shed their exoskeleton through a process called molting that is controlled by a fine tuned consequential series of events involving the hormones ecdysone, juvenile and the prothroacicotropic hormone (PTTH) (**Figure 3**) that control animal growth. With each molt, the larva will enter a new instar stage, progressing through three instars before a final molt to form a pupa. Each instar stage is regulated by the level of PTTH that rises to control the release of ecdysone allowing larvae to grow [9]. PTTH is released from neurosecretory cells in the brain, a process that is remotely controlled by growth of the larval organs and imaginal discs, sac-like structures of monoepithelial cells that become organs after metamorphosis, with the release of dilp8, a member of the secreted insulin-like peptide family, in a negative feed-back loop that controls PTTH production [10–12]. PTTH stimulates the release

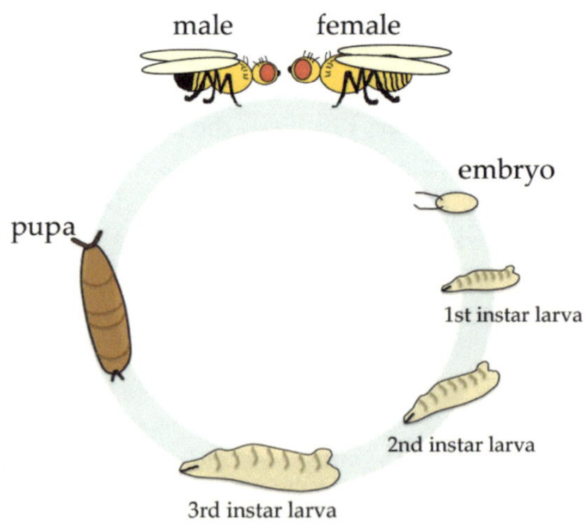


**Figure 2.** *Drosophila melanogaster* life cycle. The entire life cycle lasts approximately 10 days at 25°C. Flies complete embryonic development as eggs before hatching as first instar larvae. The larvae eat, grow, and molt though three instar stages before pupariating. Flies undergo metamorphosis during the pupal stage and adult structures are formed. Upon completing metamorphosis, an adult fly hatches.

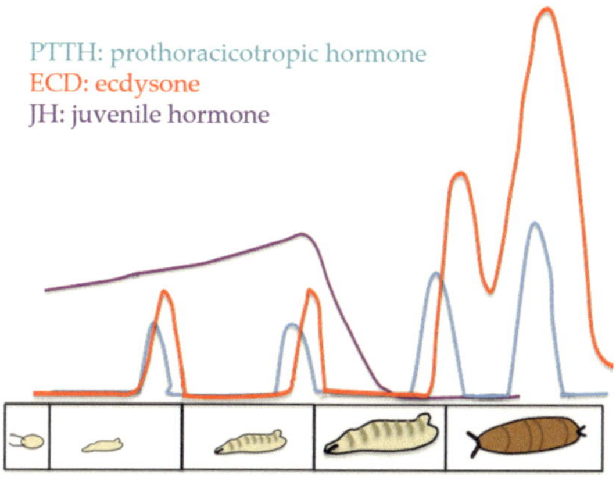


**Figure 3.** Hormonal control of molting and metamorphosis. This chart illustrates the levels in the hemolymph of hormones involved in the regulation of molting and metamorphosis. Before each molt, there is a burst of PTTH, triggering the release of ecdysone from the prothoracic gland. JH is present through most of larval development and instructs the larva to maintain the larval stage during each molt. When the critical weight is reached JH levels begin to fall, which occurs during the third instar stage. A large amount of growth occurs during the third instar stage until feeding stops and larvae begin to wander to pupariate, a stage triggered by a burst of PTTH and, subsequently ecdysone. A second larger burst of PTTH and ecdysone starts the final molt, and due to the absence of JH, the larva molts to the pupal stage.

of the molting hormone ecdysone from the prothoracic glands into the hemolymph causing the formation of a new cuticle (exoskeleton). As ecdysone levels fall again, another hormone, eclosion hormone (EH) initiates the actual molt where the larva sheds its exoskeleton and enters a new instar stage [13]. At the end of the third instar stage, larvae begin to wander to find a place to pupate, and are appropriately referred to as "wandering larvae." As adults do not grow, their final body size is primary regulated by the growth occurring after the critical size is reached [13]. Generally, larvae must attain a specific critical size where organs must be properly developed, for pupation to occur [14], a process highly regulated by both genetic and environmental conditions (see below). Metamorphosis arises during the pupal stage and most embryonic and larval structures degrade during this time while the imaginal discs, which consist mostly of undifferentiated epithelial cells, differentiate into the adult organs. During this transition, the animal cannot feed from the external environment and in order to survive activates a process called developmental- autophagy, a "self-eating" signaling mechanism that converts the stored nutrients (fats and sugars) from the fat body into nutrients and macromolecules necessary to produce the energy required for the animal survival [15, 16].

### 2.2. Regulation of body size the interplay between hormones and growth factors

Larval growth is regulated by the interplay of the function of different organs (**Figure 4**), among which the fat body works as a hub to regulate several important processes. First, by sensing the amino acid concentration in the hemolymph, the fat body remotely controls the release of Dilps, in particular dilp2, from the Insulin Producing Cells (IPCs) in the brain [17]. This mechanism depends on the release into the hemolymph of secreted factors, like the Growth-Blocking Peptide-1 (GBP1) and CG11395 (GBP2) [18] and Stunted [19] with a mechanism that is dependent on the activation of the Target of Rapamycin (TOR) pathway in

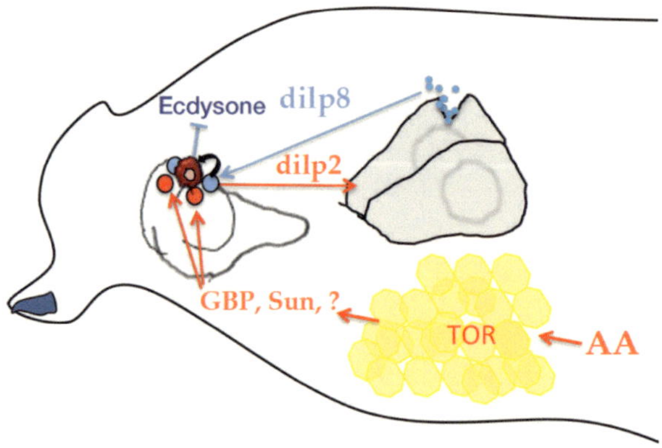


**Figure 4.** Integration of signals to control growth and metamorphosis. This schematic illustrates the coordination between nutritional status and developmental status to control the timing of metamorphosis and growth. The fat body senses amino acid concentration through TOR signaling and releases factors (in red) including growth-blocking peptides (GBPs), stunted (Sun), and other unknown factors (X?) into the hemolymph. These are sensed by the insulin producing cells of the brain, which then release dilp2 to cause growth. To ensure all organs are properly developed before metamorphosis, dilp8 is released from damaged or regenerating tissues to inhibit the production of PTTH in the brain, thereby blocking ecdysone levels to delay metamorphosis.

the fat cells. Second, the fat body controls animal survival with the activation of autophagy, consuming the fats and sugars that accumulated during the feeding phase. Third, the fat body responds to reduced ecdysone signaling from the brain by restraining metabolism and protein synthesis cell-autonomously before each molt by controlling the expression of the growth regulator Myc [20], which was shown to also regulate growth and Dilp2 secretion [21] constituting a regulatory loop that controls animal growth. Insulin signaling is the foremost important growth signal that in flies controls both growth/development and metabolism, with a unique and conserved pathway [22]. Dilps are produced by different organs and activate the Insulin Receptor (InR). Among Dilps (1-8), Dilp2, 3, 5 are produced by the IPCs in the brain and control animal growth and development [22, 23] while Dilp6, produced by the fat body and regulated by FOXO, functions to indirectly restrain Dilp2 secretion from the IPCs and to regulate longevity in the adult flies, a function similar to mammalian InR in aging [24]. A novel and exciting function was recently identified for dilp8, the last member of the Dilp family, to indirectly control ecdysone levels [25, 26]. *Dilp8* is a gene that encodes for a protein in the insulin/relaxin like family of peptides, originally identified for its control of bilateral symmetry [25, 26]. It is also produced by damaged or regenerating tissues, and is released into the hemolymph to remotely bind the *Drosophila* Leucine-rich repeat containing G protein-coupled receptor 3 (Lgr3), a member of the relaxin receptor expressed by specific neuronal cells located in a cluster in both sides of the brain [10–12]. These cells are part of a newly identified neuroendocrine circuit that ultimately acts on the hormone PTTH to reduce the levels of ecdysone during development. The growth of the imaginal discs times pupariation; indeed the time of pupariation is delayed with a reduction of ecdysone levels if a disc is injured. Thus, dilp8 activity seems to coordinate other signals to control the timing of pupariation and growth of the discs to ensure proper development of the animal [11]. These data on the function of dilp8 are described as part of the neurosecretory cell-rich *pars intercerebralis*, an axis that mediates the larval to pupal transition, highlighting the presence in *Drosophila* of a mechanism similar to the hypothalamic pituitary axis in vertebrates to control development.

## 3. Fly genetics

### 3.1. Generation of transgenes

*Drosophila melanogaster* have four pairs of chromosomes, the first pair (X or Y) are also the sex chromosomes. Generally, having two X chromosomes designates the fly as female, whereas an X and Y will designate a male; however, the Y chromosome is extremely small and contains very few genes. Of the 2nd, 3rd, and 4th pairs, the 4th is the smallest and less commonly noticed because of the difficulty to insert transgenes and to obtain balancer lines [27]. A huge array of genetic techniques exists to allow researchers to manipulate the fly genome to overexpress, knock-down, mutate, tag, or alter the expression of a gene or genes [28]. Nearly all techniques are based on the ability to insert a foreign piece of DNA into the fly genome, generating what are referred to as transgenes. This foreign DNA can consist of an entire gene, a promoter region, gene fragments, mutated genes or almost any DNA sequence a researcher desires.

#### *3.1.1. P-element transposons*

Several commonly used techniques exist to integrate DNA into the genome. Transposon mediated integration, first utilized by Rubin and Spradling in 1982 [29] is one of the most commonly employed methods [28]. This technique capitalizes on the action of the P-element transposon. Transposons are pieces of DNA with specific sequence characteristics that have the ability to cut themselves out of the genome and reintegrate in another location through the action of the transposase enzyme [30]. The transposons can be modified though cloning to contain a desired piece of DNA Plasmids containing the modified P-element constructs are injected into the *Drosophila* embryo germline cells together with plasmids encoding for the P transposase, the latter of which will subsequently chop out the P-element backbone from the first plasmid and insert the entire segment into a random point in the germ cell genome [31]. When these animals reach adulthood, they will produce a number of offspring that contain the modified transgene containing chromosomes in every cell. To determine whether the foreign DNA has been integrated, it is often necessary to link a marker gene to the desired transgene. This marker gene is usually inserted in the plasmid linked to the desired transgene and produces an easily observable phenotype, such as eye or body color, to clearly identify which flies contain the transgene (**Figure 5**). Marker genes are also used with many of the other techniques described below.

#### *3.1.2. Homologous recombination*

P-element transposon mediated transgenesis has several drawbacks, including that the location of the insertion cannot be selected and sometimes the transgene may be inserted within the regulatory or coding region of another gene and disrupt its function [32]. Rong and Golic in 2000 pioneered a procedure that can target specific genes in the *Drosophila* genome using

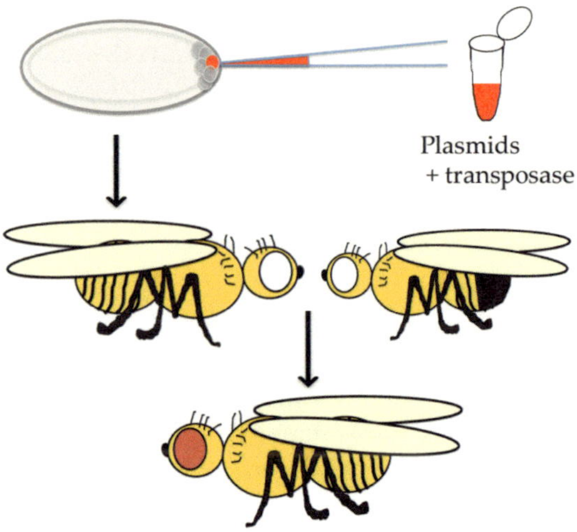


**Figure 5.** Microinjection of *Drosophila* embryos to generate transgenic flies. In this schematic, the embryo of a white eyed fly is injected with two types of plasmids: plasmids containing a P-element transposon with a desired transgene linked to a marker gene that will produce red eye pigment, and a plasmid containing the P-transposase enzyme. The injection occurs in the location of the embryo that will become the germline cells. Once both constructs are injected, the transposase enzyme will be produced and chop the entire transposon construction out of the other plasmid then, hopefully, insert it into the genome of the developing germ cells. The embryos will undergo development and the adult flies will now potentially contain sperm or eggs with the desired transgene. The flies can be mated, and if their offspring contain the transgene, they will have red eyes thanks to the expression of the marker gene.

homologous recombination [33]. This technique uses the cell's own DNA repair machinery and homologous recombination to swap out one allele or piece of DNA for another and can even be used to knockout genes [34]. Fly lines must first be generated, often by using P-element transposons, that express: 1- the site-specific recombinase flippase (FLP), an enzyme that recognizes specific sequences known as FLP recombination targets (FRT) and will excise DNA contained between FRT sequences to catalyze recombination; 2- a site-specific endonuclease (I-Sce1), which is an enzyme that generates a double stranded break in DNA at a specific sequence; and 3- a transgenic donor construct that contains the FRT sites on either end, a recognition site for the endonuclease, and some sequence similarity to the gene or area that is to be targeted [33]. When all three of these elements are brought together in a single fly, generally by mating, the FLP recombinase can excise the DNA contained between the two FRT sites creating a circular exogenous piece of DNA that the endonuclease will then cleave at its recognition site to linearize it. The broken piece of exogenous DNA will then recombine with the genomic area that it has been designed to target, effectively inserting itself into the genome at the desired area. In this way, an endogenous gene can be removed (knockout) or modified [33, 35]. While this method is effective at targeting specific genes for modification, it is very labor intensive, often requiring the generation of at least three different transgenic fly lines to target just one gene.

Several other methods that generate double stranded breaks to trigger homologous recombination have been developed [28]. These include using zinc-finger nucleases (ZFNs) and transcription activator-like effector nucleases (TALEN), which are enzymes that can target specific

DNA sequences and cause double stranded breaks, however each gene requires generating a new specific enzyme and can be challenging [36].

### *3.1.3. phiC31 integrase: site-specific integrase insertion*

Another method to target specific locations in the genome uses the bacteriophage ϕC31 integrase which can insert a transgene at a specific recognition site in the genome [37]. Bacteriophages are viruses that target bacteria. The ϕC31 integrase is an enzyme that recognizes specific attachment sites in both the bacteriophage genome (designated *attP*) and in its bacterial host's genome (*attB*) and catalyzes recombination between the two to insert itself into the bacteria's genome [38]. *Drosophila* lines have been generated with *attP* sites inserted in their genome using P-element transposons, allowing the ϕC31 integrase to effectively insert a large segment of DNA from a plasmid containing the *attB* site via recombination at the *attP* site in the *Drosophila* genome [37]. This technique is an effective method for inserting large pieces of DNA, something that is problematic for P-element transposons, and allows researchers to insert their transgene at a specific site [28].

### *3.1.4. Bacterial artificial chromosomes*

Bacterial artificial chromosomes (BACs) and recombination engineering (recombineering) are gaining traction in the *Drosophila* community because they allow the insertion of very large pieces of DNA [28]. Systems been developed that use specially modified BACs that are easy to amplify [39]. Genes or other DNA fragments are inserted into these constructs using recombineering, a process much easier than cutting DNA with restriction enzymes and reattaching it with ligases [40]. One BAC in particular combines several technological platforms and contains P-element transposase recognition sites and ϕC31-mediated integration sites [40].

### *3.1.5. CRISPR/Cas9*

The CRISPR/*Cas9* system is another rapidly expanding technology for altering the genome [41]. First coined in 2002, Clustered Regularly Interspersed Short Palindromic Repeats (CRISPR) arrays are sets of repetitive nucleotide sequences with interspaced non-repetitive sequences that function in bacterial immunity against viruses [42, 43]. There are also a number of CRISPR-associated genes (*Cas* genes). Parts of the CRISPR arrays are transcribed and processed into shorter CRISPR RNA's (crRNA) and noncoding trans-activating crRNA (tracrRNA) [44, 45]. After being infected with a bacteriophage, cas enzymes (encoded by several *Cas* genes) allow bacteria to keep copies of viral DNA within their genome in the form of CRISPR arrays. When the virus attacks again, these copies are transcribed and processed into crRNA and tracrRNA that together target the virus to be cleaved by other cas enzymes like Cas9 [42, 45–47]. In 2013, researchers began to exploit this technology to modify the mammalian genome [48, 49] and the *Drosophila* genome [50]. By taking advantage of the fact that Cas9 is guided by crRNA and tracrRNA to cleave a specific DNA sequence, artificial constructs can be designed to target any desired gene. The double strand break generated by CAs9 is then repaired either by homologous recombination with an exogenous piece of DNA containing a mutated form of the gene or any other desired piece of DNA, or through error-prone non homologous end joining [48–50]. In this way, any gene or part of the genome can be easily modified.

### 3.2. Generation of mutants

In addition to ectopically expressing or reducing the function of a gene using the classic binary system (see last section in methods), another useful way to study gene function is to generate mutations in the genome and observe the resulting phenotypes, and then work backwards to figure out what gene was modified. The function of this gene can be inferred by the phenotype that occurred when the gene was destroyed. These studies involve mutating a large number of genes in many flies, then screening though the phenotypes and determining what genes were altered. There are a number of ways to generate mutants, including using P-element transposons and chemical mutagens like Ethyl Methanesulfonate (EMS).

#### *3.2.1. P-element mutagenesis*

This technique utilizes P-elements (usually containing gene markers as described above), or other transposable elements, to move around in the genome to disrupt gene function. This is possible either by inserting themselves in a new position that could interfere with a gene or removing it from a gene and degrading a little bit more of the DNA sequence from where it was removed [51, 52]. Though P-elements show certain preferences for where they reinsert, they cannot be directed to a specific location and have no precise recognition sequence [30]. It is therefore necessary after P-element mobilization to screen the flies that show an altered phenotype to determine which gene or genes were disrupted and use PCR to identify where the insertion occurred [51].

#### *3.2.2. EMS mutagenesis*

This method uses the chemical EMS to generate random mutations in the genome [53]. EMS produces a form of guanine, $O^6$-methylguanine, that incorrectly base pairs with thymine during DNA replication, usually generating GC to AT transitions that potentially alter codons or destroy splice sites, which will damage the function of a gene product. These mutations are generated at random and while some create visible phenotypes, or even lethality, others show no obvious changes, so extensive screening is needed to determine which mutation or mutations caused the observed phenotypes [53].

### 3.3. Genetic screens

The large use of *Drosophila* as a model organism is also due to the ability to perform genetic screens aimed at the identification of novel genes important for understanding biological process. In order to do so, Drosophilists have developed balancer chromosomes that are essential to maintain mutant fly stocks and for mating design. These chromosomes carry multiple inversions and gene rearrangement that firstly suppress recombination, and secondly contain a dominant phenotypic marker visible in the larvae or adults. Because they segregate normally during meiosis, they can be followed using the dominant phenotype. All the information on balancer stocks are available from *Drosophila* stock centers.

A genetic screen can follow two main strategies: Forward or Reverse Genetics [54]. A Forward Genetics approach is based on random, genome-wide mutagenesis to generate a large progeny with aberrant phenotypes and allows the identification of individual genes involved in any given process. Traditional forward genetic screens in *Drosophila* use X-rays, chemical EMS and transposon mutagenesis [53]. Reverse Genetics instead is a targeted mutagenesis applied to a gene of a known visible phenotype and is normally used to understand the gene's biological function. The goal is to find new mutations that enhance the preexisting mutant phenotype or that suppress it [55, 56], but these genetic screens are often laborious and time consuming. The best strategy is to start with a strong phenotype. This type of mutagenesis can be accomplished via numerous mechanisms, such as classical loss of function mutant alleles, transposable P-element insertions existing for virtually all gene loci, knock-down using RNA interference strategies, or more recently using the CRISPR/*Cas9* techniques (see previous section).

### 3.4. Most common techniques in *Drosophila*

A huge step forward on the feasibility of genetic screens was improved by the generation of the UAS/GAL4 system [57] that that allowed the expression of transgenes within specific tissues of interest.

#### *3.4.1. The UAS/GAL4 system and its modifications*

This system requires the use of lines that are generated and maintained in separate stocks and targeted gene expression will be visible only in the progeny of the cross. Using the yeast transcription factor GAL4 cloned into a P-element vector, a tissue specific promoter is cloned upstream of the GAL4 gene. In parallel, a line is generated that includes a P-element vector containing the upstream activating sequences (UAS) to which GAL4 protein can bind [58]. This binary expression system is used to drive the expression of a gene of interest in any tissues where the promoter GAL4 is expressed (**Figure 6**). Because experimental design may demand expression in a more limited time window (i.e. in adult only or if the expression of the gene of interest is detrimental), the UAS/GAL4 system is often accompanied by the use of the yeast *GAL80*, a gene that encodes for a protein that physically binds to GAL4 and represses its activity [59]. This strategy was improved by the use of the ubiquitously expressed temperature-sensitive allele of GAL80 ($GAL80^{ts}$), that is active and binds GAL4 at the permissive temperature of 18°C, while at the restrictive temperature of 29°C GAL80 is degraded [60]. Another method to modulate the activity of GAL4 is to use the hormone inducible variant of GAL4, either the GAL-ER, where GAL4 is fused with the domain of the human estrogen receptor and activated by estradiol [61], or the GeneSwitch system [62] where GAL4 is fused with the domain that binds the human progesterone receptor and is activated by RU486. Finally, a more sophisticated system is the Split GAl4 [58], which allows a better control of the timing of the expression. This method is based on the use the DNA binding domain of GAL4 and its transcriptional activation domain fused separately to a promoter or hemi drivers. Only when their expression domain overlaps is the active GAL4 reconstituted and activation occurs [63].

Nowadays, the number of GAL4 lines available is constantly growing. There are UAS lines both for overexpression or RNAi interference targeting almost for all the genes in the fly. UAS- lines

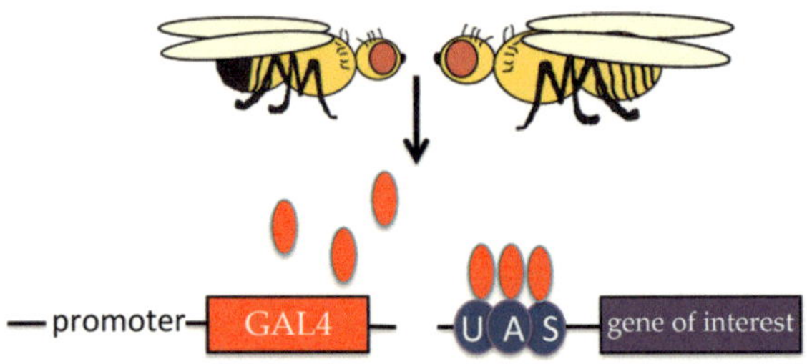


**Figure 6.** Gene expression using the GAL4/UAS system. This diagram illustrates how to drive expression of a gene of interest in a specific tissue using the GAL4/UAS system. Here a female fly is carrying the transgene for the GAL4 transcription factor (indicated in red) downstream of a tissue specific promoter region. This female fly is mated with a male fly carrying another transgene in which a UAS (upstream activating sequence) is upstream of a "gene of interest," indicated in purple. The UAS sequence is the binding site for the GAL4 transcription factor. The offspring of these flies will now have both constructs. In these flies, a tissue specific transcription factor will bind to the promoter region of the GAL4 transcription factor and that GAL4 will be transcribed then translated (indicated in red) and subsequently bind to the UAS (indicated in blue) upstream of the gene of interest and activate the transcription of this gene. In this way, the gene of interest will only be produced in a designated tissue because of the specificity of the promoter placed upstream of the GAL4 transcription factor.

with more applied specific modifications, like the enhancer-trap GFP vectors, include those from the Janelia Farm Fly light project that created more than 7000 driver lines with an intergene overlapping sequence of 3 kb fragment near the gene of interest. These lines have been characterized for their expression pattern in embryos, larvae brain and adult CNS [64] and in the larval imaginal discs [65], available at the BDRC stock center. The use of binary systems is continually evolving to provide even more inter-exchangeable systems. Indeed the recent design of the LexA/lexAop [66] and the Q system [67], both inducible systems that can be used in combination with GAL4/UAS gene expression, allows researchers to perform screens in a tissue using the UAS/GAL4, with the specific patterns of expression determined by the LexA/LexAop or Q system. They can be used simultaneously in the same animal because neither of these systems cross-react to each-other.

### *3.4.2. The FLP/FRT system and Mosaic Analysis with a Repressible Cell Marker*

To characterize the role of a gene in a small group of cells and not in the whole compartment, or to analyze the role of a mutation, it is possible to create mosaics that have homozygous mutant cells (clones) in an otherwise heterozygous animal via mitotic recombination. These studies were made possible with the combined used of the UAS/GAL4 system with the *Saccharomyces cerevisiae* enzyme FLP, a recombinase that recognizes 34 bp recombination target sequences (FRTs) on DNA [68]. FRTs on chromosomes enable mitotic recombination between homologous chromosomes in the presence of the FLP recombinase (FLP/FRT system) [68, 69]. Ubiquitous promoters like actin5C or tubulin1$\alpha$ were cloned separated from GAL4 by a FLP-out cassette containing an inert gene between the two FRTs [70]. When flies carrying the cassette are crossed with flies expressing the FLP recombinase under control of the hsp70 heat shock promoter, a heat shock temporally activates the recombination event in their progeny; FRTs remove the cassette allowing the expression of GAL4 and results in random clone-induction in all tissues of the animal. The timing and duration of the heat shock determines the number of cells in which the recombination event occurs and can be recombined with GAL80ts to restrict GAL4 expression. The FLP/FRT system can be used to manipulate gene expression when associated with UAS lines (overexpression or RNAi) and to generate loss-of-function clones in heterozygous mutant tissues [71]. Mosaic Analysis with a Repressible Cell Marker (MARCM) [72] is a technique that allows the expression of a marker or of a gene of interest in mutant clones, and is based on the ability of GAL80 to inhibit GAL4 activity and to produce positively labeled clones. In this case, GAL4 and GAL80 need to be expressed ubiquitously using the tubulin 1$\alpha$ promoter. The *tubGal80* transgene is in trans with the mutation and distal to a FRT site. The event of recombination results in two populations of daughter cells, in which one inherits two copies of tubGAL80 and is wild-type, while the other one that loses the tubGAL80 and is homozygous mutant. Loss of GAL80 de-represses GAL4 allowing the expression of a marker or of a gene of interest in the mutant clone.

## 4. Conclusions

As illustrated in Part I of this chapter, *Drosophila melanogaster* provides a very useful platform for studying a variety of diseases and conditions. The basic lifecycle and biology of *Drosophila* combined with a wide array of genetic tools allows researchers to easily and quickly manipulate the function of any gene or genes of interest. In Part II of this chapter, it will be shown how *Drosophila* is employed as a model to study a variety of pathologies and to uncover potential therapies.

## Conflict of interest

The authors declare no conflict of interest.

## Author details

Mariateresa Allocca, Sheri Zola and Paola Bellosta*

*Address all correspondence to: paola.bellosta@unitn.it

Center of Integrated Biology, University of Trento, Trento, Italy

## References

[1] Adams MD, Celniker SE, Holt RA, Evans CA, Gocayne JD, Amanatides PG, Scherer SE, Li PW, Hoskins RA, Galle RF, et al. The genome sequence of *Drosophila melanogaster*. Science. 2000;**287**(5461):2185-2195

[2] Pandey UB, Nichols CD. Human disease models in *Drosophila melanogaster* and the role of the fly in therapeutic drug discovery. Pharmacological Reviews. 2011;**63**(2):411-436

[3] Millburn GH, Crosby MA, Gramates LS, Tweedie S, FlyBase C. FlyBase portals to human disease research using Drosophila models. Disease Models & Mechanisms. 2016;**9**(3):245-252

[4] Kenney DE, Borisy GG. Thomas Hunt Morgan at the marine biological laboratory: Naturalist and experimentalist. Genetics. 2009;**181**(3):841-846

[5] Bilder D, Irvine KD. Taking stock of the Drosophila research ecosystem. Genetics. 2017;**206**(3):1227-1236

[6] Patel S, Prokop A. The Manchester Fly Facility: Implementing an objective-driven long-term science communication initiative. Seminars in Cell & Developmental Biology. 2017;**70**:38-48

[7] Nusslein-Volhard C, Frohnhofer HG, Lehmann R. Determination of anteroposterior polarity in Drosophila. Science. 1987;**238**(4834):1675-1681

[8] Gebelein B, McKay DJ, Mann RS. Direct integration of Hox and segmentation gene inputs during Drosophila development. Nature. 2004;**431**(7009):653-659

[9] Nijhout HF, Williams CM. Control of moulting and metamorphosis in the tobacco hornworm, *Manduca sexta* (L.): Growth of the last-instar larva and the decision to pupate. The Journal of Experimental Biology. 1974;**61**(2):481-491

[10] Colombani J, Andersen DS, Boulan L, Boone E, Romero N, Virolle V, Texada M, Leopold P. Drosophila Lgr3 couples organ growth with maturation and ensures developmental stability. Current Biology : CB. 2015;**25**(20):2723-2729

[11] Garelli A, Heredia F, Casimiro AP, Macedo A, Nunes C, Garcez M, Dias AR, Volonte YA, Uhlmann T, Caparros E, et al. Dilp8 requires the neuronal relaxin receptor Lgr3 to couple growth to developmental timing. Nature Communications. 2015;**6**:8732

[12] Vallejo DM, Juarez-Carreno S, Bolivar J, Morante J, Dominguez M. A brain circuit that synchronizes growth and maturation revealed through Dilp8 binding to Lgr3. Science. 2015;**350**(6262):aac6767

[13] Nijhout HF, Riddiford LM, Mirth C, Shingleton AW, Suzuki Y, Callier V. The developmental control of size in insects. Wiley Interdisciplinary Reviews Developmental Biology. 2014;**3**(1):113-134

[14] Edgar BA. How flies get their size: Genetics meets physiology. Nature Reviews Genetics. 2006;**7**(12):907-916

[15] Neufeld TP, Baehrecke EH. Eating on the fly: Function and regulation of autophagy during cell growth, survival and death in Drosophila. Autophagy. 2008;**4**(5):557-562

[16] Anding AL, Baehrecke EH. Autophagy in cell life and cell death. Current Topics in Developmental Biology. 2015;**114**:67-91

[17] Geminard C, Rulifson EJ, Leopold P. Remote control of insulin secretion by fat cells in Drosophila. Cell Metabolism. 2009;**10**(3):199-207

[18] Koyama T, Mirth CK. Growth-blocking peptides as nutrition-sensitive signals for insulin secretion and body size regulation. PLoS Biology. 2016;**14**(2):e1002392

[19] Delanoue R, Meschi E, Agrawal N, Mauri A, Tsatskis Y, McNeill H, Leopold P. Drosophila insulin release is triggered by adipose Stunted ligand to brain Methuselah receptor. Science. 2016;**353**(6307):1553-1556

[20] Delanoue R, Slaidina M, Leopold P. The steroid hormone ecdysone controls systemic growth by repressing dMyc function in Drosophila fat cells. Developmental Cell. 2010;**18**(6):1012-1021

[21] Parisi F, Riccardo S, Zola S, Lora C, Grifoni D, Brown LM, Bellosta P. dMyc expression in the fat body affects DILP2 release and increases the expression of the fat desaturase Desat1 resulting in organismal growth. Developmental Biology. 2013;**379**(1):64-75

[22] Partridge L, Alic N, Bjedov I, Piper MD. Ageing in Drosophila: The role of the insulin/Igf and TOR signalling network. Experimental Gerontology. 2011;**46**(5):376-381

[23] Gronke S, Clarke DF, Broughton S, Andrews TD, Partridge L. Molecular evolution and functional characterization of Drosophila insulin-like peptides. PLoS Genetics. 2010;**6**(2):e1000857

[24] Bai H, Kang P, Tatar M. Drosophila insulin-like peptide-6 (dilp6) expression from fat body extends lifespan and represses secretion of Drosophila insulin-like peptide-2 from the brain. Aging Cell. 2012;**11**(6):978-985

[25] Colombani J, Andersen DS, Leopold P. Secreted peptide Dilp8 coordinates Drosophila tissue growth with developmental timing. Science. 2012;**336**(6081):582-585

[26] Garelli A, Gontijo AM, Miguela V, Caparros E, Dominguez M. Imaginal discs secrete insulin-like peptide 8 to mediate plasticity of growth and maturation. Science. 2012;**336**(6081):579-582

[27] Stocker H, Gallant P. Getting started: An overview on raising and handling Drosophila. Methods in Molecular Biology. 2008;**420**:27-44

[28] Venken KJ, Bellen HJ. Transgenesis upgrades for *Drosophila melanogaster*. Development. 2007;**134**(20):3571-3584

[29] Rubin GM, Spradling AC. Genetic transformation of Drosophila with transposable element vectors. Science. 1982;**218**(4570):348-353

[30] O'Hare K, Rubin GM. Structures of P transposable elements and their sites of insertion and excision in the *Drosophila melanogaster* genome. Cell. 1983;**34**(1):25-35

[31] Karess RE, Rubin GM. Analysis of P transposable element functions in Drosophila. Cell. 1984;**38**(1):135-146

[32] Bellen HJ. Hugo J. Bellen. Current Biology: CB. 2004;**14**(6):R218

[33] Rong YS, Golic KG. Gene targeting by homologous recombination in Drosophila. Science. 2000;**288**(5473):2013-2018

[34] Rong YS, Golic KG. A targeted gene knockout in Drosophila. Genetics. 2001;**157**(3): 1307-1312

[35] Rong YS, Titen SW, Xie HB, Golic MM, Bastiani M, Bandyopadhyay P, Olivera BM, Brodsky M, Rubin GM, Golic KG. Targeted mutagenesis by homologous recombination in *D. melanogaster*. Genes & Development. 2002;**16**(12):1568-1581

[36] Gaj T, Gersbach CA, Barbas CF 3rd. ZFN, TALEN, and CRISPR/Cas-based methods for genome engineering. Trends in Biotechnology. 2013;**31**(7):397-405

[37] Groth AC, Fish M, Nusse R, Calos MP. Construction of transgenic Drosophila by using the site-specific integrase from phage phiC31. Genetics. 2004;**166**(4):1775-1782

[38] Thorpe HM, Smith MC. In vitro site-specific integration of bacteriophage DNA catalyzed by a recombinase of the resolvase/invertase family. Proceedings of the National Academy of Sciences of the United States of America. 1998;**95**(10):5505-5510

[39] Wild J, Hradecna Z, Szybalski W. Conditionally amplifiable BACs: Switching from single-copy to high-copy vectors and genomic clones. Genome Research. 2002;**12**(9):1434-1444

[40] Venken KJ, He Y, Hoskins RA, Bellen HJ. P[acman]: A BAC transgenic platform for targeted insertion of large DNA fragments in *D. melanogaster*. Science. 2006;**314**(5806):1747-1751

[41] Hsu PD, Lander ES, Zhang F. Development and applications of CRISPR-Cas9 for genome engineering. Cell. 2014;**157**(6):1262-1278

[42] Barrangou R, Fremaux C, Deveau H, Richards M, Boyaval P, Moineau S, Romero DA, Horvath P. CRISPR provides acquired resistance against viruses in prokaryotes. Science. 2007;**315**(5819):1709-1712

[43] Jansen R, Embden JD, Gaastra W, Schouls LM. Identification of genes that are associated with DNA repeats in prokaryotes. Molecular Microbiology. 2002;**43**(6):1565-1575

[44] Brouns SJ, Jore MM, Lundgren M, Westra ER, Slijkhuis RJ, Snijders AP, Dickman MJ, Makarova KS, Koonin EV, van der Oost J. Small CRISPR RNAs guide antiviral defense in prokaryotes. Science. 2008;**321**(5891):960-964

[45] Deltcheva E, Chylinski K, Sharma CM, Gonzales K, Chao Y, Pirzada ZA, Eckert MR, Vogel J, Charpentier E. CRISPR RNA maturation by trans-encoded small RNA and host factor RNase III. Nature. 2011;**471**(7340):602-607

[46] Deveau H, Barrangou R, Garneau JE, Labonte J, Fremaux C, Boyaval P, Romero DA, Horvath P, Moineau S. Phage response to CRISPR-encoded resistance in *Streptococcus thermophilus*. Journal of Bacteriology. 2008;**190**(4):1390-1400

[47] Jinek M, Chylinski K, Fonfara I, Hauer M, Doudna JA, Charpentier E. A programmable dual-RNA-guided DNA endonuclease in adaptive bacterial immunity. Science. 2012;**337**(6096):816-821

[48] Cong L, Ran FA, Cox D, Lin S, Barretto R, Habib N, Hsu PD, Wu X, Jiang W, Marraffini LA, et al. Multiplex genome engineering using CRISPR/Cas systems. Science. 2013;**339**(6121): 819-823

[49] Andersen DS, Colombani J, Palmerini V, Chakrabandhu K, Boone E, Rothlisberger M, Toggweiler J, Basler K, Mapelli M, Hueber AO, et al. The Drosophila TNF receptor Grindelwald couples loss of cell polarity and neoplastic growth. Nature. 2015;**522**(7557): 482-486

[50] Gratz SJ, Cummings AM, Nguyen JN, Hamm DC, Donohue LK, Harrison MM, Wildonger J, O'Connor-Giles KM. Genome engineering of Drosophila with the CRISPR RNA-guided Cas9 nuclease. Genetics. 2013;**194**(4):1029-1035

[51] Hummel T, Klambt C. P-element mutagenesis. Methods in Molecular Biology. 2008;**420**:97-117

[52] Cooley L, Berg C, Spradling A. Controlling P element insertional mutagenesis. Trends in Genetics : TIG. 1988;**4**(9):254-258

[53] Bokel C. EMS screens : From mutagenesis to screening and mapping. Methods in Molecular Biology. 2008;**420**:119-138

[54] St Johnston D. The art and design of genetic screens: *Drosophila melanogaster*. Nature Reviews Genetics. 2002;**3**(3):176-188

[55] Moulton MJ, Letsou A. Modeling congenital disease and inborn errors of development in *Drosophila melanogaster*. Disease Models & Mechanisms. 2016;**9**(3):253-269

[56] Hales KG, Korey CA, Larracuente AM, Roberts DM. Genetics on the fly: A primer on the Drosophila model system. Genetics. 2015;**201**(3):815-842

[57] Brand AH, Perrimon N. Targeted gene expression as a means of altering cell fates and generating dominant phenotypes. Development. 1993;**118**(2):401-415

[58] Elliott DA, Brand AH. The GAL4 system: A versatile system for the expression of genes. Methods in Molecular Biology. 2008;**420**:79-95

[59] Nogi Y, Shimada H, Matsuzaki Y, Hashimoto H, Fukasawa T. Regulation of expression of the galactose gene cluster in *Saccharomyces cerevisiae*. II. The isolation and dosage effect of the regulatory gene GAL80. Molecular & General Genetics : MGG. 1984;**195**(1-2):29-34

[60] Zeidler MP, Tan C, Bellaiche Y, Cherry S, Hader S, Gayko U, Perrimon N. Temperature-sensitive control of protein activity by conditionally splicing inteins. Nature Biotechnology. 2004;**22**(7):871-876

[61] Mondal K, Dastidar AG, Singh G, Madhusudhanan S, Gande SL, VijayRaghavan K, Varadarajan R. Design and isolation of temperature-sensitive mutants of Gal4 in yeast and Drosophila. Journal of Molecular Biology. 2007;**370**(5):939-950

[62] Han DD, Stein D, Stevens LM. Investigating the function of follicular subpopulations during Drosophila oogenesis through hormone-dependent enhancer-targeted cell ablation. Development. 2000;**127**(3):573-583

[63] Luan H, Peabody NC, Vinson CR, White BH. Refined spatial manipulation of neuronal function by combinatorial restriction of transgene expression. Neuron. 2006;**52**(3):425-436

[64] Manning L, Heckscher ES, Purice MD, Roberts J, Bennett AL, Kroll JR, Pollard JL, Strader ME, Lupton JR, Dyukareva AV, et al. A resource for manipulating gene expression and analyzing cis-regulatory modules in the Drosophila CNS. Cell Reports. 2012;**2**(4):1002-1013

[65] Jory A, Estella C, Giorgianni MW, Slattery M, Laverty TR, Rubin GM, Mann RS. A survey of 6,300 genomic fragments for cis-regulatory activity in the imaginal discs of *Drosophila melanogaster*. Cell Reports. 2012;**2**(4):1014-1024

[66] Lai SL, Lee T. Genetic mosaic with dual binary transcriptional systems in Drosophila. Nature Neuroscience. 2006;**9**(5):703-709

[67] Potter CJ, Tasic B, Russler EV, Liang L, Luo L. The Q system: A repressible binary system for transgene expression, lineage tracing, and mosaic analysis. Cell. 2010;**141**(3):536-548

[68] Golic KG, Lindquist S. The FLP recombinase of yeast catalyzes site-specific recombination in the Drosophila genome. Cell. 1989;**59**(3):499-509

[69] Andrews BJ, Proteau GA, Beatty LG, Sadowski PD. The FLP recombinase of the 2 micron circle DNA of yeast: Interaction with its target sequences. Cell. 1985;**40**(4):795-803

[70] Struhl G, Basler K. Organizing activity of wingless protein in Drosophila. Cell. 1993;**72**(4):527-540

[71] Xu T, Rubin GM. Analysis of genetic mosaics in developing and adult Drosophila tissues. Development. 1993;**117**(4):1223-1237

[72] Lee T, Luo L. Mosaic analysis with a repressible cell marker for studies of gene function in neuronal morphogenesis. Neuron. 1999;**22**(3):451-461

Chapter 7

# The Fruit Fly, *Drosophila melanogaster*: Modeling of Human Diseases (Part II)

Mariateresa Allocca, Sheri Zola and Paola Bellosta

Additional information is available at the end of the chapter

http://dx.doi.org/10.5772/intechopen.73199

**Abstract**

The fruit fly, *Drosophila melanogaster* (Meigen, 1830) has been established as a key model organism thanks in part to their considerable biological similarity to mammals and an abundance of available genetic tools. *Drosophila* have been used to model many human disease states and have been critical in elucidating the genetic mechanisms contributing to them. Part I of this chapter covered basic *Drosophila* biology and relevant genetic tools available to *Drosophila* researchers. Here in part II, we review the use of *Drosophila* as a model organism to study neurodegenerative disorders, cardiovascular diseases, kidney diseases, cancer, metabolic disorders, and immune disorders, as well as key findings made in those fields thanks to *Drosophila* research.

**Keywords:** animal model, cancer, diseases, *Drosophila*, genetic techniques, heart, immunology, kidney, metabolic disorders, neurodegeneration

## 1. Introduction

Please refer to the Introduction of Part I, The fruit fly, *Drosophila* melanogaster: The Making of a Model.

In this two-part chapter, some of the many aspects that make *Drosophila* such a fundamental model organism are covered.

Part I covered the basic fly biology and key genetic tools.

Here, Part II provides an overview of important disease states that *Drosophila* is used to model and some significant advances made in those fields.

## 2. *Drosophila melanogaster* as model to study human diseases

*Drosophila melanogaster* is a widely used model organism to understand many molecular and developmental processes common to higher eukaryotes. A prerogative for a good model system is to share higher physiology within the molecular pathways with humans, and it is remarkable that approximately 75% of genes associated with human diseases have *Drosophila* homologs and share similarities in their functions, which is of particular interest for medical purposes [2]. Based on this genetic similarity, the fly is a valid tool for understanding the function of genes involved in human disorders. Clearly, *Drosophila* has the limitation of being an invertebrate system, as some biological processes evolved only within the vertebrate lineage. Despite this, *Drosophila* exhibits complex behaviors, and each phenotype observed must be contextualized considering that mammalian physiology is not very different from that of the tiny fly. It is not easy to choose an appropriate organism to model a disease due to the higher complexity of humans, and it is necessary to evaluate the nature of the pathology before choosing. *Drosophila* provides a good background for genetic and biological studies of different pathological conditions such as neurological, cardiac, and metabolic disorders (**Table 1**).

| Organ system | Diseases |
|---|---|
| **Brain and nervous system**<br>The Drosophila brain is two-lobed and contains approximately 100,000 neurons. It is organized into several main structures including: supraesophageal ganglion (optic lobes and cerebrum) and a subesophageal ganglion. Flies also have a segmented nerve cord similar to a mammalian spinal cord (FLYBRAIN neuron Database) | • Neurodegeneration<br>o Huntington's disease<br>o Amyotrophic lateral sclerosis<br>o Spinocerebellar ataxia<br>o Alzheimer's disease<br>o Parkinson's disease<br>• Cancer |
| **Immune system**<br>Circulating immune cells called hemocytes (consisting of plasmatocytes, lamellocytes, and crystal cells) fight pathogens by encapsulating them, generating ROS, and/or producing antimicrobial peptides (AMPs). Many tissues are also capable of generating AMPs including the gut and fat body [72] | • Wound healing<br>• Cancers, including acute myeloid leukemia<br>• Autoimmune diseases<br>• Allergies |
| **Digestive system**<br>Consists of mouth parts for chewing, salivary glands to produce saliva, a crop (similar to a stomach), the proventriculus for grinding food, and a gut (midgut and hindgut) for digestion and nutrient and water absorption | • Intestinal infections<br>• Intestinal inflammation<br>• Cancer |
| **Excretory system**<br>Structures called Malpighian tubules and nephrocytes function similar to kidneys and filter nitrogenous waste from hemolymph. The tubules connect to the hindgut and excretory waste is eliminated along with digestive waste in the form of uric acid | • Nephrotic syndrome<br>• Polycystic kidney disease<br>• Kidney stones |

| Organ system | Diseases |
|---|---|
| **Circulatory system**<br>*Drosophila* has an open circulatory system. The tube-like heart (consisting of the dorsal vessel and the aortic arches) circulates hemolymph (insect blood) around the body cavity | • Congenital heart defects<br>• Cardiomyopathies<br>• Arrhythmias<br>• Channelopathies<br>• Heart failure |
| **Respiratory system**<br>*Drosophila*, like many other insects, does not carry oxygen in their hemolymph. Instead, a system of trachea connects directly with organs for gas exchange. Trachea open to the environment though tiny holes in the exoskeleton called spiracles | • Viral infection<br>• Respiratory disorders, including asthma and COPD (not discussed here) |
| **Energy storage**<br>Flies store glycogen and triglycerides in a specialized structure called the fat body. The fat body has functions similar to the mammalian liver and adipose tissue and is heavily involved in regulating growth, metabolism, and the immune system [16, 73–75]. | • Metabolic disorders<br>• Non-alcoholic fatty liver disease<br>• Diabetes<br>• Cancer |
| **Reproductive system**<br>Flies have ovaries for egg production in females, and testes for sperm production in males. These structures develop from imaginal discs in the larva. A fertile female fly can lay hundreds of eggs | • Female reproductively<br>• Cancer<br>• Aging<br>• Epigenetics |
| **Musculoskeletal system**<br>Flies have an exoskeleton composed mostly of a chitinous cuticle and an outer waxy coating. The cuticle is produced by epithelial cells and can be hard like bone or softer (as in the case of larvae). Muscles attach to points inside the exoskeleton and allow the fly to move | • Parkinson's disease (and other neurodegenerative diseases affecting movement)<br>• Musculoskeletal disorders (not discussed here) |

**Table 1.** The "Organ-Disease".

*Drosophila* has certain characteristics unique to insects such as an open circulatory system, exoskeleton, and tracheal system for gas exchange; however, they also share many similar organs and biological processes with mammals. The following summarizes the major organ and physiological systems in *Drosophila* and their comparative function to human diseases.

## 2.1. Neurodegenerative disorders

The *Drosophila* central nervous system (CNS) is composed of a bilaterally symmetrical brain with two cell types, neurons and glia, both originating from neural progenitors named neuroglioblasts. The fly CNS is considerably simpler than that of vertebrates and the neurodevelopment pattern is conserved among the organisms. Wnt, the mammalian homolog of the *Drosophila wingless* plays an important function during neuronal development [7] and Notch signaling, which plays a pivotal role during neurogenesis and neuronal differentiation, is also evolutionary conserved [8]. Neurons attend to neurotransmission while glia sustain the neurons during development and adult life mainly by providing trophic factors [9, 10]. When studying neuropathies, it is relevant to consider the interaction between neurons and glia, and research in *Drosophila* is contributing to this. In fact, the power of the neurodegenerative fly model is in the ability to explore the disease in a physiological context. While glia support neuronal survival and promote recovery in cases of neuronal damage, impairment of glial function induces non-autonomous neuronal death. Glial anti-neurodegenerative functions suggest using them as targets in human neurodegeneration [11]. The *Drosophila* brain, in particular the visual system, is widely employed for research related to neurodegenerative diseases [12]. The nervous system of people suffering from these debilitating conditions exhibits the progressive loss of neurons. The origins are disparate, and in many cases, they are unknown so it is necessary to intensify the research, aiming to understand how to treat them. Interestingly, insects lack the human hematoencephalic barrier allowing for pharmacological screening directed at the central nervous system. Depending on the mechanisms inducing the disorder and the symptomatology, we can differentiate several types of human neurodegeneration. Most neurodegenerative disorders are characterized by the presence of protein aggregates in the neurons that are different for the various classes of diseases. Despite identifying many causative factors, it remains to be determined how these proteins become neurotoxic. Thanks to the precious genetic tools available, the fly is an excellent model to explore the function of the genes coding for the proteins involved. In addition, the molecular pathways are remarkably conserved allowing for parallels with humans [13]. The simpler fruit fly CNS allows for a better understanding of the function of a gene involved in a disease and its relationship with the other neuronal patterns.

In order to characterize neuronal dysfunction in *Drosophila,* several approaches can be used including testing motility, individual and social behaviors, hearing, learning, and memory [14–16]. A histological method based on measuring the vacuoles in adult fly brains allows for the quantification of neuronal degeneration [17]; moreover, electrophysiological assays enable the analysis of synapse functionality [18]. Fruit flies affected by neurodegeneration share behavioral defects and reduced lifespans.

*Drosophila* is already used to investigate proteinopathies (protein misfolding diseases) such as Huntington's disease, amyotrophic lateral sclerosis, and spinocerebellar ataxia [19–21]. The cause of Huntington's disease (HD) is the expansion of CAG repeats in the *huntingtin* gene, leading to a polyglutamine (poliQ) repeats in the huntingtin (htt) protein. Htt is required for axonal transport and synapsis, and the fly homolog shares the same expression pattern and function [22]. The poliQ expansion is toxic also for *Drosophila* neurons; in fact, the fly gradually loses photoreceptors when human htt is expressed in the eye compartment. When human mutated genes encoding for polyQ are expressed in *Drosophila,* there is a phenotype comparable to the human disease, for instance late onset, progressive loss of neurons and motility, and premature death, and the formation of large protein aggregates of mutant Htt visible also in neurons of *Drosophila* (**Figure 1**). The *Drosophila* HD model has contributed to some findings, for instance it uncovered that the histone deacetylase (HDAC) controls the level of neurodegeneration, making it an important achievement for

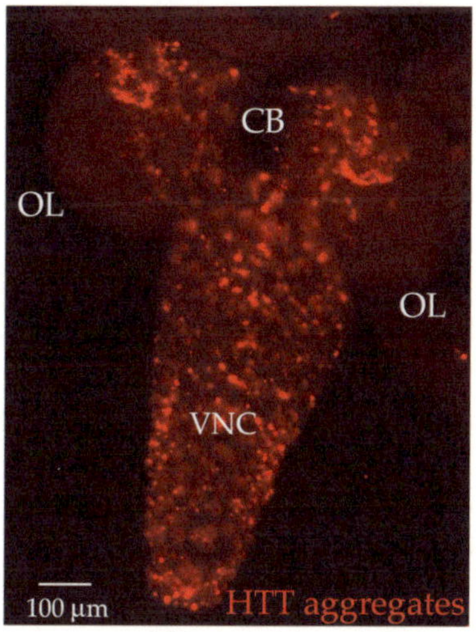


**Figure 1.** Human huntingtin aggregates in neurons. Photograph of a larval brain showing the formation of aggregates of mutant human huntingtin (HTT) with 93-polyQ repeats (red) in neurons using *Elav-Gal4* to express *UAS-HTTQ93*. HTT aggregates are visualized by immunofluorescence with anti-HTT antibodies. OP: optical lobe, CB: central brain, and VNC: ventral nerve cord.

human poliQ diseases [23]. In the fly, as in humans, the neurodegeneration rate is related to polyQ repeat length [24]. Spinocerebellar ataxia (SCA) is another disorder originating from abnormal CAG repeats. Humans can be affected by several types of SCA and *ataxin* is the mutated gene. Autophagy is a fundamental process to limit the poliQ aggregation, and in a fly model of SCA3, autophagy proteins are overexpressed allowing for a rescue of the toxicity [25]. Amyotrophic lateral sclerosis (ALS) is a disease characterized by loss of cortical and spinal motor neurons [26]. Several genes are involved in ALS and most of them can be expressed in *Drosophila* to assess their contribution to neurodegeneration. A causative factor of ASL is a mutation in superoxide dismutase SOD1 [27], and interestingly, loss of *Drosophila* SOD1 causes neuronal death while human SOD1 expression increases the fly lifespan [28, 29].

Tauopathies, including Alzheimer's, Parkinson's, and others, refer to disorders caused by aberrant accumulation of the microtubule-associated protein tau [30]. *Drosophila* has a tau homolog and the pathways involved in tau neurotoxicity such as wnt, JNK, and TOR are shared with humans [31–33]. More than 30 transgenic fly models have been established that express various forms of human wild-type and mutant tau and have uncovered many potential mechanisms for tau toxicity in a variety of neurodegenerative diseases [34]. Alzheimer's disease (AD) is one of the most common neurodegenerative disorders and yet its pathogenesis is still unclear. The tiny fly is once again a good organism to model this affliction because the AD-associated genes, such as APP and *presenilins*, are evolutionarily conserved. The brains of Alzheimer's patients are marked by aggregation of beta-amyloid protein and neurofibrillary tangles (NFTs) originating from hyperphosphorylation of Tau [35]. Tau expression induces learning and memory deficits in *Drosophila,* mimicking AD in humans [36]. Some recent advances uncovered by *Drosophila* Alzheimer's models include: explaining the mechanisms behind the phosphorylation of tau and its toxicity [37–40] along with ways to reverse it [41, 42], as well as linking DNA damage and oxidative stress triggered by tau phosphorylation in causing neurotoxicity [33, 43]. Moreover, *Drosophila* models are helping researchers to uncover the interaction between beta-amyloid proteins and tau and how they cause neuronal death [34]. Parkinson's disease (PD) is characterized by the progressive loss of dopamine neurons in the substantia nigra, a part of the brain responsible for motor control, as well as the formation of protein accumulations known as Lewy bodies, which are composed primarily of alpha synuclein [44]. Many mechanisms have been proposed for the cause of this neuronal death including disruptions in protein degradation, oxidative stress, mitochondrial dysfunction, autophagy and lysosomal dysfunction, and problems with calcium homeostasis [45] Furthermore, phosphorylated tau has been found to be associated with alpha synuclein in Lewy bodies [46, 47] and the two may function together to destabilize microtubules and damage axonal transport, also contributing to cell death [48]. Many fly models exist to study Parkinson's disease [49]. The fly dopamine neurotransmitter is similar to the human version and its function in movement is conserved [50]. Homologs of several PD-related genes are present in *Drosophila,* allowing researchers to model this neurodegenerative disease [51]. *Drosophila* models are currently being used to test a variety of potential therapeutic approaches, including boosting antioxidant mechanisms, reducing the oxidative stress caused by dopamine metabolites, and using inhibitors for members of the TOR pathway to improve Parkinson's symptoms [49].

## 2.2. Cardiovascular diseases

*Drosophila melanogaster* and humans share some aspects of heart development and function making the fly a good model for studying cardiovascular diseases, which are the leading causes of death worldwide. The heart precursors of *Drosophila* originate in the lateral mesoderm and converge on the dorsal midline to form a linear tubular structure comparable to the early vertebrate embryo heart. In *Drosophila,* a simple contractile tube pumps the hemolymph through the larval body cavity in an open cardiovascular system and regulates cardiac rhythm (**Figure 2**). The cardiovascular system has an anteroposterior polarity and it consists of the posterior portion named the dorsal vessel, corresponding to the heart, and the narrow anterior portion named the aorta, which facilitates the transport of hemolymph to the head [52]. The dorsal vessel is made up of two cell types: the cardiomyocytes, which are the inner contractile muscle cells, and the pericardial non-contractile cells, which flank the cardiomyocytes. The human heart has four distinct chambers, likewise the fly heart is divided into four chambers, each one consisting of six myocardial cells [53] that have a sarcomere structure similar to mammalian cardiac cells. The hemolymph flow moves nutrients, immune cells, and molecules required for homeostasis; however, oxygen is transported through diffusion from spiracles that invaginate from the cuticle into the interior of the animal. Despite the fly dorsal vessel being much simpler than the mammalian looped heart, the signaling

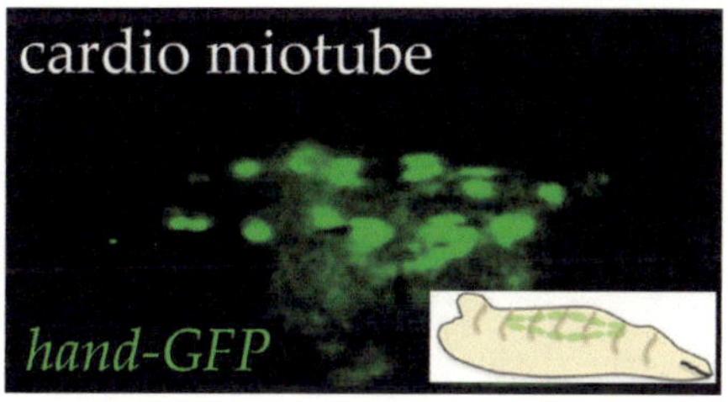


**Figure 2.** Cardiomyotube. Photograph of larval cardiomyotube with the cardiomyocytes visualized by the expression of the reporter *hand-GFP*.

pathways involved are remarkably conserved [54]. Cardiogenic genes required for the proper development of the *Drosophila* embryonic heart were identified through genome wide screens [55] showing that many molecules important for heart development and morphogenesis are conserved in humans [56]. *Tinman* is a homeobox transcription factor discovered in *Drosophila* and it is a master gene of cardiac development conserved in higher organisms [57, 58]. In addition, *pannier* and *hand,* which play crucial roles for heart specification as well as *neuromancer,* have counterparts in humans [59–61]. Moreover, these signaling pathways are required for some adult function both in *Drosophila* and in mammals suggesting that they have a conserved physiology [62].

Even if most studies are based on the embryonic development of the fly heart, nowadays the focus is shifting to the function and structure of the *Drosophila* adult heart as a model of human heart defects. Indeed, the great availability of genetic tools in *Drosophila* allows for the identification of elements important for heart functions and facilitates the analysis of mutant isoforms associated with congenital heart defects [63]. The physiological mechanisms are conserved among *Drosophila* and vertebrates supporting the utility of the fly to investigate cardiomyopathies and arrhythmias [52]. The improvement of techniques for the measurement of cardiac performance in *Drosophila* also permits the analysis of the effect of aging and the stress response on the heart [64]. Cardiac dysfunction can occur naturally in *Drosophila,* and this phenotype depends on age, just like in humans [64]. Some strategies allow heart rate monitoring in response to externally applied electric pacing in order to understand the effects of aging in adult flies. Insulin-IGF receptor (InR) and TOR signaling play an important role in regulation of age-dependent cardiac performance [65]. *Drosophila* is also one of the most efficient model organism used to discern the mechanism underlying channelopathies and cardiomyopathies as many impaired pathways are evolutionarily conserved [66]. Cardiomyopathies affecting *Drosophila* resemble those of humans both in terms of the genes responsible and the resulting effect. Such a similarity among the fly and humans is also found in the case of channelopathies and arrhythmias.

Several assay systems are helpful in characterizing *Drosophila* heart function, such as optical coherence tomography (OCT), an imaging of the *Drosophila* heart tube to observe contraction *in vivo* similar to clinical echocardiography [62]. In addition, semi-automated measurements allow researchers to record heart function to quantify cardiac impairment in *Drosophila*.

## 2.3. Kidney diseases

Despite millions of people suffering from kidney disorders, there is a disconcerting lack of therapies available to patients because the primary causes of kidney disorders are not completely characterized. *Drosophila* is advantageous to model renal disorders since many genes, proteins, and even some functions of the vertebrate kidney have parallels with the fruit fly. Despite many differences due the greater complexity of the human kidney, several orthologous genes have an important role in renal development and function, both in humans and in *Drosophila* [67]. For example, many genes encoding for electrolyte transporting proteins affected in congenital renal disorders have fly counterparts [68, 69].

The insect Malpighian tubules and the nephrocytes are functionally analogous to the vertebrate kidney; in fact, these two organs in *Drosophila* guide the metabolite homeostasis and the excretory process (**Figure 3**). Nephrocytes, which surround the heart and esophagus, are responsible for filtering the hemolymph, similar to the podocytes in the human glomerulus. In addition, nephrocytes have filtration diaphragms similar to the podocyte slit diaphragms that work as a filtration barrier in higher organisms [70, 71]. The Malpighian tubules, corresponding to the tubular part of nephrons, are two pairs of elongated and thin tubes connected to the hindgut that secrete urine after absorption of water, ions, solutes, and organic metabolites from the hemolymph. The principal cells and the stellate cells are the two main cell types in Malpighian tubules involved in excretion [72].

Nephrotic syndrome refers to ultrafiltration dysfunction leading mostly to extra protein in the urine and deficiency of protein in blood [73]. Given the evolutionary conservation of the diaphragms and their regulative mechanisms, *Drosophila* is a good option to look into this kind of disease. Some events during the renal development are shared

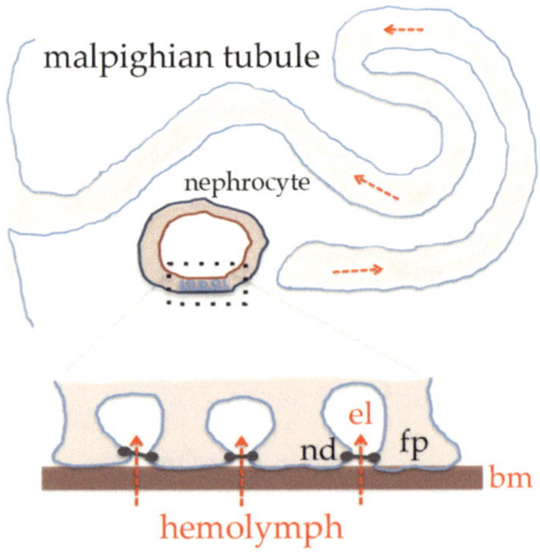


**Figure 3.** Excretory system in larvae. Malpighian tubule and nephrocyte are composing the filtration barrier; hemolymph is filtrated by nephrocyte. Nd: nephrocyte diaphragm, fp: foot process, bm: basal membrane, and el: extracellular lacunae.

between the fly and humans and the molecular pathways are conserved. All the genes playing a pivotal role in renal development, such as *Kruppel* and *Cut* involved in cell specification, *Dwnt* in tubulogenesis, and *Sns*, a nephrin-like protein, in cell differentiation, have a counterpart in mammals. One of the fundamental phases of Malpighian tubules formation is a mesenchymal-to-epithelial transition that resembles the steps of kidney development [74]. This makes the fruit fly organ able to provide insights on disorders affecting the tubular nephrons such as polycystic kidney disease and renal agenesis [75, 76]. *Drosophila* is also useful to study nephrolithiasis, also known as kidney stones, since insects also produce stone formations like calcium phosphate and calcium oxalate [77]. A simple method exists to score the filtration and the uptake of a secreted fluorescently tagged protein (ANFRFP) that accumulates in nephrocytes to assess the renal function in *Drosophila* [75].

The similarities among the species definitely allow the use of the *Drosophila* renal structure as a model to better understand the basis of human kidney impairments and consequently to develop personalized therapeutic agents. Furthermore, immune and inflammatory responses are trigger factors of kidney diseases so they should be taken into account when analyzing these pathologies [78].

### 2.4. Cancer and growth

The fly is a simple model to improve the understanding of tumor biology and progression [79–83] as the available genetic tools support the analysis of the mechanisms underlying growth regulation in an intact epithelium rather than in cell cultures. The advantage is remarkable since cell-cell and cell-environment interactions contribute to tissue size regulation. The *Drosophila* cell cycle can escape the normal control system leading to the typical cancer hyperproliferation. Reproducing human tumors in *Drosophila* allowed for the identification of many oncosuppressor genes that regulate cell division and differentiation [84]. In the fly, the tumor hallmarks mimic the human ones: autonomous proliferation signals and overgrowth, irregular cell morphology, bypassed apoptosis, and metastasis [85]. In spite of these similarities, there are several limitations including lack in flies of processes such as telomere maintenance and angiogenesis that participate in cancer development.

A great conservation across species is detected in regards to the signaling pathways affecting growth. Initial studies using activated proto-oncogenes such as the receptor tyrosine kinase (*ret*), a gene responsible for medullary thyroid carcinoma (MTC), allowed researchers to perform genetic screens for suppressors or enhancers of the rough eye phenotype, which indicates an overproliferation of cells in the eye [86]. These studies evolved to include tumors that were induced by the activation of growth signaling pathways, such as PI3K and EGFR in glia, which resemble human glioma [87], or studies involving tuberous sclerosis, an autosomal dominant disorder characterized by benign tumors in multiple organs induced by the loss of function activity of the TSC1 and 2 tumor suppressor genes [88]. A large number of studies also demonstrated how the Hippo pathway, which regulates growth through the activation of Yki, is highly conserved and required for cellular proliferation as well as for apoptosis, has a human counterpart that retains sequence and function, and is mutated within the context of cancer [89, 90]. The same goes for *Salvador*, a gene promoting apoptosis, and *Archipelago* [91–94]. The two organisms also share PTEN, a tumor suppressor that plays a crucial role in carcinogenesis both in humans and in flies [95].

New studies defined how the loss of cell polarity could be considered a hallmark of malignancy [96]. Members of discs large (dlg) and lethal giant larvae (lgl) were identified as tumor suppressors in the fly by promoting cell invasion if mutated, with a similar role also seen in human neoplasm [97]. The role of proteins involved in cellular adhesion, such as Rho1 and E-cadherin, was also shown to be conserved and relevant for migration and invasion helping the study of the metastatic process [98, 99]. Other well-studied oncogenes in *Drosophila* that promote overgrowth and cell survival are *Ras* and *Notch* and were also shown to play a role in cellular polarity [100]. Dpp, the homolog of human bone morphogenetic protein/transforming growth factor-beta (BMP/TGF beta), is also responsible for epithelial integrity [101] and implicated in a model for cancer in *Drosophila*. All these parallelisms provide the potential to dissect in vivo the interacting patterns causing the tumor growth.

As anticipated, the communication between neighboring cells must be taken into consideration when analyzing a tumor tissue. Competitive interactions occur among cells with different growth rates in a process known as cell competition, which was first described in *Drosophila* using ribosomal proteins [102, 103] and then characterized using dMyc, the fly homolog of human cMyc [104]. Cells expressing higher levels of Myc behave as supercompetitors: they survive and acquire a proliferative advantage inducing apoptosis in the weaker nearby cells, termed losers [105–107]. The mechanisms controlling overproliferation and metastasis are comparable to those involved in cell competition since in human cancer, cells overexpressing Myc acquire the capacity to grow more than normal and to invade the neighboring normal cells. Since then, a few additional oncogenes and tumor suppressor genes have been associated with a competitive behavior, and cell competition is now thought to have an important role in human cancer [108–112]. This similitude underscores the utility of using flies for studying how cells compete for survival.

More studies are arising on the connection between the insurgence of tumors and diet or obesity. Recent studies linked the growth of prostate tumors and the status of obesity [113]. Caloric restriction reduces the growth of tumor cells in rodent models through reduced systemic insulin and IGF-1 signaling [114], while the activation of PI3K induces tumors to be resistant to diet restriction [115] suggesting an important relationship between PI3K signaling in tumors and the nutrients in the tumor environment. The exact link between obesity and cancer has not yet been established and the fly may facilitate this research thanks to the ability to combine obesity and tumor models in *Drosophila.* Insulin signaling is the main regulator of metabolic homeostasis, and it is also involved in cancer development and progression [116] but we have yet to understand how hyperinsulinemia promotes tumor formation. Interestingly, the oncogenes Src and Ras were overexpressed in a *Drosophila* model of obesity and increasing the level of insulin exacerbates the malignant phenotype due to *wingless* activity [117]. The interplay between obesity and cancer is an important area of study to understand the relevance of fat to tumor growth, since fatty acids are unable to penetrate the biological membranes and need to be cleaved by lipases (lipolysis). Recent studies indicate that in the peritumoural area, an increase in adipose triglyceride lipase

(ATGL) that mediates lipolysis results in tumor survival [118, 119]. The ability to manipulate flies genetically and the possibility to change the composition of lipids or nutrients in their food will likely put *Drosophila* as a key model to investigate the relationship between obesity and cancer and the mechanisms that control cellular overgrowth. Cancer research can only benefit from the ability to create specific disease models in *Drosophila.* This approach lets researchers detect oncogenes and tumor suppressors, allowing a detailed in vivo analysis of the mechanisms triggering cancer. From these findings, drug therapy compounds can then be developed and tested.

### 2.5. Metabolic disorders

Hepatic diseases affect a large proportion of the population worldwide making it crucial to investigate the underlying pathogenic mechanisms that still remain unclear. Identification of the molecular defects underlying liver disease requires studies in model organisms, and recently *Drosophila* has been proposed for this purpose [120].

The use of the fruit fly in the study of hepatic disorders is partially restricted due to the absence of a homologous organ for the liver. The fat body in *Drosophila* acts as storage for sugar and fat and also performs metabolic functions similar to those of the mammalian hepatocytes, regulated by insulin through an evolutionarily conserved mechanism [121, 122]. During starvation, triglycerides are transported from the fat body into the hemolymph where they are captured by the oenocytes, clusters of hepatocyte-like cells that are important for lipid metabolism [123]. Therefore, some functions of hepatocytes are performed by oenocytes, which are located near the body wall surface and play a prominent role in the fly lipid processing. *Drosophila* homologs of genes specifically expressed in human hepatocytes are expressed in larval oenocytes and the fat metabolism pathway is similar among the organisms [123]. An interesting aspect regarding lipid metabolism is the interaction between oenocytes and fat cells, as oenocytes control lipolysis in fat cells through a feedback similar to that in mammals [123]. Underfed *Drosophila* stores many fat droplets resulting in the accumulation of triacylglycerols in the liver, a condition called steatosis, and forms an excellent model for understanding human non-alcoholic fatty liver disease (NAFLD) [124]. Moreover, the relationship between oenocytes and fat cells needs to be elucidated because it contributes to the pathogenesis of metabolic syndrome [125], and fly modeling can be useful for this purpose.

It is necessary to improve assays examining the function of the fat body and oenocytes to solidify *Drosophila* as a liver disease model. To date, the analyses are based on evaluating lipid accumulation depending on different nutritional conditions. Fly lipid homeostasis can be monitored by Raman scattering microscopy that allows for the visualization of the lipid content in larval oenocytes and in the fat body by *in vivo* imaging [126]. Oil Red-O and BODIPY are dyes permitting the assessment of lipid content [123, 127].

Several proteins that contribute to lipid metabolism in *Drosophila,* including proteins responsible for lipid storage, transport, and utilization, have counterparts in higher organisms [128, 129]. This similitude makes the fruit fly helpful in describing the main pathways controlling homeostasis and provides an opportunity to examine metabolic disorders affecting humans such as diabetes and obesity [122]. For example, the main regulator of sugar and fat metabolism is the nutrient-sensing target of rapamycin (TOR) both in *Drosophila* and in mammals [130]. Flies are able to regulate carbohydrate metabolism by cellular storage of excess nutrients. The hormone insulin controls hemolymph sugar levels and maintains carbohydrate homeostasis through a phylogenetically conserved signaling pathway [122, 131]. *Drosophila* insulin induces an increase in fat cell mass, just as in mammals, because insulin acts on triglyceride storage and on fat body cell number. Shaggy is a serine/threonine protein kinase orthologous to glycogen synthase kinase 3 (GSK3), and it is responsible for the lipid accumulation in *Drosophila* fat cells while the transcription factor *Drosophila* FOXO (dFOXO) influences the adipocyte cell number [121]. Both of these key factors are regulated by the conserved insulin pathway [121]. Dilp2, 3 and 5, members of the *Drosophila* insulin-like peptides (Dilps) are expressed in the insulin-producing cells (IPCs), a cluster of cells in the brain that function similarly to human pancreatic β cells [132]. Additionally, the adipokinetic hormone participates in fly glucose regulation with a glucagon-like function [55]. Functional changes to these metabolic regulators in *Drosophila* cause a phenotype similar to metabolic impairment as well as affecting body size [132, 133]. The resemblance between *Drosophila* and mammals helps to elucidate the main mechanisms of metabolic homeostasis involved in common pathologies such as type 2 diabetes mellitus (T2DM), which is characterized by insulin resistance, hyperglycemia, and defects in lipid metabolism [134]. High-glycemic diets promote obesity, a disorder characterized by excessive fat storage. *Drosophila* fed a high fat diet store fat in the fat body and in the midgut [135]. This condition changes the animal physiology and lifespan mainly due to insulin resistance [136, 137]. Moreover, obesity is considered among the risk factors for diabetes, cardiac diseases, and several types of cancer [138, 139]. Insulin resistance is also related to NFALD, the most frequent chronic hepatic disorder [140]. NAFLD originates from metabolic impairment highlighting the strong relationship between the liver and metabolism and the subsequent need to examine the pathways linking them [124].

*Drosophila* has facilitated the study of metabolic pathways thanks to the availability of several assays of metabolic function, including some that are available for use only in *Drosophila,* which allow for the quantification of lipids, sugars, ATP, and mitochondria. In spite of the anatomical differences between flies and humans, the identification of novel genes and pathways in the fruit fly could arrange for new therapies to treat metabolic disease in humans.

### 2.6. Immunological diseases

The mechanism of the innate immune system is fairly conserved across species, and *Drosophila* is a leading organism for elucidating the process of defense from pathogens and its evolution [141]. Since the adaptive immune response of vertebrates could hide some aspects of the innate immunity, it is beneficial to use *Drosophila* to detail the regulation of innate immunity because this organism does not have an adaptive one [141]. Pathogenic microorganisms, such as bacteria, fungi, nematodes, and viruses, can infect *Drosophila,* priming an immune reaction. Despite the greater refinement of mammalian immunity, *Drosophila* and humans share general defense strategies like epithelial barriers, phagocytosis, and antimicrobial peptides. The fly's first line of defense against to pathogens is a physical barrier represented by the epithelia of the epidermis, trachea, and gut. Clotting factors in the hemolymph provide a second barrier because they can entrap invaders by means of their protein filaments [142]. Epithelia then release antimicrobial peptides (AMPs) and reactive oxygen species (ROS), triggering a local immune response [143, 144]. Beside their toxic activity, ROS are involved in wound healing and tissue repair both in *Drosophila* and mammals [145]. In addition to epithelia, blood cells and the fat body are also required for *Drosophila* immunity. The external agents are phagocytized by hemocytes; the circulating blood cells and different types of hemocytes are involved in this reaction. Plasmatocytes are monocyte-like cells, which able to phagocytose pathogens, apoptotic bodies, and other foreign particles. Crystal cells, another type of hemocyte, are involved in the production of melanin, a protein involved in both encapsulating and killing microorganisms as well as being implicated in wound healing. Hemocytes differentiate into lamellocytes if a more specialized response is required, and lamellocytes can trap larger parasites, producing a cellular capsule around it in a process named encapsulation [146, 147]. In *Drosophila*, the majority of blood cells have phagocytic activity.

Some fly macrophages originate via self-renewing and others from progenitor cells that are located in the lymph gland, a specialized hematopoietic organ. The great importance of the lymph gland in controlling the blood cell homeostasis makes this *Drosophila* organ comparable with the hematopoietic stem cell niche in the bone marrow [148, 149]. ROS levels have a crucial role in the regulation of *Drosophila* hematopoiesis [150]. Moreover, the signaling pathways regulating blood cell differentiation are conserved from *Drosophila* to humans [151, 152]. These similarities with vertebrate hematopoiesis underscore the utility of the fly to elucidate the basis of hematopoietic injury, necessary because an impairment in hematopoietic differentiation and homeostasis causes several diseases such as leukemia. *Drosophila* has already been used to study acute myeloid leukemia, a widespread form of leukemia, in particular to identify the genes promoting the disease. AML1 is one of the transcription factors activating myeloid differentiation and it has a counterpart in the fly [153]. When AML1 is fused with the repressor ETO, the differentiation is inhibited while the proliferation of multilineage progenitors is activated, leading to acute myeloid leukemia. AML1-ETO expression in *Drosophila* causes the same effect, confirming the fly as a good genetic model for leukemia [153, 154].

The great availability of genetic tools in the fly contributed to defining the innate immune system and to establishing that it is a specific mechanism. In fact, *Drosophila* can respond specifically to pathogens, discriminating between classes of surface molecules on different intruders. AMPs have different targets, for instance drosomycin acts on fungi, defensin on Gram-positive bacteria, and drosocin on Gram-negative bacteria [155]. Moreover, the sequences of AMPs are conserved between humans and insects [156]. Not only is the defense mechanism evolutionarily conserved, but also is the molecular pattern promoting innate immune reactions. *Toll* and *Imd* are the two master genes of *Drosophila* immunity, but *FoxO, JAK/STAT, and JNK* transduction also play a part [157]. After pathogen detection, *Toll* and *Imd* induce a cascade of events that finally release the antimicrobial peptides in fat body cells through the activation of the NF-κB transcription factors Dif, homolog of Dorsal, and Relish, respectively [155]. *Toll* encodes an interleukin 1 receptor-like protein that in *Drosophila* acts in parallel during two different processes: the dorsoventral specification and the immune response regulation [158]. *Toll* is activated by fungi and most Gram-positive bacteria and has a pivotal function both in the humoral response and in phagocytosis. Dissecting *Toll* signaling in *Drosophila* helped to understand toll-like receptors that play an important role in inflammatory responses [159–161]. The Immune deficiency (*Imd*) signaling is mainly involved in the *Drosophila* reaction to Gram-negative bacterial infection [162]. The flies are also helpful in examining the defense against viral infection as they share with humans some proteins, named restriction factors, involved in the reaction to viral infection. Restriction factors, for instance Pastrel in *Drosophila,* are induced in host cells by virus infection and they can recognize specific viral elements, but the mechanism by which they act in insects is not very clear yet [163].

In order to examine immunity in the fly, an efficient and simple procedure has been developed to elucidate the physiological effect after infection and to quantify the pathogen load. It consists in scoring bacterial load, fly mortality, and also evaluating the effect on immune transcription factors after the direct introduction of bacteria in the fly body cavity, eluding the epithelial barrier [164].

The innate immunity contributes to *Drosophila* homeostasis and it is regulated by endocrine and metabolic systems. Since immune dysfunction leads to several human diseases, including autoimmune disorders, allergy, and intestinal infections, it is fruitful to use this model organism to better understand how all these systems are regulated. The fruit fly is also used to investigate the association between the microbiome and host, trying to characterize the resistance and tolerance mechanisms that are conserved in humans [165–167]. Circadian rhythms also participate in immune regulation both in *Drosophila* and in humans providing another similarity between organisms [168].

## 3. Conclusions

As illustrated throughout these two chapters, *Drosophila melanogaster* has been an invaluable tool for unlocking mechanisms contributing to the pathogenesis of many diseases such as cancer, diabetes, obesity, neurodegenerative disorders, kidney disease, immunological impairments, and many others. Given the advances in the field of genetics, new tools and techniques are continually being developed that will keep flies at the forefront of biomedical research.

## Acknowledgements

We thanks the Confocal facility at IFOM-Milan, Matteo Cascinelli, Matteo Frattaroli, Valeria Lupi, and John Benedict Pollard from Liceo Scientifico "A. Einstein Milano" and Zhasmine Mirzoyan University of Milan for helping with the images. Funding from Cariplo Foundation and EHDN to PB and from CiBio to MTC.

## Conflict of interest

The authors declare no conflict of interest.

## Author details

Mariateresa Allocca, Sheri Zola and Paola Bellosta*

*Address all correspondence to: paola.bellosta@unitn.it

Center of Integrated Biology, University of Trento, Trento, Italy

## References

[1] Adams MD, Celniker SE, Holt RA, Evans CA, Gocayne JD, Amanatides PG, Scherer SE, Li PW, Hoskins RA, Galle RF, et al. The genome sequence of *Drosophila melanogaster*. Science. 2000;**287**(5461):2185-2195

[2] Pandey UB, Nichols CD. Human disease models in *Drosophila melanogaster* and the role of the fly in therapeutic drug discovery. Pharmacological Reviews. 2011;**63**(2):411-436

[3] Millburn GH, Crosby MA, Gramates LS, Tweedie S, FlyBase C. FlyBase portals to human disease research using *Drosophila* models. Disease Models & Mechanisms. 2016; **9**(3):245-252

[4] Kenney DE, Borisy GG. Thomas hunt Morgan at the marine biological laboratory: Naturalist and experimentalist. Genetics. 2009;**181**(3):841-846

[5] Bilder D, Irvine KD. Taking stock of the *Drosophila* research ecosystem. Genetics. 2017; **206**(3):1227-1236

[6] Patel S, Prokop A. The Manchester fly facility: Implementing an objective-driven long-term science communication initiative. Seminars in Cell & Developmental Biology. 2017; **70**:38-48

[7] Bejsovec A. Wnt pathway activation: New relations and locations. Cell. 2005;**120**(1):11-14

[8] Kopan R, Ilagan MX. The canonical notch signaling pathway: Unfolding the activation mechanism. Cell. 2009;**137**(2):216-233

[9] Bergmann A, Tugentman M, Shilo BZ, Steller H. Regulation of cell number by MAPK-dependent control of apoptosis: A mechanism for trophic survival signaling. Developmental Cell. 2002;**2**(2):159-170

[10] Chotard C, Salecker I. Glial cell development and function in the *Drosophila* visual system. Neuron Glia Biology. 2007;**3**(1):17-25

[11] Evans JR, Barker RA. Neurotrophic factors as a therapeutic target for Parkinson's disease. Expert Opinion on Therapeutic Targets. 2008;**12**(4):437-447

[12] Marsh JL, Thompson LM. *Drosophila* in the study of neurodegenerative disease. Neuron. 2006;**52**(1):169-178

[13] Bellen HJ, Tong C, Tsuda H. 100 years of *Drosophila* research and its impact on vertebrate neuroscience: A history lesson for the future. Nature Reviews Neuroscience. 2010; **11**(7):514-522

[14] Branson K, Robie AA, Bender J, Perona P, Dickinson MH. High-throughput ethomics in large groups of *Drosophila*. Nature Methods. 2009;**6**(6):451-457

[15] Inagaki HK, Kamikouchi A, Ito K. Protocol for quantifying sound-sensing ability of *Drosophila melanogaster*. Nature Protocols. 2010;**5**(1):26-30

[16] McGuire SE, Deshazer M, Davis RL. Thirty years of olfactory learning and memory research in *Drosophila melanogaster*. Progress in Neurobiology. 2005;**76**(5):328-347

[17] Sunderhaus ER, Kretzschmar D. Mass histology to quantify Neurodegeneration in *Drosophila*. Journal of Visualized Experiments: JoVE. 2016;**118**

[18] Broadie K, Bate M. Activity-dependent development of the neuromuscular synapse during *Drosophila* embryogenesis. Neuron. 1993;**11**(4):607-619

[19] Jackson GR, Salecker I, Dong X, Yao X, Arnheim N, Faber PW, MacDonald ME, Zipursky SL. Polyglutamine-expanded human huntingtin transgenes induce degeneration of *Drosophila* photoreceptor neurons. Neuron. 1998;**21**(3):633-642

[20] Watson MR, Lagow RD, Xu K, Zhang B, Bonini NM. A *drosophila* model for amyotrophic lateral sclerosis reveals motor neuron damage by human SOD1. The Journal of Biological Chemistry. 2008;**283**(36):24972-24981

[21] Casci I, Pandey UB. A fruitful endeavor: Modeling ALS in the fruit fly. Brain Research. 2015;**1607**:47-74

[22] Li Z, Karlovich CA, Fish MP, Scott MP, Myers RM. A putative *Drosophila* homolog of the Huntington's disease gene. Human Molecular Genetics. 1999;**8**(9):1807-1815

[23] Steffan JS, Bodai L, Pallos J, Poelman M, McCampbell A, Apostol BL, Kazantsev A, Schmidt E, Zhu YZ, Greenwald M, et al. Histone deacetylase inhibitors arrest polyglutamine-dependent neurodegeneration in *Drosophila*. Nature. 2001;**413**(6857):739-743

[24] Marsh JL, Walker H, Theisen H, Zhu YZ, Fielder T, Purcell J, Thompson LM. Expanded polyglutamine peptides alone are intrinsically cytotoxic and cause neurodegeneration in *Drosophila*. Human Molecular Genetics. 2000;**9**(1):13-25

[25] Bilen J, Bonini NM. *Drosophila* as a model for human neurodegenerative disease. Annual Review of Genetics. 2005;**39**:153-171

[26] Boillee S, Vande Velde C, Cleveland DW. ALS: A disease of motor neurons and their nonneuronal neighbors. Neuron. 2006;**52**(1):39-59

[27] Rothstein JD. Current hypotheses for the underlying biology of amyotrophic lateral sclerosis. Annals of Neurology. 2009;**65**(Suppl 1):S3-S9

[28] Phillips JP, Tainer JA, Getzoff ED, Boulianne GL, Kirby K, Hilliker AJ. Subunit-destabilizing mutations in *Drosophila* copper/zinc superoxide dismutase: Neuropathology

and a model of dimer dysequilibrium. Proceedings of the National Academy of Sciences of the United States of America. 1995;**92**(19):8574-8578

[29] Parkes TL, Elia AJ, Dickinson D, Hilliker AJ, Phillips JP, Boulianne GL. Extension of *Drosophila* lifespan by overexpression of human SOD1 in motorneurons. Nature Genetics. 1998;**19**(2):171-174

[30] Lee VM, Goedert M, Trojanowski JQ. Neurodegenerative tauopathies. Annual Review of Neuroscience. 2001;**24**:1121-1159

[31] Jackson GR, Wiedau-Pazos M, Sang TK, Wagle N, Brown CA, Massachi S, Geschwind DH. Human wild-type tau interacts with wingless pathway components and produces neurofibrillary pathology in *Drosophila*. Neuron. 2002;**34**(4):509-519

[32] Dias-Santagata D, Fulga TA, Duttaroy A, Feany MB. Oxidative stress mediates tau-induced neurodegeneration in *Drosophila*. The Journal of Clinical Investigation. 2007; **117**(1):236-245

[33] Khurana S, Robinson BG, Wang Z, Shropshire WC, Zhong AC, Garcia LE, Corpuz J, Chow J, Hatch MM, Precise EF, et al. Olfactory conditioning in the third instar larvae of *Drosophila melanogaster* using heat shock reinforcement. Behavior Genetics. 2012; **42**(1):151-161

[34] Fernandez-Funez P, de Mena L, Rincon-Limas DE. Modeling the complex pathology of Alzheimer's disease in *Drosophila*. Experimental Neurology. 2015;**274**(Pt A):58-71

[35] Ballatore C, Lee VM, Trojanowski JQ. Tau-mediated neurodegeneration in Alzheimer's disease and related disorders. Nature Reviews Neuroscience. 2007;**8**(9):663-672

[36] Mershin A, Pavlopoulos E, Fitch O, Braden BC, Nanopoulos DV, Skoulakis EM. Learning and memory deficits upon TAU accumulation in *Drosophila* mushroom body neurons. Learning & Memory. 2004;**11**(3):277-287

[37] Kosmidis S, Grammenoudi S, Papanikolopoulou K, Skoulakis EM. Differential effects of tau on the integrity and function of neurons essential for learning in *Drosophila*. The Journal of Neuroscience. 2010;**30**(2):464-477

[38] Nishimura I, Yang Y, Lu B. PAR-1 kinase plays an initiator role in a temporally ordered phosphorylation process that confers tau toxicity in *Drosophila*. Cell. 2004;**116**(5):671-682

[39] Steinhilb ML, Dias-Santagata D, Mulkearns EE, Shulman JM, Biernat J, Mandelkow EM, Feany MB. S/P and T/P phosphorylation is critical for tau neurotoxicity in *Drosophila*. Journal of Neuroscience Research. 2007;**85**(6):1271-1278

[40] Chatterjee S, Sang TK, Lawless GM, Jackson GR. Dissociation of tau toxicity and phosphorylation: Role of GSK-3beta, MARK and Cdk5 in a *Drosophila* model. Human Molecular Genetics. 2009;**18**(1):164-177

[41] Cowan CM, Bossing T, Page A, Shepherd D, Mudher A. Soluble hyper-phosphorylated tau causes microtubule breakdown and functionally compromises normal tau in vivo. Acta Neuropathologica. 2010;**120**(5):593-604

[42] Mudher A, Shepherd D, Newman TA, Mildren P, Jukes JP, Squire A, Mears A, Drummond JA, Berg S, MacKay D, et al. GSK-3beta inhibition reverses axonal transport defects and behavioural phenotypes in *Drosophila*. Molecular Psychiatry. 2004;**9**(5):522-530

[43] Frost B, Hemberg M, Lewis J, Feany MB. Tau promotes neurodegeneration through global chromatin relaxation. Nature Neuroscience. 2014;**17**(3):357-366

[44] Dauer W, Przedborski S. Parkinson's disease: Mechanisms and models. Neuron. 2003; **39**(6):889-909

[45] Michel PP, Hirsch EC, Hunot S. Understanding dopaminergic cell death pathways in Parkinson disease. Neuron. 2016;**90**(4):675-691

[46] Arima K, Hirai S, Sunohara N, Aoto K, Izumiyama Y, Ueda K, Ikeda K, Kawai M. Cellular co-localization of phosphorylated tau- and NACP/alpha-synuclein-epitopes in lewy bodies in sporadic Parkinson's disease and in dementia with Lewy bodies. Brain Research. 1999;**843**(1-2):53-61

[47] Ishizawa T, Mattila P, Davies P, Wang D, Dickson DW. Colocalization of tau and alpha-synuclein epitopes in Lewy bodies. Journal of Neuropathology and Experimental Neurology. 2003;**62**(4):389-397

[48] Lei P, Ayton S, Finkelstein DI, Adlard PA, Masters CL, Bush AI. Tau protein: Relevance to Parkinson's disease. The International Journal of Biochemistry & Cell Biology. 2010; **42**(11):1775-1778

[49] Hewitt VL, Whitworth AJ. Mechanisms of Parkinson's disease: Lessons from *Drosophila*. Current Topics in Developmental Biology. 2017;**121**:173-200

[50] Monastirioti M. Biogenic amine systems in the fruit fly *Drosophila melanogaster*. Microscopy Research and Technique. 1999;**45**(2):106-121

[51] Shulman JM, De Jager PL. Evidence for a common pathway linking neurodegenerative diseases. Nature Genetics. 2009;**41**(12):1261-1262

[52] Bier E, Bodmer R. Drosophila, an emerging model for cardiac disease. Gene. 2004; **342**(1):1-11

[53] Lehmacher C, Abeln B, Paululat A. The ultrastructure of *Drosophila* heart cells. Arthropod Structure & Development. 2012;**41**(5):459-474

[54] Ahmad SM. Conserved signaling mechanisms in *Drosophila* heart development. Developmental Dynamics: An Official Publication of the American Association of Anatomists. 2017;**246**(9):641-656

[55] Kim SK, Rulifson EJ. Conserved mechanisms of glucose sensing and regulation by *Drosophila* corpora cardiaca cells. Nature. 2004;**431**(7006):316-320

[56] Neely GG, Kuba K, Cammarato A, Isobe K, Amann S, Zhang L, Murata M, Elmen L, Gupta V, Arora S, et al. A global in vivo *Drosophila* RNAi screen identifies NOT3 as a conserved regulator of heart function. Cell. 2010;**141**(1):142-153

[57] Bodmer R. The gene tinman is required for specification of the heart and visceral muscles in *Drosophila*. Development. 1993;**118**(3):719-729

[58] Olson EN. Gene regulatory networks in the evolution and development of the heart. Science. 2006;**313**(5795):1922-1927

[59] Qian L, Bodmer R. Partial loss of GATA factor pannier impairs adult heart function in *Drosophila*. Human Molecular Genetics. 2009;**18**(17):3153-3163

[60] Qian L, Mohapatra B, Akasaka T, Liu J, Ocorr K, Towbin JA, Bodmer R. Transcription factor neuromancer/TBX20 is required for cardiac function in *Drosophila* with implications for human heart disease. Proceedings of the National Academy of Sciences of the United States of America. 2008;**105**(50):19833-19838

[61] Hallier B, Hoffmann J, Roeder T, Togel M, Meyer H, Paululat A. The bHLH transcription factor hand regulates the expression of genes critical to heart and muscle function in *Drosophila melanogaster*. PLoS One. 2015;**10**(8):e0134204

[62] Choma MA, Suter MJ, Vakoc BJ, Bouma BE, Tearney GJ. Physiological homology between *Drosophila melanogaster* and vertebrate cardiovascular systems. Disease Models & Mechanisms. 2011;**4**(3):411-420

[63] Amodio V, Tevy MF, Traina C, Ghosh TK, Capovilla M. Transactivation in *Drosophila* of human enhancers by human transcription factors involved in congenital heart diseases. Developmental Dynamics: An Official Publication of the American Association of Anatomists. 2012;**241**(1):190-199

[64] Ocorr KA, Crawley T, Gibson G, Bodmer R. Genetic variation for cardiac dysfunction in *Drosophila*. PLoS One. 2007;**2**(7):e601

[65] Wessells R, Fitzgerald E, Piazza N, Ocorr K, Morley S, Davies C, Lim HY, Mitchell L, Hayes M, Oldham S, et al. d4eBP acts downstream of both dTOR and dFoxo to modulate cardiac functional aging in *Drosophila*. Aging Cell. 2009;(5):542-552

[66] Wolf MJ, Amrein H, Izatt JA, Choma MA, Reedy MC, Rockman HA. *Drosophila* as a model for the identification of genes causing adult human heart disease. Proceedings of the National Academy of Sciences of the United States of America. 2006;**103**(5):1394-1399

[67] Dow JA, Romero MF. *Drosophila* provides rapid modeling of renal development, function, and disease. American Journal of Physiology Renal Physiology. 2010;**299**(6):F1237-F1244

[68] Wang J, Kean L, Yang J, Allan AK, Davies SA, Herzyk P, Dow JA. Function-informed transcriptome analysis of *Drosophila* renal tubule. Genome Biology. 2004;**5**(9):R69

[69] Hatton-Ellis E, Ainsworth C, Sushama Y, Wan S, VijayRaghavan K, Skaer H. Genetic regulation of patterned tubular branching in *Drosophila*. Proceedings of the National Academy of Sciences of the United States of America. 2007;**104**(1):169-174

[70] Weavers H, Prieto-Sanchez S, Grawe F, Garcia-Lopez A, Artero R, Wilsch-Brauninger M, Ruiz-Gomez M, Skaer H, Denholm B. The insect nephrocyte is a podocyte-like cell with a filtration slit diaphragm. Nature. 2009;**457**(7227):322-326

[71] Zhuang S, Shao H, Guo F, Trimble R, Pearce E, Abmayr SM. Sns and Kirre, the *Drosophila* orthologs of Nephrin and Neph1, direct adhesion, fusion and formation of a slit diaphragm-like structure in insect nephrocytes. Development. 2009;**136**(14):2335-2344

[72] Sozen MA, Armstrong JD, Yang M, Kaiser K, Dow JA. Functional domains are specified to single-cell resolution in a *Drosophila epithelium*. Proceedings of the National Academy of Sciences of the United States of America. 1997;**94**(10):5207-5212

[73] Nielsen R, Christensen EI. Proteinuria and events beyond the slit. Pediatric Nephrology. 2010;**25**(5):813-822

[74] Singh SR, Hou SX. Lessons learned about adult kidney stem cells from the malpighian tubules of *Drosophila*. Journal of the American Society of Nephrology: JASN. 2008;**19**(4):660-666

[75] Zhang Q, Taulman PD, Yoder BK. Cystic kidney diseases: All roads lead to the cilium. Physiology. 2004;**19**:225-230

[76] Perin L, Giuliani S, Sedrakyan S, Das S, De Filippo RE. Stem cell and regenerative science applications in the development of bioengineering of renal tissue. Pediatric Research. 2008;**63**(5):467-471

[77] Miller J, Chi T, Kapahi P, Kahn AJ, Kim MS, Hirata T, Romero MF, Dow JA, Stoller ML. *Drosophila melanogaster* as an emerging translational model of human nephrolithiasis. The Journal of Urology. 2013;**190**(5):1648-1656

[78] Imig JD, Ryan MJ. Immune and inflammatory role in renal disease. Comprehensive Physiology. 2013;**3**(2):957-976

[79] Hariharan IK, Bilder D. Regulation of imaginal disc growth by tumor-suppressor genes in *Drosophila*. Annual Review of Genetics. 2006;**40**:335-361

[80] Brumby AM, Richardson HE. Using *Drosophila melanogaster* to map human cancer pathways. Nature Reviews Cancer. 2005;**5**(8):626-639

[81] Gonzalez C. *Drosophila melanogaster*: A model and a tool to investigate malignancy and identify new therapeutics. Nature Reviews Cancer. 2013;**13**(3):172-183

[82] Miles WO, Dyson NJ, Walker JA. Modeling tumor invasion and metastasis in *Drosophila*. Disease Models & Mechanisms. 2011;**4**(6):753-761

[83] Vidal M, Larson DE, Cagan RL. Csk-deficient boundary cells are eliminated from normal *Drosophila* epithelia by exclusion, migration, and apoptosis. Developmental Cell. 2006; **10**(1):33-44

[84] Gateff E. Tumor suppressor and overgrowth suppressor genes of *Drosophila melanogaster*: Developmental aspects. The International Journal of Developmental Biology. 1994;**38**(4):565-590

[85] Mechler BM, Strand D. Tumor suppression in *Drosophila*. Immunology Series. 1990; **51**:123-144

[86] Das TK, Cagan RL. A *Drosophila* approach to thyroid cancer therapeutics. Drug Discovery Today Technologies. 2013;**10**(1):e65-e71

[87] Read RD, Cavenee WK, Furnari FB, Thomas JB. A *drosophila* model for EGFR-Ras and PI3K-dependent human glioma. PLoS Genetics. 2009;**5**(2):e1000374

[88] Pan D, Dong J, Zhang Y, Gao X. Tuberous sclerosis complex: From *Drosophila* to human disease. Trends in Cell Biology. 2004;**14**(2):78-85

[89] Yu J, Zheng Y, Dong J, Klusza S, Deng WM, Pan D. Kibra functions as a tumor suppressor protein that regulates hippo signaling in conjunction with Merlin and Expanded. Developmental Cell. 2010;**18**(2):288-299

[90] Pan D. The hippo signaling pathway in development and cancer. Developmental Cell. 2010;**19**(4):491-505

[91] Tapon N, Harvey KF, Bell DW, Wahrer DC, Schiripo TA, Haber DA, Hariharan IK. Salvador promotes both cell cycle exit and apoptosis in *Drosophila* and is mutated in human cancer cell lines. Cell. 2002;**110**(4):467-478

[92] Udan RS, Kango-Singh M, Nolo R, Tao C, Halder G. Hippo promotes proliferation arrest and apoptosis in the Salvador/Warts pathway. Nature Cell Biology. 2003;**5**(10):914-920

[93] Wu S, Huang J, Dong J, Pan D. Hippo encodes a Ste-20 family protein kinase that restricts cell proliferation and promotes apoptosis in conjunction with salvador and warts. Cell. 2003;**114**(4):445-456

[94] Moberg KH, Bell DW, Wahrer DC, Haber DA, Hariharan IK. Archipelago regulates Cyclin E levels in *Drosophila* and is mutated in human cancer cell lines. Nature. 2001;**413**(6853):311-316

[95] Hafen E. Cancer, type 2 diabetes, and ageing: News from flies and worms. Swiss Medical Weekly. 2004;**134**(49-50):711-719

[96] Grzeschik NA, Amin N, Secombe J, Brumby AM, Richardson HE. Abnormalities in cell proliferation and apico-basal cell polarity are separable in *Drosophila* lgl mutant clones in the developing eye. Developmental Biology. 2007;**311**(1):106-123

[97] Naora H, Montell DJ. Ovarian cancer metastasis: Integrating insights from disparate model organisms. Nature Reviews Cancer. 2005;**5**(5):355-366

[98] Brumby AM, Goulding KR, Schlosser T, Loi S, Galea R, Khoo P, Bolden JE, Aigaki T, Humbert PO, Richardson HE. Identification of novel Ras-cooperating oncogenes in *Drosophila melanogaster*: A RhoGEF/rho-family/JNK pathway is a central driver of tumorigenesis. Genetics. 2011;**188**(1):105-125

[99] Navarro C, Nola S, Audebert S, Santoni MJ, Arsanto JP, Ginestier C, Marchetto S, Jacquemier J, Isnardon D, Le Bivic A, et al. Junctional recruitment of mammalian scribble relies on E-cadherin engagement. Oncogene. 2005;**24**(27):4330-4339

[100] Pagliarini RA, Xu T. A genetic screen in *Drosophila* for metastatic behavior. Science. 2003;**302**(5648):1227-1231

[101] Cordero JB, Larson DE, Craig CR, Hays R, Cagan R. Dynamic decapentaplegic signaling regulates patterning and adhesion in the *Drosophila* pupal retina. Development. 2007;**134**(10):1861-1871

[102] Morata G, Ripoll P. Minutes: Mutants of *drosophila* autonomously affecting cell division rate. Developmental Biology. 1975;**42**(2):211-221

[103] Simpson P, Morata G. Differential mitotic rates and patterns of growth in compartments in the *Drosophila* wing. Developmental Biology. 1981;**85**(2):299-308

[104] Gallant P, Shiio Y, Cheng PF, Parkhurst SM, Eisenman RN. Myc and max homologs in *Drosophila*. Science. 1996;**274**(5292):1523-1527

[105] Johnston LA, Prober DA, Edgar BA, Eisenman RN, Gallant P. *Drosophila* myc regulates cellular growth during development. Cell. 1999;**98**(6):779-790

[106] de la Cova C, Abril M, Bellosta P, Gallant P, Johnston LA. *Drosophila* myc regulates organ size by inducing cell competition. Cell. 2004;**117**(1):107-116

[107] Moreno E, Basler K. dMyc transforms cells into super-competitors. Cell. 2004;**117**(1):117-129

[108] Tamori Y, Deng WM. Cell competition and its implications for development and cancer. Journal of Genetics and Genomics. 2011;**38**(10):483-495

[109] Wagstaff L, Kolahgar G, Piddini E. Competitive cell interactions in cancer: A cellular tug of war. Trends in Cell Biology. 2013;**23**(4):160-167

[110] Baker NE, Li W. Cell competition and its possible relation to cancer. Cancer Research. 2008;**68**(14):5505-5507

[111] Moreno E. Is cell competition relevant to cancer? Nature Reviews Cancer. 2008;**8**(2):141-147

[112] Di Giacomo S, Sollazzo M, de Biase D, Ragazzi M, Bellosta P, Pession A, Grifoni D. Human cancer cells signal their competitive fitness through MYC activity. Scientific Reports. 2017;**7**(1):12568

[113] Pettersson A, Lis RT, Meisner A, Flavin R, Stack EC, Fiorentino M, Finn S, Graff RE, Penney KL, Rider JR, et al. Modification of the association between obesity and lethal prostate cancer by TMPRSS2:ERG. Journal of the National Cancer Institute. 2013;**105**(24):1881-1890

[114] Breese CR, Ingram RL, Sonntag WE. Influence of age and long-term dietary restriction on plasma insulin-like growth factor-1 (IGF-1), IGF-1 gene expression, and IGF-1 binding proteins. Journal of Gerontology. 1991;**46**(5):B180-B187

[115] Kalaany NY, Sabatini DM. Tumours with PI3K activation are resistant to dietary restriction. Nature. 2009;**458**(7239):725-731

[116] Donadon V, Balbi M, Zanette G. Hyperinsulinemia and risk for hepatocellular carcinoma in patients with chronic liver diseases and Type 2 diabetes mellitus. Expert Review of Gastroenterology & Hepatology. 2009;**3**(5):465-467

[117] Hirabayashi S, Baranski TJ, Cagan RL. Transformed *Drosophila* cells evade diet-mediated insulin resistance through wingless signaling. Cell. 2013;**154**(3):664-675

[118] Gnerlich JL, Yao KA, Fitchev PS, Goldschmidt RA, Bond MC, Cornwell M, Crawford SE. Peritumoral expression of adipokines and fatty acids in breast cancer. Annals of Surgical Oncology. 2013;**20**(Suppl 3):S731-S738

[119] Martinez-Outschoorn UE, Pestell RG, Howell A, Tykocinski ML, Nagajyothi F, Machado FS, Tanowitz HB, Sotgia F, Lisanti MP. Energy transfer in "parasitic" cancer metabolism: Mitochondria are the powerhouse and Achilles' heel of tumor cells. Cell Cycle. 2011;**10**(24):4208-4216

[120] Ugur B, Chen K, Bellen HJ. *Drosophila* tools and assays for the study of human diseases. Disease Models & Mechanisms. 2016;**9**(3):235-244

[121] Diangelo JR, Birnbaum MJ. The regulation of fat cell mass by insulin in *Drosophila melanogaster*. Molecular and Cellular Biology. 2009;(24):6341-6352

[122] Baker KD, Thummel CS. Diabetic larvae and obese flies-emerging studies of metabolism in *Drosophila*. Cell Metabolism. 2007;**6**(4):257-266

[123] Gutierrez E, Wiggins D, Fielding B, Gould AP. Specialized hepatocyte-like cells regulate *Drosophila* lipid metabolism. Nature. 2007;**445**(7125):275-280

[124] Samuel VT, Shulman GI. Nonalcoholic fatty liver disease as a nexus of metabolic and hepatic diseases. Cell Metabolism. 2017;(17):30487-30494

[125] Oike Y, Akao M, Kubota Y, Suda T. Angiopoietin-like proteins: Potential new targets for metabolic syndrome therapy. Trends in Molecular Medicine. 2005;**11**(10):473-479

[126] Chien CH, Chen WW, JT W, Chang TC. Investigation of lipid homeostasis in living *Drosophila* by coherent anti-stokes Raman scattering microscopy. Journal of Biomedical Optics. 2012;**17**(12):126001

[127] Kohyama-Koganeya A, Kim YJ, Miura M, Hirabayashi Y, *Drosophila* A. Orphan G protein-coupled receptor BOSS functions as a glucose-responding receptor: Loss of boss causes abnormal energy metabolism. Proceedings of the National Academy of Sciences of the United States of America. 2008;**105**(40):15328-15333

[128] Canavoso LE, Jouni ZE, Karnas KJ, Pennington JE, Wells MA. Fat metabolism in insects. Annual Review of Nutrition. 2001;**21**:23-46

[129] Gronke S, Muller G, Hirsch J, Fellert S, Andreou A, Haase T, Jackle H, Kuhnlein RP. Dual lipolytic control of body fat storage and mobilization in *Drosophila*. PLoS Biology. 2007;**5**(6):e137

[130] Wullschleger S, Loewith R, Hall MN. TOR signaling in growth and metabolism. Cell. 2006;**124**(3):471-484

[131] Garofalo RS. Genetic analysis of insulin signaling in *Drosophila*. Trends in Endocrinology and Metabolism: TEM. 2002;**13**(4):156-162

[132] Rulifson EJ, Kim SK, Nusse R. Ablation of insulin-producing neurons in flies: Growth and diabetic phenotypes. Science. 2002;**296**(5570):1118-1120

[133] Rajan A, Perrimon N. *Drosophila* cytokine unpaired 2 regulates physiological homeostasis by remotely controlling insulin secretion. Cell. 2012;**151**(1):123-137

[134] Musselman LP, Fink JL, Narzinski K, Ramachandran PV, Hathiramani SS, Cagan RL, Baranski TJ. A high-sugar diet produces obesity and insulin resistance in wild-type *Drosophila*. Disease Models & Mechanisms. 2011;**4**(6):842-849

[135] Birse RT, Choi J, Reardon K, Rodriguez J, Graham S, Diop S, Ocorr K, Bodmer R, Oldham S. High-fat-diet-induced obesity and heart dysfunction are regulated by the TOR pathway in *Drosophila*. Cell Metabolism. 2010;**12**(5):533-544

[136] Skorupa DA, Dervisefendic A, Zwiener J, Pletcher SD. Dietary composition specifies consumption, obesity, and lifespan in *Drosophila melanogaster*. Aging Cell. 2008;**7**(4):478-490

[137] Woodcock KJ, Kierdorf K, Pouchelon CA, Vivancos V, Dionne MS, Geissmann F. Macrophage-derived upd3 cytokine causes impaired glucose homeostasis and reduced lifespan in *Drosophila* fed a lipid-rich diet. Immunity. 2015;**42**(1):133-144

[138] Na J, Cagan R. The *Drosophila* nephrocyte: Back on stage. Journal of the American Society of Nephrology: JASN. 2013;**24**(2):161-163

[139] Renehan AG, Roberts DL, Dive C. Obesity and cancer: Pathophysiological and biological mechanisms. Archives of Physiology and Biochemistry. 2008;**114**(1):71-83

[140] Younossi Z, Henry L. Contribution of alcoholic and nonalcoholic fatty liver disease to the burden of liver-related morbidity and mortality. Gastroenterology. 2016; **150**(8):1778-1785

[141] Hoffmann JA, Kafatos FC, Janeway CA, Ezekowitz RA. Phylogenetic perspectives in innate immunity. Science. 1999;**284**(5418):1313-1318

[142] Scherfer C, Karlsson C, Loseva O, Bidla G, Goto A, Havemann J, Dushay MS, Theopold U. Isolation and characterization of hemolymph clotting factors in *Drosophila melanogaster* by a pullout method. Current Biology: CB. 2004;**14**(7):625-629

[143] Bulet P, Hetru C, Dimarcq JL, Hoffmann D. Antimicrobial peptides in insects; structure and function. Developmental and Comparative Immunology. 1999;**23**(4-5):329-344

[144] Ha EM, CT O, Ryu JH, Bae YS, Kang SW, Jang IH, Brey PT, Lee WJ. An antioxidant system required for host protection against gut infection in *Drosophila*. Developmental Cell. 2005;**8**(1):125-132

[145] Juarez MT, Patterson RA, Sandoval-Guillen E, McGinnis W. Duox, Flotillin-2, and Src42A are required to activate or delimit the spread of the transcriptional response to epidermal wounds in *Drosophila*. PLoS Genetics. 2011;**7**(12):e1002424

[146] Carton Y, Nappi AJ. Immunogenetic aspects of the cellular immune response of Drosophilia against parasitoids. Immunogenetics. 2001;**52**(3-4):157-164

[147] Gold KS, Bruckner K. Macrophages and cellular immunity in *Drosophila melanogaster*. Seminars in Immunology. 2015;**27**(6):357-368

[148] Krzemien J, Dubois L, Makki R, Meister M, Vincent A, Crozatier M. Control of blood cell homeostasis in *Drosophila* larvae by the posterior signalling centre. Nature. 2007; **446**(7133):325-328

[149] Mandal L, Martinez-Agosto JA, Evans CJ, Hartenstein V, Banerjee U. A hedgehog- and Antennapedia-dependent niche maintains *Drosophila* haematopoietic precursors. Nature. 2007;**446**(7133):320-324

[150] Owusu-Ansah E, Banerjee U. Reactive oxygen species prime *Drosophila* haematopoietic progenitors for differentiation. Nature. 2009;**461**(7263):537-541

[151] Lebestky T, Jung SH, Banerjee U. A serrate-expressing signaling center controls *Drosophila* hematopoiesis. Genes & Development. 2003;**17**(3):348-353

[152] Jung SH, Evans CJ, Uemura C, Banerjee U. The *Drosophila* lymph gland as a developmental model of hematopoiesis. Development. 2005;**132**(11):2521-2533

[153] Sinenko SA, Hung T, Moroz T, Tran QM, Sidhu S, Cheney MD, Speck NA, Banerjee U. Genetic manipulation of AML1-ETO-induced expansion of hematopoietic precursors in a *Drosophila* model. Blood. 2010;**116**(22):4612-4620

[154] Osman D, Gobert V, Ponthan F, Heidenreich O, Haenlin M, Waltzer L, *Drosophila* A. Model identifies calpains as modulators of the human leukemogenic fusion protein AML1-ETO. Proceedings of the National Academy of Sciences of the United States of America. 2009;**106**(29):12043-12048

[155] Lemaitre B, Hoffmann J. The host defense of *Drosophila melanogaster*. Annual Review of Immunology. 2007;**25**:697-743

[156] Hoffmann JA, Reichhart JM. *Drosophila* innate immunity: An evolutionary perspective. Nature Immunology. 2002;**3**(2):121-126

[157] Varma H, Cheng R, Voisine C, Hart AC, Stockwell BR. Inhibitors of metabolism rescue cell death in Huntington's disease models. Proceedings of the National Academy of Sciences of the United States of America. 2007;**104**(36):14525-14530

[158] Lemaitre B. The road to toll. Nature Reviews Immunology. 2004;**4**(7):521-527

[159] Manfruelli P, Reichhart JM, Steward R, Hoffmann JA, Lemaitre B. A mosaic analysis in *Drosophila* fat body cells of the control of antimicrobial peptide genes by the Rel proteins Dorsal and DIF. The EMBO Journal. 1999;**18**(12):3380-3391

[160] Tauszig S, Jouanguy E, Hoffmann JA, Imler JL. Toll-related receptors and the control of antimicrobial peptide expression in *Drosophila*. Proceedings of the National Academy of Sciences of the United States of America. 2000;**97**(19):10520-10525

[161] Zettervall CJ, Anderl I, Williams MJ, Palmer R, Kurucz E, Ando I, Hultmark D. A directed screen for genes involved in *Drosophila* blood cell activation. Proceedings of the National Academy of Sciences of the United States of America. 2004;**101**(39):14192-14197

[162] Lemaitre B, Kromer-Metzger E, Michaut L, Nicolas E, Meister M, Georgel P, Reichhart JM, Hoffmann JA. A recessive mutation, immune deficiency (imd), defines two distinct control pathways in the *Drosophila* host defense. Proceedings of the National Academy of Sciences of the United States of America. 1995;**92**(21):9465-9469

[163] Cogni R, Cao C, Day JP, Bridson C, Jiggins FM. The genetic architecture of resistance to virus infection in *Drosophila*. Molecular Ecology. 2016;**25**(20):5228-5241

[164] Khalil S, Jacobson E, Chambers MC, Lazzaro BP. Systemic bacterial infection and immune defense phenotypes in *Drosophila melanogaster*. Journal of Visualized Experiments: JoVE. 2015;**99**:e52613

[165] Buchon N, Broderick NA, Poidevin M, Pradervand S, Lemaitre B. *Drosophila* intestinal response to bacterial infection: Activation of host defense and stem cell proliferation. Cell Host & Microbe. 2009;**5**(2):200-211

[166] Storelli G, Defaye A, Erkosar B, Hols P, Royet J, Leulier F. Lactobacillus plantarum promotes *Drosophila* systemic growth by modulating hormonal signals through TOR-dependent nutrient sensing. Cell Metabolism. 2011;**14**(3):403-414

[167] Ferrandon D. The complementary facets of epithelial host defenses in the genetic model organism *Drosophila melanogaster*: From resistance to resilience. Current Opinion in Immunology. 2013;**25**(1):59-70

[168] Lee JE, Edery I. Circadian regulation in the ability of *Drosophila* to combat pathogenic infections. Current Biology: CB. 2008;**18**(3):195-199

Chapter 8

# Parkinson's Disease: Insights from *Drosophila* Model

Mohamad Ayajuddin, Abhik Das,
Limamanen Phom, Priyanka Modi, Rahul Chaurasia,
Zevelou Koza, Abuno Thepa, Nukshimenla Jamir,
Pukhrambam Rajesh Singh,
Sentinungla Longkumer, Pardeshi Lal and
Sarat Chandra Yenisetti

Additional information is available at the end of the chapter

http://dx.doi.org/10.5772/intechopen.72021

## Abstract

Parkinson's disease (PD) is a medical condition that has been known since ancient times. It is the second most common neurodegenerative disorder affecting approximately 1% of the population over 50 years. It is characterized by both motor and non-motor symptoms. Most of PD cases are sporadic while 5–10% cases are familial. Environment factors such as exposure to pesticides, herbicides and other heavy metals are expected to be the main cause of sporadic form of the disease. Mutation of the susceptible genes such as SNCA, PINK1, PARKIN, DJ1, and others are considered to be the main cause of the familial form of disease. *Drosophila* offers many advantages for studying human neurodegenerative diseases and their underlying molecular and cellular pathology. Shorter life span; large number of progeny; conserved molecular mechanism(s) among fly, mice and human; availability of many techniques, and tools to manipulate gene expression makes drosophila a potential model system to understand the pathology associated with PD and to unravel underlying molecular mechanism(s) responsible for dopaminergic neurodegeneration in PD—understanding of which will be of potential assistance to develop therapeutic strategies to PD. In the present review, we made an effort to discuss the contribution of fly model to understand pathophysiology of PD, in understanding the biological functions of genes implicated in PD; to understand the gene-environment interaction in PD; and validation of clues that are generated through genome-wide association studies (GWAS) in human through fly; further to screen and develop potential therapeutic molecules for PD. In nutshell, fly has been a great model system which has immensely contributed to the biomedical research relating to understand and addressing the pathology of human neurological diseases in general and PD in particular.

**Keywords:** dopamine, *Drosophila*, Parkinson's disease, mitochondrial dysfunction, neuroprotective therapeutics, pathophysiology and translational research

## 1. Introduction

Parkinson's disease (PD) is a medical condition that has been known about since ancient times in Indian and Chinese civilization [1, 2]. It is referred to as *Kampavata* in the ancient Indian medical system of Ayurveda ("*kampa*" means tremor in Sanskrit). An Egyptian papyrus from the twelfth century BC mentions a king drooling with age [3]. In Western medical literature, the tremor symptom was described by the physician Galen in 175 AD [4]. In 1817, James Parkinson wrote an essay on "shaking palsy" based on six cases that he had observed in his own practice and on walks around his neighborhood. The essay was intended to encourage others to study the disease. This established the disease as a recognized medical condition. He termed this medical condition as "*shaking palsy* or *paralysis agitans*". He published a detailed medical essay entitled "An Essay on the Shaking Palsy" where he described shaking palsy as "involuntary tremulous motion, with lessened muscular power, in parts not in action and even when supported; with a propensity to bend the trunk forwards, and to pass from a walking to a running pace: the senses and intellects being uninjured" [5]. The term "Parkinson's disease" was coined in 1865 by William Sanders and later popularized by French neurologist Jean Martin Charcot [6]. Charcot and colleagues described the clinical symptoms of this disease, noting two prototypes: the tremorous and the rigid or akinetic form. They described the detailed arthritic changes, dysautonomia, and pain that can accompany Parkinson's disease. He recognized that PD patients are not markedly weak and do not necessarily have tremor [7]. All these observations instigated the curiosity among the clinicians to understand this condition better with an aim to improve the patient's quality of life.

## 2. Pathophysiology

Parkinson's disease is the second most common neurodegenerative disorder affecting approximately 1% of the population over 50 years [8]. A central pathological hallmark of PD is the selective loss of dopamine (DA) neurons in the *substantia nigra pars compacta* (SN). These dopaminergic neurons are required for proper motor function, and their deficiency manifests its characteristic features: bradykinesia, tremors, and rigidity [9]. A second neuropathological hallmark of PD is the Lewy body (LB), which is a cytoplasmic spherical proteinaceous inclusion. LBs have been reported to contain various proteins including $\alpha$-synuclein, ubiquitin, parkin, and neurofilaments [10]. The mechanisms by which $\alpha$-synuclein and other proteins aggregate to form Lewy pathology are uncertain, but may involve oxidative modifications and/or cross-linking. Although, the neurodegeneration of PD was considered to be confined to dopaminergic cell loss in the SN, cell loss, and neuropathology is found to occur in other parts as well including the locus coeruleus, raphe, nucleus basalis of Meynert, dorsal motor nucleus of the vagus, cerebral cortex, olfactory bulb, and autonomic nervous system [11]. Several non-motor symptoms such as sleep disturbances, constipation, cognitive decline, depression, fear, anxiety, bladder problems, weight changes, fatigue and loss of energy, hypotension, and sexual problems can be dominant and debilitating in a sizeable number of patients, affecting the quality of their life [12]. Till date, treatments address only the symptoms but they fail to stop the progression of the disease and PD patients continue to experience a higher mortality rate compared to the general population [13].

### 2.1. Sporadic Parkinson's disease

A sporadic PD has unknown cause with implication of environmental influence coupled with genetic factors. The pathology of PD therefore may be multifactorial involving gene and environment interactions. Studies indicate role of neurotoxicants or neuroprotective compounds in pathogenesis of nigrostriatal degeneration, supporting the concept of association between the environment and PD [14]. Additionally, the identification of the mutated $\alpha$-synuclein (SNCA) gene that cause familial PD [15] as a risk factor for sporadic disease [16] provides a genetic background for the disease. Studies suggest that rural people, well water consumption, pesticide use, and occupations like rural farming, mining, and welding have an increased risk of PD [17, 18]. Epidemiological studies suggest association of PD with environmental toxic factors, primarily the mitochondrial complex I inhibitors such as rotenone [19]. Some other findings suggest that exposure to pesticide such as bipyridyl, organochlorine, and carbamate derivatives could contribute to PD [20].

### 2.2. 1-Methyl-4-phenyl-1,2,3,6-tetrahydropyridine (MPTP)

Exploring the contribution of environmental exposure markedly advanced our understanding of the mechanisms involved in the development of PD. Initial evidence came from findings that subjects exposed to 1-methyl-4-phenyl-1,2,3,6-tetrahydropyridine (MPTP) developed PD-like symptoms [21]. MPTP was accidentally discovered in a synthesis process that went wrong, and, although it may have caused some problems in certain circles, today it represents the most important and most frequently used parkinsonian toxin applied in animal models. 1-Methyl-4-phenyl-1,2,3,6-tetrahydropyridine (MPTP) is a representative strong neurotoxin that has been recognized from several young drug addicts from Northern California developed severe parkinsonism [22]. Since then, environmental exposure to pesticides [23], polychlorinated biphenyls [24], organic solvents [25], metals [26], and air pollutants [27] has been proposed to increase risk for PD.

### 2.3. Mechanism of MPTP neurotoxicity

Though the exact mechanism regarding the mode of MPTP toxicity is not known, it has been postulated that $MPP^+$ entry into dopaminergic neurons is dependent on selective uptake by dopamine transporter localizing it and interfering into mitochondrial activity. MPTP is not toxic per se, but becomes toxic once it is converted to 1-methyl-4-phenyl-2,3-dihydropyridinium (MPDP+) by action of monoamine oxidase B (MAO-B) in glial cells and serotoninergic neurons [28] followed by oxidation into 1-methyl-4-phenylpyridinium (MPP+), which is a highly toxic compound [29]. Then the dopamine transporter (DAT) carries it to dopaminergic neurons leading to its accumulation in cytoplasm and into synaptic vesicles by the vesicular monoamine transporter (VMAT). The driving force of mitochondrial membrane potential lets MPP+ enter these organelles, where it blocks complex I [30]. This leads to abnormally increased concentrations of the toxin to interfere with mitochondrial respiration by blocking the mitochondrial oxidation. Thus it results in impairment of ATP synthesis and involving in the generation of oxidative stress.

### 2.4. Paraquat

Paraquat (1,1-dimethyl-4,4′-bipyridinium dichloride) is a quaternary nitrogen herbicide widely used for broadleaf weed control. It is a fast-acting, non-selective compound which destroys tissues of green plants on contact and by translocation with the plant. Significant damage to the brain was seen in individuals who died from paraquat intoxication [31]. For many years, experimental studies using paraquat were focusing on its effects on lung, liver, and kidney, probably because the toxicity induced by this herbicide in these organs is responsible for death after acute exposure [32]. Epidemiological studies in agricultural communities have suggested an increased risk for PD due to paraquat use, raising the possibility that paraquat could be an environmental parkinsonian toxin. This chemical causes extensive oxidative stress in mitochondria of the cell, resulting in the perturbation of biochemical processes, cell death, multi-organ failure, and neurodegenerative diseases [33].

It is still vague about how the molecular mechanism of paraquat leads to cell death. However, studies have shown that paraquat can trigger the sequential phosphorylation of c-Jun N-terminal kinase (JNK), c-Jun, and the activation of caspase-3 consequently leading to neuronal death both in vitro and in vivo [34], suggesting that JNK signaling pathway contributes to paraquat-induced neurodegeneration.

### 2.5. Rotenone

Rotenone is a commonly used natural pesticide prepared from the roots of certain tropical plants, such as *Derris elliptica*. It is a classical high-affinity-specific inhibitor of mitochondrial complex I. This lipophilic compound crosses the blood-brain barrier rapidly and accumulates in subcellular organelles such as mitochondria where it impairs oxidative phosphorylation by inhibiting complex I of the electron transport chain [35]. Postmortem studies implicated mitochondrial impairments [36], and epidemiological studies suggested an association with environmental toxins, in particular mitochondrial complex I inhibitors such as rotenone [37]. *In vitro*, rotenone has been shown to produce cell apoptosis, accumulation, and aggregation of $\alpha$-synuclein and ubiquitin, oxidative damage, and endoplasmic reticulum stress [38, 39]. In a study in post mortem idiopathic PD brain, the substantia nigra is seen to comprise of a strong inhibition of complex I activity [40] suggesting this could be the cause of degeneration of dopaminergic neurons.

## 3. Familial/genetic Parkinson's disease

Till date, 15 known genes and 21 loci have been identified for familial PD. Some of the genes are discussed further.

### 3.1. SNCA

In a study involving a large Italian family, Polymeropoulos and colleagues identified the missense mutations in the *SNCA* gene. Through a traditional linkage approach, they managed to track the underlying genetic injury to an area located in the long arm of human chromosome number 4 [41]. This discovery has been a FRAME shift in the genetic research of PD. A separate study showed that the $\alpha$-synuclein protein is the main constituent of the Lewy body which is the pathological hallmark of PD [42]. These two vital research findings brought about a link between sporadic and familial forms of PD.

Five different missense mutations in *SNCA* have been implicated in PD namely A53T, A30P, E46K, H50Q, and G51D mutation. Clinically, early age onset of parkinsonism with a positive initial response to levodopa treatment are seen among patients with missense mutations but later on the disease progresses rapidly with dementia as a common feature. Eventually, appearance of cognitive decline, hallucinations, and fluctuations of consciousness in patients becomes clear. Histopathological study reveals an abundant LB pathology [43].

### 3.2. LRRK2

An association between apparent autosomal-dominant parkinsonism and chromosome number 12 was suggested by findings on a study involving a large Japanese family [44]. Later it was established that mutations in the gene *LRRK2* lead to the basic genetic cause of chromosome number 12 linked PD [45]. The most frequent *LRRK2* mutation is G2019S, detected in approximately 1% of sporadic and about 3–6% of familial PD cases [46]. R1441G is the second most common mutation after G2019S [47]. Most of the *LRRK2* cases described demonstrate LB in the brainstem accompanied by loss of neurons in the SN. Only a minority of cases exhibit neurofibrillary tangle pathology, glial cytoplasmic inclusions, or neuronal nigral loss without LB [45].

### 3.3. PARKIN

An uncommon syndrome characterized by early dystonia and problems from levodopa treatment, osteotendinous hyperreflexia and comparatively slow motor progression was described in Japan in 1973, which is now known to be an autosomal-recessive juvenile parkinsonism (AR-JP) [48]. Mutations in parkin gene was identified as a cause of this condition [49]. AR-JP maps to the long arm of chromosome number 6 and linked to the markers *D6S305* and *D6S253* [50]. It was found that the former is deleted in an AR-JP Japanese patient [51]. By positional cloning within this microdeletion, Kitada and colleagues isolated a cDNA clone of 2960 bp with a 1395 bp open reading frame which code a 465 amino acid protein consisting of an N-terminal ubiquitin-like domain and RING domain having two RING finger motifs. The gene spans more than 500 kb and has 12 exons of which 5 exons (3–7) are found to be deleted in the patient. Also four other AR-JP unrelated patients have a deletion that affect only the exon 4. A 4.5 kb transcript expressed in various human tissues abundantly in the brain, including the substantia nigra, is shorter in brain tissue from one of the groups of exon 4 deleted patients. Therefore inferring that the mutations in this newly identified gene is responsible for the pathogenesis of AR-JP. In a number of families, PD is related with heterozygous *parkin* mutations through an apparently dominant way of transmission, suggesting that the carriers of a sole *parkin* mutation might be at risk to develop the PD [52].

### 3.4. DJ1

A homozygous deletion and a missense mutation in the Daisuke-Junko-1 (*DJ-1*) gene as a cause of autosomal-recessive early onset PD was identified [53]. A number of novel *DJ-1* mutations have been discovered in patients with early onset PD. However, these mutations are rare and can be found in only ~ about 1% of early onset PD cases [54]. At the clinical level, the phenotype of *DJ-1* subjects is the same as that of *parkin* and *PINK1*-related parkinsonism, with age at onset (AAO) usually around the mid-30s, good response to levodopa treatment, slow disease progression, and often focal dystonia such as blepharospasm [55]. However, the neuropathology of *DJ-1*-linked parkinsonism still remains unidentified.

### 3.5. PINK1

Mutations in the phosphate and tensin homolog-induced Putative kinase 1 (*PINK1*) were initially identified in a Sicilian family with autosomal-recessive parkinsonism [56]. In most of *PINK1* mutations, the type of mutation seen is missense. However, mutations of copy number, genetic, and exonic rearrangements have been described [57]. In both the cases of familial and sporadic PD, investigation of mutation has recognized homozygous and compound heterozygous type *PINK1* mutations. This raised the potential role on a single *PINK1* heterozygous mutation to be a possible risk factor for PD [58]. In a number of families, *PINK1* mutations have been associated with early onset PD and *PINK1* mutations association in sporadic cases is about 2–4%. In a clinical phenotype, this type of parkinsonism is quite comparable to those seen among patients with *parkin* and *DJ-1* mutations. They display progressive levodopa-responsive disease gradually [57].

### 3.6. Vacuolar protein sorting-associated protein 35 (VPS35)

Mutation in VPS35 causes monogenic form of PD was described by using exome sequencing in two separate studies with identification of p.D620N mutation in VPS35 among the members of a Swiss kindred with a late onset, autosomal-dominant PD [59]. An independent study published the identification of the p.D620N mutation in a large multi-generation Austrian PD family and in two additional families screened for VPS35 mutations [60]. The VPS35-linked families reportedly fulfill the London Brain Bank criteria for PD, but there are fairly limited clinical and pathological details on these cases.

### 3.7. ATPase (P-type) 13A2 (ATP13A2)

In a neuronal P-type ATPase gene, ATP13A2, a loss of function was initially described in a consanguineous Jordanian family [61]. Clinically, the subjects showed a very early onset of the disease accompanied by rigid-akinetic phenotype with reduced resting tremor, pyramidal syndrome, progressive cognitive impairment, vertical gaze palsy, mini myoclonus, insomnia, and levodopa responsive [61]. Mutations of this gene mapping on chromosome 1p36 cause PD in atypical form which is known as Kufor-Rakeb syndrome [62]. Clinical phenotype of this early onset pallido-pyramidal syndrome varies in severity but only a handful of cases and families have been reported [63, 64].

### 3.8. PLA2G6

Homozygous mutations in phospholipase A2 gene (PLA2G6) located on chromosome 22q13.1 was reported in three patients of two inbred Pakistani families. The phenotypes were associated with cognitive and psychiatric problems and dystonic features, pyramidal syndrome [65]. In an Asian group with early onset recessive parkinsonism caused by compound heterozygous mutations in PLA2G6, the phenotype reported were frontotemporal lobar atrophy and dementia [66]. The PLA2G6 gene encodes a protein of phospholipase A2 group VI (PLA2G6), which act as catalysis of fatty acids elimination from phospholipids and involved in maintaining membrane phospholipids homeostasis [67].

### 3.9. UCH-L1

Missense mutations of gene coding for the ubiquitin carboxyl-terminal hydrolase L1 (UCH-L1), an ubiquitin recycling enzyme located on chromosome 4p14 was reported in a German family with an autosomal-dominant transmission PD [68]. The phenotype of affected individuals was consistent with that of idiopathic PD. An epidemiological study suggested an association between the UCH-L1 gene S18Y variant and PD [69]. Overexpression of UCH-L1 gene upregulated aggresomes formation through dysfunction of proteasome system [70].

### 3.10. HtrA2

Various studies indicate risk factor for parkinsonism due to loss of function of the gene encoding Omi/high temperature requiring A2 mitochondrial protein (HtrA2) in German [71] and Belgians PD patients [72]. The serine protease Omi/HtrA2 is released from mitochondria and promotes apoptosis [73] and mutations of Omi/HtrA2 gene affect its protease activity linked to mitochondrial dysfunction [71, 72]. Although it acts independently of parkin, Omi/HtrA2 functions in the PINK1/parkin pathway downstream of PINK1 [74]. These findings were not confirmed in Omi/HtrA2 knockout mutants in contrast to PINK1 or parkin null mutants [75].

### 3.11. EIF4G1

Mutations in the eukaryotic translation initiation factor 4-gamma (*EIF4G1*) was identified as a risk factor in a study of a northern French family with autosomal-dominant late onset parkinsonism on the chromosome 3q26-q28. Genomic analysis identified a heterozygous mutation in EIF4G1 which was confirmed subsequently in 2 PD patients and 2 Lewy body disease patients among 225 more patients [76]. Further, in all the affected members of another multiplex unrelated family, a pathogenic mutation was detected including in one unaffected 86-year-old family member suggesting an incomplete penetrance [77].

## 4. Animal models of Parkinson's disease

### 4.1. Criteria for modeling PD in animals

Being a neurodegenerative disorder, the prominent hall mark of PD is progressive loss of dopaminergic neurons in the *Substantia niagra parse compacta* (SNpc) [78]. Together with DA neurons, there is a loss of cholinergic neurons, serotonergic neurons, and nor adrenergic neurons in the brain [79]. The prominent biomarker being aggregation of Lewy bodies in intracytoplasmic space [80]. A combination of all these features shows the hallmark motor and non-motor symptom of PD.

A suitable model for PD should show histopathologically characterizable progressive loss of DA neurons together with other neurons and significant reduction in DA level. Moreover, the onset of the disease should be in adulthood this should manifest in such a way that it would mimic the PD-affected human motor symptom such as bradykinesia, rigidity, postural instability, and resting tremor, with motor features being responsive to L-DOPA or any anti-PD drug therapy. Even though non human primate and mouse has been the traditional model of PD, because of low cost of maintenance, shorter life cycle, and defined neuropathological profile is making zebrafish and *Drosophila* are among the emerging and more interesting model of PD [81].

### 4.2. Advantages and limitation of modeling PD using drosophila

#### *4.2.1. Advantages*

*Drosophila* offers many advantages for studying human neurodegenerative diseases and their underlying molecular and cellular pathology. Benefits include a faster time frame due to the shorter life span of the fly (10–14 day from pupae to adult), large number of progeny, availability of many techniques, and tools to manipulate gene expression [82]. Also *Drosophila* has been studied for longer than any other model out there which makes its anatomy and phenotypes very well known to experimental biologists [83]. A well developed CNS and prominent number of DA neurons [84] combined with well characterized behaviors which are conserved among strains in 90% cases [85] makes fly very cost effective and efficient model. Genetically, it has been estimated that nearly 75% of disease-related genes in humans have functional orthologs in the fly [86]. While overall similarity at nucleotide or protein level is 40% but in conserved domain it is 80–90% or higher [87].

#### *4.2.2. Disadvantages*

The major difference being an invertebrate there will be some prominent difference in physiology with human (e.g., brain anatomy, cardio vascular system, and respiratory system) which relating complex motor behavior with human a difficult task [88]. For CNS-related studies, there is also an issue on blood-brain permeability difference [89]. The metabolic differences are also to be considered when studying drug efficacy and toxin-induced disease phenotype.

### 4.3. Relevance of study of PD in flies

#### *4.3.1. Drosophila mimicking human PD symptoms*

Although the physiological difference between human and flies are very prominent core pathology observed in human PD can be produced in a very accurate extent by toxin-induced or transgenic modification. It is as accurate as area specific and age-dependent DA neuron loss as observed in PD condition and hallmark PD biomarker the LB and LN inclusions are visible in transgenic system [90]. *Drosophila* also performs complex behavior such as mating, aggression, conditioning to fear, learning and motor behaviors such as flying, climbing, and walking [85] which like human are affected by the PD onset and progression. These multitudes of behavior are very much helpful in characterizing different aspects of PD.

### 4.4. Physiological attributes of fly brain

*Drosophila* model of PD has two principle phenotypes: the specific loss of DAergic neurons with aging brain and defects in motor behavior. In fly brain, the DA neurons are distributed as a group of clusters throughout the brain and project their effect on different behavioral patterns of the fly by different functional areas of the adult brain. Target areas include:

- **The mushroom bodies**: involved in memory formation and motivation.
- **The Central Complex**: controls the motor behavior.

The phenomenon of different part of brain controlling different behavior pattern highly resembles the mammalian brain (**Figure 1**) [84].

### 4.5. Genetic similarity between *Drosophila* and human

Drsophila shares 61% homology with human genetically. In fact all the familial or sporadic genes reported so far in human associated with parkinsonism are available in *Drosophila* as a

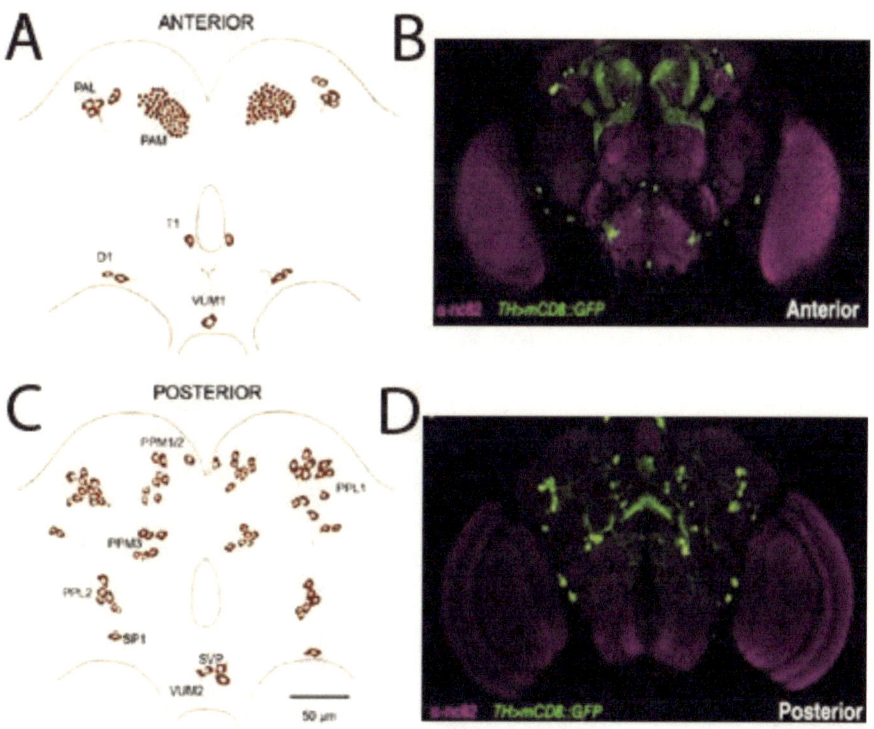


**Figure 1.** (**A** and **C**) Schematic representation of an adult fly brain with the distribution of DA neurons grouped in clusters and arranged with bilateral symmetry (image adapted from Botella et al. [91]). (**B** and **D**) Confocal Z-stack of *TH > mCD8::GFP* brain; anti-nc82 immunoreactivity together with GFP labeling reveals dopaminergic neurons in the anterior and posterior brain (image adapted from White et al. [84]).

| Gene/protein | Inheritance | Fly homolog | Protein function |
|---|---|---|---|
| $\alpha$-Synuclein | AD | None | Pre-synaptic protein |
| Parkin | AR | Parkin/CG10523 | E3 ubiquitin ligase |
| UCH-L1 | unclear | Uch/CG4265 | E3 ubiquitin hydrolase/ligase |
| PINK1 | AR | Pink1/CG4523 | Mitochondrial kinase |
| DJ-1 | AR | DJ-1a/CG6646<br>DJ-1b/CG1349 | Redox sensor/Chaperone |
| LRRK2 | AD | lrrk2/CG5483 | Kinase/GTPase |
| HtrA2 | AD | HtrA2/CG8486 | Mitochondrial pro-apoptotic protease |
| GBA | unclear | CG33090 | Lysosomal enzyme |
| POLG | unclear | tamas/CG8987 | Mitochondrial DNA polymerase |
| Tau | unclear | tau/CG31057 | Microtubule stabilization |

Notes: UCH-L1 = ubiquitin carboxyl-terminal esterase L1; PINK1 = PTEN-induced putative kinase 1; LRRK2 = leucine-rich repeat kinase 2; HtrA2 = high temperature requirement protein A2; GBA = glucocerebrosidase; POLG = polymerase gamma; AD = autosomal dominant; AR = autosomal recessive.

**Table 1.** Showing parkinsonian genes and their fly homologs [92].

homolog. But there is no homolog for the gene $\alpha$-synuclein which produces Lewy bodies and Lewy neuritis at the extracellular matrix of brain a hallmark biomarker under PD condition in mammalian brain. Nevertheless $\alpha$-synuclein transgenic models respond very well under PD conditions and recapitulate the PD phenotypes. Given (**Table 1**) are the list of parkinsonian genes and their homologs in flies.

## 5. Gene-environment interaction studies in PD

The identification of PD symptoms subsequent to ingestion of MPTP led to the idea that environmental factors could be related to the causes of pathogenesis of the disease. When ingested, MPTP is metabolized to neurotoxin MPP+ which was originally developed as an herbicide, cyperquat. The chemical structure of MPP+ is similar to that of the widely used herbicide paraquat. This finding suggested that exposure to environmental factors such as pesticide, herbicides may contribute to human sporadic PD. Over the years, environmental factors, including pesticides and herbicides, metals, tobacco and caffeine, head injuries, etc. have been largely considered as possible PD risk factors.

Over the last few decades either through genetic or toxin challenges many animal models have been developed to study PD. Using *Drosophila melanogaster* as a model has proved to be of great value and has contributed significantly toward understanding the mechanism underlying PD pathogenesis. **Table 2** illustrates the interaction studies between environmental toxins and PD genes using *Drosophila* model.

| Gene/protein | Toxin(s) | Observations | Reference |
|---|---|---|---|
| $\alpha$-Synuclein | Rotenone | *Drosophila* larvae expressing mutant human $\alpha$-synuclein showed significant loss of DA neurons and deficit in locomotion | [93] |
| DJ-1 | Paraquat<br>Rotenone | DJ-1 KO flies selectively sensitive to oxidative stress induced by toxins | [94]<br>[95] |
| LRRK2 | Rotenone | hLRRk2 transgenic flies showed loss of DA neurons and sensitivity to environmental toxin | [96] |
| Parkin | Rotenone | Transgenic flies expressing parkin mutation show age-dependent DA neuron degeneration and locomotor deficits | [97] |
| PINK1 | Rotenone | Flies expressing PINK1 mutations show increased rates of mortality in the mutants | [98] |

**Table 2.** Gene-environment interaction studies in drosophila model of PD.

Environmental toxins such as paraquat (herbicide) increases the risk factors for PD and this susceptibility is influenced by the genetic background. Fly model exposed to paraquat shows neurodegenerative symptoms induced by oxidative stress which are similar with most behavioral characters of PD. As seen from epidemiological studies, male flies are more susceptibility to paraquat toxicity than female flies. *Drosophila* mutant for dopamine regulating genes show variable susceptibility to paraquat such as mutations which increase DA pathway function helps in reducing paraquat neurotoxicity while loss of function mutations increases susceptibility to paraquat by decreasing dopamine levels. The loss of function mutation in negative regulator of DA production (Catecholamines-up (Catsup)) acts by delaying the onset of PD symptoms and loss of DA neurons and confers protection against paraquat exposure [99]. *Drosophila* DJ-1 mutants developed motor deficits when exposed to paraquat [100]. Loss of function of DJ-1$\beta$ mutants confers resistance to paraquat and lowers mortality of DA neuron and overexpression of DJ-1$\alpha$ in DA neurons, protects against paraquat toxicity [95]. Acute paraquat exposure in *Drosophila* showed elevated levels of oxidative stress markers and mitochondrial dysfunction [101]. Role of ubiquitin proteosome system (UPS) to sporadic PD is not very clear though it is a potential target for PD risk associated with pesticide [102]. Data from epidemiological studies show that paraquat in addition with maneb or ziram increases PD risk [103]. *Drosophila* knockdown of E1 ligase when exposed to paraquat + maneb showed significant DA neuron loss thus imply synergistic effects of the pesticides on risk for PD [104].

Rotenone exposure suppresses proteasomal activity through ATP depletion thus inhibiting mitochondrial function [105]. Parkin through its E3 ligase function offers neuroprotection against neuronal insults including rotenone [106]. Rotenone alters parkin solubility increasing intracellular aggregation and S-nitrosylation of parkin [107]. Loss of parkin increases rotenone-induced DA cell death in mice [108]. *Drosophila* shows negative geotaxis characteristics and rotenone exposure has shown to cause mortality and locomotor defects affecting the climbing ability of flies [109]. Rotenone also inhibits learning and memory functions in fly and the damage caused shows severe effect than those in $\alpha$-synuclein A30P mutant [110].

Exposure of adult *Drosophila* to sublethal doses of rotenone causes concentration-dependent locomotor deficiencies, specific dopaminergic neuronal loss, and reduction in the DA levels in flies [111]. Non-motor symptoms such as circadian rhythms in *Drosophila* are also altered following exposure to rotenone [112]. All these studies highlight the utility of drosophila model to understand the gene-environment interaction in PD.

### 5.1. *Drosophila* PD models and associated tools

*Drosophila* is proved to be one of the important genetic models to study the disease mechanism in PD. Even though with limitations, the fly model enables rapid and elaborate genetic study, providing in depth cell and molecular studies which offers a unique look into the mechanisms and pathways underlying PD pathogenesis [113]. Currently 14 genes for PD have been identified of which 12 genes have homology with Drosphila. Loss of function and gain of function analysis using fly model would provide insights into the mechanism of action of these genes associated with PD (**Table 3**).

### 5.2. Mitochondrial genetics of PD: insights from drosophila model

*Drosophila* model has made major contribution in the research area of mitochondrial genetics. The early hints of PINK1/parkin on mitochondrial homeostasis came from studies using *Drosophila* model [98, 114, 115]. The most compelling evidence for a mitochondrial etiology of

| Gene/symbol | Drosophila homolog | (Over)Expression construct | (Over)Expression construct with point mutation | Loss of function mutants |
|---|---|---|---|---|
| SNCA/ PARK1 | No | YES | YES | NO |
| PARK2 encoding parkin/ PARK2 | Parkin | YES | YES | YES |
| UCHL1/PARK5 | Uch | NO | NO | NO |
| PINK1/PARK6 | Pink1 | YES | YES | YES |
| PARK7 encoding DJ1/PARK7 | Dj-1α and Dj-1β | YES | NO | YES |
| LRRK2/PARK8 | Lrrk | YES | YES | YES |
| ATP13A2/PARK9 | CG32000 | YES | NO | NO |
| HTRA2/PARK13 | Htra2 | YES | YES | YES |
| PLA2G6/PARK14 | iPLA2-VIA | NO | NO | YES |
| FBOX7/PARK15 | No homolog | YES | YES | NO |
| VPS35/PARK17 | Vps35 | YES | Yes | YES |
| EIG4G1/PARK18 | eIF4G | No | NO | NO |
| DNAJC6/PARK19 | auxillin | YES | NO | YES |
| SYNJ1/PARK20 | Synj | YES | YES | NO |

**Table 3.** *Drosophila* model of Parkinson's disease (genetic) (adapted from Vanhauwaert and Verstreken [113]).

PD was derived from the study of genes mediating familial forms of the disease in fly model [116, 117]. Mutations in *Pink1* (*PARK6*), which encodes a serine-threonine kinase localized to mitochondria and *parkin* (*PARK2*), which encodes a RING finger-containing E3 ubiquitin ligase have been found in recessively inherited and sporadic PD cases [56, 118]. *Drosophila* PINK1 and parkin function in the same genetic pathway, with *pink1* acting upstream of parkin, to regulate mitochondrial integrity in testes, muscle, and dopaminergic neurons [98, 115, 119]. Flies lacking PINK1 or parkin function are viable and show muscle degeneration and TUNEL staining, indicative of cell death [98, 115, 119]. Parkin and PINK1 mutant flies show locomotor defects, abnormal wing position, and dented thoraces [98, 115, 119]. Mitochondrial defects seen in parkin or PINK1 mutant flies are majorly found in the muscle cells though other cell types like DA neurons in the fly brain also show mitochondrial defects which suggests that loss of PINK1 or parkin would result in systemic mitochondrial defects but not extended to all tissues with similar extent [98, 115, 119]. PINK1 and parkin mutant flies show very similar phenotypes suggesting that these genes act together in protecting mitochondria from damage. *Drosophila* expressing wild-type parkin in a PINK1 mutant partially rescues PINK1-associated phenotypes. Alternatively, PINK1 wild type expression in a parkin mutant does not rescue the parkin-associated phenotypes which shows that parkin acts downstream of PINK1 [98, 115]. Parkin is also associated with degradation of mitochondria by autophagy and proteasomal degradation of mitochondrial components [113, 120].

Mitochondrial DNA mutations were first associated with different sporadic or maternally inherited neuromuscular disorders [121]. These disorders were characterized by either the accumulation of multiple mtDNA deletions in post-mitotic tissues [122] or tissue-specific mtDNA depletion [123], and a genetic defect affecting nuclear genes involved in mtDNA replication and maintenance [124]. Mitochondrial DNA often exists in a state of heteroplasmy (cells carrying mtDNA of different genotypes), in which mutant mtDNA co-exists in cells with wild-type mtDNA. Pathology is seen when the frequency of such a mutation reaches a threshold [125]. It accumulates throughout life and is thought to contribute to diseases of aging that include neurodegeneration, metabolic disorders, cancer, heart disease, and sarcopenia [126, 127]. A new investigation of mtDNA in the dopaminergic neurons [128] expanded the previous results showing a prevalent deletion in single neurons on a background of multiple mtDNA deletions [129]. It showed that complex I and complex II are most consistently affected in single neurons, which also displayed a reduced mtDNA copy number [128]. Stimulation of autophagy, activation of the PINK1/parkin pathway, or decreased levels of mitofusin results in a selective decrease in lethal mtDNA deletion [130]. Similar to flies lacking parkin, the flies with mtDNA deletions display striking mitochondrial abnormalities such as disrupted cristae resulting in reduced ATP levels leading to apoptosis and subsequently neurodegeneration.

These dynamic processes regulate mitochondrial function by enabling mitochondrial recruitment to critical subcellular compartments, content exchange between mitochondria, mitochondrial shape control, mitochondrial communication with the cytosol, and mitochondrial quality control making the mitochondria readily adapt to changes in cellular requirements. When mitochondrial dynamics is disrupted, cellular dysfunction follows. The view of mitochondrial dynamics has expanded from a curious phenomenon into an integral cell biological process influencing many cellular functions and survival [131].

### 5.3. Mitochondrial dynamics: fusion and fission

Mitochondria undergo frequent fission and fusion events contributing to the filamentous and interconnected morphology of the organelles and serve crucial physiological functions [132]. The first mitofusin was discovered in mutant *Drosophila* characterized by male sterility [133]. During cell division, mitochondrial fission plays an important role for proper inheritance of mitochondria to the daughter cells. Moreover, organelle fission appears to be important to meet the energy demands of cells at particular subcellular locations. This is true especially for neurons, which heavily depend on energy supply, where mitochondria can travel via microtubule-associated motor proteins to serve specialized neuronal functions such as synaptic transmission. Furthermore, an interconnected network of contiguous mitochondrial organelles seems to be required for the intramitochondrial exchange of metabolic substrates as well as the maintenance of respiratory capacity and mitochondrial membrane potential in hypoxic cellular regions [134].

Mitochondrial fission and fusion events are regulated and coordinated by evolutionarily highly conserved molecular machineries. The molecules regulating mitochondrial dynamics include the homologous GTPases Mitofusin 1 (Mfn1) and Mitofusin 2 (Mfn2), which mediate fusion of the mitochondrial outer membrane, and Optic atrophy 1 (Opa1), a GTPase required for fusion of the inner membrane. Mitochondrial fission, conversely, requires Dynamin-related protein 1 (Drp1), which is also a GTPase [131, 135]. The *Drosophila melanogaster* genome encodes two homologs of Mfn, one being Fuzzy onion (Fzo) [133]. The expression of Fzo is restricted to the testes, and mutations in *fzo* causes mitochondrial fusion defects in testes and male sterility [133]. The second *Drosophila* Mfn homolog is a largely uncharacterized protein known as mitochondrial assembly regulatory factor (Marf), which is expressed in germline and somatic cells [136]. The *Drosophila* genome also encodes single homologs of *opa1* [137] and *drp1* [138], both of which have been shown to function in mitochondrial dynamics in flies.

### 5.4. Genome-wide association studies (GWAS) and genetic screens

GWAS are large-scale population-based genotyping studies that were designed to capture common genetic risk loci and searches for small variations, called single nucleotide polymorphisms or SNPs, which occur more frequently in people with particular disease than in people without the disease. These approaches lead to the identification of new disease causing genes, new biological pathways to explain disease origin or progression, and potential therapeutic targets which is much more precise than the corresponding information from linkage-based studies.

Maraganore and colleagues conducted the first GWAS for PD. It had limited sample sizes (few hundred patients) and patient-control series were sometimes mismatched, but they remain in the genetic history of PD and suggested the low heritability of PD generating a large amount of genetic data into public domain to be further examined and completed by other researchers [139]. Over the last two decades, human GWAS have begun to reveal the genetic risk factors for countless common disorders with complex genetic etiologies including most of the major causes of morbidity and mortality in the developed world [140].

Involvement of multiple genes and pathways complicate the identifying and developing effective therapeutic strategies in different geographical and ethnically divergent populations. Here lies the necessity and importance of GWAS relating to complex neurological disorders such as PD and cancers. Mutations in five genes have been identified to contribute to Mendelian forms of PD risk in fewer than 5% of those with PD which suggest that additional genes too contribute to disease risk [141]. In 2009, two GWAS papers provided unequivocal evidence for an association of the MAPT locus and SNCA variations with sporadic PD. Additionally, both papers implicated variants close to LRRK2 and at two new loci on chromosome 1 (1q32) and chromosome 4, close to the bone marrow stromal cell antigen 1 (BST1) gene [142]. Shortly after these findings, Pankratz et al. conducted the first GWAS in familial PD, confirming the previous discoveries and providing preliminary evidence for an association of a new locus containing the genes cyclin G association kinase (*GAK*) and diacylglycerol kinase theta (*DGKQ*) with PD [142, 143]. A meta-analysis on more than seven million polymorphisms originating either from GWAS datasets and/or from smaller scale PD association studies was performed, where 10 loci showed genome-wide significant association with PD risk (*BST1, CCDCC2/HIP1R, DGKQ/GAK, GBA, LRRK2, MAPT, MCCC1/LAMP3, SNCA, STK39,* and *SYT11/RAB25*) and novel evidence for genome-wide significant association with polymorphism in *STGA8* was found [144]. To date, the largest GWAS was performed in 2014 carrying out a meta-analysis in all existing European-ancestry PD GWAS data with 13,708 PD cases, 95,282 controls and a replication study using genotyping array called 'Neuro X' in an independent data set identified 26 independent SNPs which showed genome-wide risk for PD [145].The first papers about the potential impact of risk loci on age at onset (AAO) in PD were published in 2015. The results using polygenic score analysis showed that patients with an early AAO had a significantly higher polygenic score when compared to those with late AAO [146]. GWAS showed a genetic association with PD in the HLA region (chromosome 6p21.3), which was designated PARK18 and the common variant associated with late onset sporadic PD was rs3129882 in intron 1 of HLA-DRA [147] GWAS sporadic PD was performed and sterol regulatory element binding transcription factor 1 (SREBF1) was identified as risk factor for PD [148]. Later, an unbiased approach on genome-wide RNAi screen was performed in *Drosophila* cells and validated in Hela cells to identify genes regulating the PINK1/parkin pathway, which act in a common genetic pathway in mediating the autophagic degradation of mitochondria (mitophagy) and found 20 genes that have a conserved function in promoting translocation and degradation of depolarized mitochondria. But most notable genes involved were (SREBF1) Fbox and WD40 domain protein 7 (FBXW7), and other components of the sterol regulatory element binding protein (SREBP) lipogenesis pathway indicating a role of lipids in mitochondrial homeostasis, which further showed that this pathway regulate mitophagy and also share a common mechanistic link between autosomal-recessive and sporadic PD [149].

The loci currently associated with PD account for only a very small amount (3–5%) of the expected heritability of PD, suggesting that additional heritable factors (genetic or epigenetic) also play a role in transforming susceptibility to PD. While current evidence suggests that common genetic variants play a role in the etiology of typical PD. GWAS by their inherent design may not be able to detect rare variants [143, 150]. Most SNPs detected by GWAS are not likely causal variants for disease risk but rather informative markers hence it is often not productive to study their direct functional consequences. Also cases selected for GWAS may not be particularly enhanced with genetic susceptibility alleles, moreover the effect sizes identified for most variants are reduced. Thus multiple approaches including linkage analysis, sequencing, and sibpair analysis would be needed to discover additional variants/causative genes and susceptibility loci. Large-scale genome and exome sequencing in conjunction with denser genotyping in large cohorts may help to identify the loci that contribute to the "missing heritability" previously unnoticed by earlier generation technologies [151].

## 6. Utilizing *Drosophila* to understand human GWAS signals

### 6.1. GAL4-UAS system

In 1993, Brand and Perrimon developed the GAL4-UAS system for precise spatial and temporal patterns directing gene expression in *Drosophila* and has been considered as a powerful research tool. A bipartite approach in which a transcriptional activator, the GAL4 gene binds to specific cis-enhancer elements, upstream activation sequence (UAS) leading to activation of the adjacent gene, and thousands of GAL4 driver lines available from individual labs and public stock collections allow expression of desired target genes, typically cDNA transgenes under control of upstream activating sequence (UAS) sites [152]. Findings showed that several familial forms of parkinsonism result from increased gene dosage of $\alpha$-syn, based on this expression levels of the $\alpha$-syn transgene was augmented by generating an $\alpha$-syn expression construct bearing sequence alterations designed to improve the translational efficiency of this cDNA in *Drosophila* and further maximized $\alpha$-syn protein expression in the fly brain by making use of flies bearing two copies each of the UAS-$\alpha$-syn transgene and the TH-GAL4 driver; these control approximately doubled the abundance of $\alpha$-syn protein relative to flies bearing a single copy of each of these transgenes generating a more robust *Drosophila* model for studying synucleinopathies [153].

### 6.2. RNA interference (RNAi)

RNAi is an RNA-dependent gene-silencing process that is regulated by the RNA-induced silencing complex (RISC) and triggered by short double-stranded RNA (dsRNA) molecules.

Efficient silencing of gene expression by dsRNA was first discovered by Fire and Mello [154]. RNAi silencing of a specific target gene relies on the ability of small interfering RNAs (siRNAs), long double-stranded RNAs (dsRNAs), or short hairpin RNAs (shRNAs) to target mRNA molecules for degradation [155–157]. *Drosophila* in vivo RNAi techniques screen both the whole genome and subsets of genes. A total of 10,689 different genes (78% of the *Drosophila* genome) were assayed that affect susceptibility to intestinal *Serratia marcescens* infection [158]. Of these, 8.3% (885 genes) were defined as hits; the majority 89.3% (790 genes) were susceptibility candidates, and 95 genes (10.7% of hits) were negative regulators. A total of 78 and 56 genes were found to function only in the gut and hemocytes, respectively, and 79 functioned in both. A primary screened of 6923 UAS-IR strains for genes involved in the glycosylation of a neural glycoprotein and identified 171candidates [159]. These were further confirmed by knock down experiments, using in-silico analysis and a secondary set of UAS-IR strains that targeted regions distinct from those of the primary strains. A total of 2970 genes were knocked down by neuron-specific RNAi in search for genes involved in the formation, growth, and maintenance of the neuromuscular junction (NMJ) [160]. Knockdown of 158 genes in post-mitotic neurons led to abnormalities in the neuromuscular system. Genome-wide small interfering RNA (siRNA) screening yielded gene candidates involved in characterization of TOMM7 required for stabilizing PINK1 on the outer mitochondrial membrane following mitochondrial damage. Also, HSPA1L (HSP70 family member) and BAG4 found to have mutually opposing roles in the regulation of parkin translocation. RNAi screens revealed that SIAH3 localize to mitochondria, inhibits PINK1 accumulation after mitochondrial insult, and reducing parkin translocation. Overall, screens provide a rich resource to understand mitochondrial quality control [161].

Using this inducible RNAi technique, large-scale screens for various biological processes have been performed successfully in *Drosophila,* proving RNAi-based in vivo screen adequate and efficient. However, RNAi-based screens have relatively high levels of false positives and negatives. To validate the screening results, experimental, and computational analyses have been proposed which will increase the accuracy of RNAi-based screen results.

### 6.3. CRISPR/Cas9

Clustered regularly interspaced short palindromic repeat/CRISPR-associated protein 9 (CRISPR/Cas9) was first applied in mammalian cells in 2013 which has been used as an essential tool in biotechnology [162]. The CRISPR/Cas9 system is a novel genome modification method in which gRNA direct the nuclease Cas9 to selected sequences of genomic DNA, and Cas9 cuts both strands at a specific location. Non-homologous end joining (NHEJ) or homology-directed repair (HDR) repairs the genomic DNA resulting in mutations that can interrupt the open reading frame and cause gene inactivation. For example, loss in function of parkin and Pink1 genes causes PD, CRISPR/Cas9-mediated mutations can mimic knockout of the parkin and/or Pink1 gene [163]. Thus, when both alleles are mutated by CIRSPR/Cas9, the complete loss of parkin or Pink1 will mimic the genetic mutations in PD patients. Indeed, CRISPR/Cas9 has been used to generate pig models of PD by targeting the genes for parkin, Pink1, and DJ1 [164]. In addition to genome editing in germline cells, CRISPR/Cas9 can efficiently target genes in somatic tissues, such as neurons in the brain [165, 166]. In PD patients, continuous loss of dopaminergic neurons in the substantia nigra is a fundamental pathological feature. Thus, gRNAs and Cas9 can be delivered to the substantia nigra of animal brains by a viral system to examine the effect of parkin or Pink1 loss in adult brains. This approach is especially useful for investigating the age-related neuropathology in PD. Also, by gaining toxicity of mutant proteins Cas9-mediated knock-in mutations within the genome can develop animal models of those neurodegenerative diseases.

For transgenic PD animal models that express mutant a-syn, CRISPR/Cas9 can be designed to reduce the expression of mutant genes via NHEJ, which can lead to gene inactivation, in dopaminergic neurons. Besides, replacing the mutant gene by CRISPR/Cas9 via HDR with normal DNA sequences can also lead to the genetic correction of DNA mutations in PD animal models. Even though efficiency of such gene replacement is low at present, the rapid development of CRISPR/Cas9 system offers a promising attempt to produce knock-in models of human diseases [167].

### 6.4. Deciphering the pathways of therapeutic molecules: role of drosophila model

The first published study about compound treatments in a drosophila PD model reported the effects of drugs commonly used for treating PD on the locomotor phenotype of $\alpha$-synuclein expressing flies and showed that some of them were able to suppress that phenotype [168]. Subsequently, and given the ability of increased chaperone activity to counteract $\alpha$-synuclein toxicity [169], the effect of geldanamycin (GA), an antibiotic able to interfere with Hsp90 activity and activate stress response, was assayed over $\alpha$-synuclein expressing flies [170, 171]. Notably, feeding these flies with GA protected DA neurons against $\alpha$-synuclein-induced degeneration, and this protection was driven by an increase in Hsp70 levels [171]. Inhibitors of the histone deacetylase SIRT2 also showed a protective effect against $\alpha$-synuclein toxicity [172]. Other studies have been also performed in several drosophila PD models to look for potentially therapeutic compounds directed to reduce oxidative stress damage. As explained previously, the study of $\alpha$-synuclein toxicity in flies led to the identification of Phase II detoxification pathway as a possible target for therapeutic treatment [173]. In fact, feeding $\alpha$-synuclein-expressing flies or *Drosophila parkin* mutants with pharmacological inducers of that pathway like sulforaphane or allyl disulfide suppresses the neuronal loss of both PD models [173]. Besides, it has been shown that dietary supplementation with *S*-methyl-*L*-cysteine (SMLC) inhibits the locomotor and circadian rhythm defects caused by ectopic expression of human $\alpha$-synuclein in drosophila [174]. SMLC participates in the catalytic antioxidant mechanism involving methionine sulfoxide reductase A (MSRA), one of the enzymes that catalyze the oxidation of the amino acid methionine to methionine sulfoxide, a reversible reaction that has been postulated to act protecting cells from oxidative damage. Furthermore, grape extract supplementation has been shown to recover locomotor ability and lifespan in $\alpha$-synuclein-expressing flies. It is known that grape extracts contain several polyphenols, compounds with antioxidant properties [175]. Other drosophila PD models in which treatments with antioxidant compounds have been shown to be beneficial are those involving the *DJ-1*$\alpha$ and *DJ-1*$\beta$ genes [176, 177]. Compounds with antioxidant and anti-inflammatory properties such as celastrol and minocycline conferred potent DA neuroprotection in RNAi *DJ-1*$\alpha$ mutants [176]. We have also recently demonstrated that chronic treatments with antioxidant compounds are able to modify the lifespan phenotype of *DJ-1*$\beta$ mutant flies, thus suggesting that oxidative stress plays a causal role in such phenotype [177]. It is known that rapamycin is a small molecule inhibitor of TOR signaling that has been shown to lead to 4E–BP hypophosphorylation in vitro and in vivo [178, 179]. Notably administration of rapamycin was able to suppress all pathologic phenotypes in *parkin* and *PINK1* mutants. Moreover, this suppression was found to be 4E-BP dependent, since the administration of rapamycin to *parkin* and *Thor* or *PINK1* and *Thor* double mutants was completely unable to suppress these phenotypes [180]. Since 4E-BP activity can be manipulated by small molecule inhibitors such as rapamycin, this pathway represents a viable therapeutic target for PD treatment. Moreover, it has been recently suggested that *parkin* mutants, apart from the described phenotypes, also present altered zinc homeostasis. This is supported by the fact that dietary zinc supplementation in the form of zinc chloride increased lifespan as well as the percentage of *parkin* mutant flies reaching adulthood while this supplemented diet was deleterious to control flies [181]. Since most PD cases are sporadic and could be associated to different environmental agents, it is also essential the use of toxin-induced drosophila PD models to assay the beneficial effects of candidate compounds. Polyphenol administration was also found to exert a beneficial effect on flies exposed to paraquat and iron, protecting, rescuing, and restoring the impaired locomotor activity caused by exposure to those agents [182]. Other antioxidant compounds such as melatonin have also been found to rescue locomotor deficits and DA neurodegeneration in flies exposed to rotenone [19].

#### *6.4.1. LRRK2 kinase inhibitors*

Several LRRK2 kinase inhibitors, including CZC-25146, GW5074, and sorafenib, have been tested in rodents, as well as in *Caenorhabditis elegans* and drosophila models, and have been shown to protect against LRRK 2 (G2019S)-induced neurodegeneration [183]. These findings indicate that increased kinase activity of LRRK2 is neurotoxic and that inhibition of LRRK2 activity can have a disease-modifying effect.

#### *6.4.2. Molecular chaperones*

##### *6.4.2.1. HSF-1 modulators*

Endogenous molecular chaperone function can be modulated pharmacologically with compounds that augment endogenous chaperone levels. Several HSF-1 modulators including celastrol and carbenoxolone can trigger HSF-1 activation, leading to downstream induction of Hsp70 expression [184]. Celastrol has been demonstrated to be effective against protein aggregation and toxicity in various neurodegenerative disease models, including dopaminergic neuroprotection in a *Drosophila* model of PD [176]. Carbenoxolone has demonstrated the ability to attenuate $\alpha$-synuclein and ubiquitin aggregation in vitro and in vivo [185, 186]. Thus, it may have potential as a chaperone-mediated therapeutic option for PD. Hsp90 Inhibitors a naturally occurring small molecule antibiotic, geldanamycin (GA), inhibits the interaction between Hsp90 and HSF-1, leading to increased Hsp70 expression [187]. In vitro cell studies have demonstrated the capability of this compound to decrease $\alpha$-synuclein aggregation and reduce cell toxicity [188], and its neuroprotective effects have been shown in *Drosophila* and MPTP mouse models of PD [169, 189]. Other analogues of GA include 17-AAG and 17-DMAG, which similarly prevent $\alpha$-synuclein aggregation and toxicity, but are more potent and less toxic than GA [190]. Moreover, 17-AAG has poor permeability of the BBB, limiting its pharmacological usage for neurodegenerative diseases [185]. Consequently, compound library screening for small molecule Hsp90 inhibitors with improved pharmacokinetics, including BBB permeability, have led to the identification of SNX compounds [185]. These compounds are associated with an increase in Hsp70 activity in the brain and a reduction in $\alpha$-synuclein oligomerization and toxicity in vitro [190].

Insights regarding identification of pathways through which different therapeutic molecules confer neuroprotection is briefed in **Table 4**.

| Pathway/process | Compound treatment* | Drosophila model | Modified phenotype/s | References |
|---|---|---|---|---|
| Oxidative stress | Sulforaphane and allyl<br>disulfide*<br>S-methyl-Lcysteine<br>Polyphenols<br>α-tocopherol<br>SOD<br>Melatonin<br>*Bacopa monieri*<br>leaf extract | *Parkin*<br>*α-synuclein*<br>*α-synuclein*<br>*α-synuclein*<br>Paraquat and iron<br>*DJ-1β*<br>*PINK1*<br>*PINK1*<br>*DJ-1β*<br>Paraquat<br>Rotenone<br>Paraquat | DA neuron number<br>DA neuron number<br>Locomotor activity<br>Lifespan, locomotor<br>activity<br>Locomotor activity<br>Lifespan<br>Ommatidial<br>degeneration<br>Ommatidial<br>degeneration<br>Lifespan<br>Locomotor activity<br>Locomotor activity, DA<br>neuron<br>Oxidative markers levels | [173]<br>[173]<br>[174]<br>[175]<br>[182]<br>[177]<br>[191]<br>[191]<br>[177]<br>[99]<br>[99]<br>[192] |
| Oxidative stress/<br>inflammatory process | Minocycline*<br>Celastrol | *DJ-1α* | DA neuron number,<br>dopamine Levels<br>DA neuron number,<br>dopamine levels,<br>locomotor activity, and<br>survival rate under<br>oxidativestress condition | [176] |
| TOR signaling | Rapamycin* | *parkin/PINK1* | Thoracic indentations,<br>locomotors activity, DA<br>neuron number, and<br>muscle integrity | [180] |
| Removal of excess or toxic<br>protein forms | Geldanamycin* | *α-synuclein* | DA neuron number | [170]<br>[171] |
| Zinc homeostasis | Zinc chloride* | *parkin* | Lifespan, locomotor<br>activity, and percentage<br>of adulthood<br>survivors | [181] |
| Chaperone therapies (HSF-1<br>modulators)<br>Trigger HSF-1 activation<br>Induces downstream Hsp70<br>expression | Celastrol<br>Carbenoxolone | *α-synuclein*<br>*α-synuclein* | dopaminergic<br>neuroprotection | [184] |
| Hsp90 inhibitors<br>Inhibits the interaction<br>between Hsp90 and HSF-1,<br>leading to increased Hsp70<br>expression and activity | Geldanamycin<br>17-AAG<br>17-DMAG<br>SNX-2112 | *α-synuclein* | decrease α- synuclein<br>aggregation and reduce<br>cell toxicity | [187]<br>[190] |
| mTOR-dependent<br>pathways/AMPK | Metformin<br>AICAR | *Drosophila<br>melanogaster*<br>mutated for LRRK2 | Reduced cell death | [193] |
| mTORC1 | Rapamycin and Rp<br>analogues<br>(CCI-779, RAD001 and<br>AP23573) | *Drosophila<br>melanogaster*<br>mutated for PINK-1<br>and<br>Parkin | Reduced mitochondrial<br>Dysfunction | [180] |

| Pathway/process | Compound treatment* | Drosophila model | Modified phenotype/s | References |
|---|---|---|---|---|
| mTor-independent pathways/unknown | Spermidine | *α-synuclein* | Reduced motor dysfunction, increased lifespan; Reduced neuronal cell loss | [194] |
| LRKK2 kinase inhibitors | GW5074, and sorafenib | *α-synuclein* | Protect again DA neuron degeneration locomotor activity | [183] |
| Histone Deacetylase inhibitors | | *α-synuclein* | Protect again DA locomotor activity | [195] |
| Antitumor agents | Geldanamycin | *α-synuclein* | Protect again DA Mobilized the stress response and increase levels of chaperon HSP70 | [195] |

*All treatments were administered as dietary complement.

**Table 4.** Listing of pathways through which neuroprotective compounds confer neuroprotection: lessons from drosophila model of PD.

## 7. Conclusion

**Bench to bedside**: role of drosophila in translational research.

Bench to bedside is a term used to describe the process by which the results of research done in the laboratory are directly used to develop new ways to treat patients. Taking advantage of studies from animal models such as drosophila certain pharmacotherapies and non-pharmacotherapies have been developed which are in different stages in clinical trials to validate their efficacy, safety, and tolerability. Pharmacotherapies include adenosine A2A receptor antagonists [196], glutamate AMPA receptor antagonists [197], monoamine oxidase inhibitors [198], anti-apoptotic agents [199], and antioxidants [200]. Non-pharmacotherapies also offer alternative approaches for treatment of the disease which include the use of viral vector gene therapy [201], stem cell transplants [202], and microRNAs [203]. Nevertheless, additional trials enrolling larger numbers of PD patients are still needed to better understand the neuroprotective effects of these therapies.

## Acknowledgements

This work is partly supported by Department of Biotechnology (DBT), India grant numbers (BT/405/NE/U-Excel/2013 and BT/PR16868/NER/95/328/2015) and DST-SERB (Department of Science and Technology-Science and Engineering Research Board), India grant number (EMR/2016/002375).

## Conflict of interest

Authors declares no conflict of interest.

## Author details

Mohamad Ayajuddin†, Abhik Das†, Limamanen Phom†, Priyanka Modi†, Rahul Chaurasia†, Zevelou Koza†, Abuno Thepa†, Nukshimenla Jamir, Pukhrambam Rajesh Singh, Sentinungla Longkumer, Pardeshi Lal and Sarat Chandra Yenisetti*

*Address all correspondence to: yschandrays@rediffmail.com

Department of Zoology, Drosophila Neurobiology Laboratory, Nagaland University (Central), India

† These authors contributed equally.

## References

[1] Manyam BV. Paralysis agitans and levodopa in "Ayurveda": Ancient Indian medical treatise. Movement Disorders. 1990;**5**:47-48

[2] Zhang ZX, Dong ZH, Roman GC. Early descriptions of Parkinson's disease in ancient China. Archives of Neurology. 2006;**63**:782-784

[3] Garcia-Ruiz PJ. Prehistoria de la enfermedad de Parkinson (Prehistory of Parkinson's disease). Neurologia (in Spanish). 2004;**19**(10):735-737

[4] Available from: http://www.parkinsons.org/parkinsons-history.html [Accessed: Oct 2, 2017]

[5] Parkinson J. An Essay on the Shaking Palsy. Needly and Jones, London: Whittingham and Rowland for Sherwood; 1817

[6] Lees AJ. Unresolved issues relating to the shaking palsy on the celebration of James Parkinson's 250th birthday. Movement Disorders. 2007;**22**(17):S327-S334

[7] Charcot J-M. De la paralysie agitante. In: Oeuvres Comple'tes (t 1) Lec¸ons sur les maladies du syste'me nerveux. Paris: A Delahaye; 1877. pp. 155-188. [In English: Charcot, J-M. On Parkinson's disease. In: Lectures on Diseases of the Nervous System Delivered at the Salpeˆtrie're (transl. Sigerson G). London: New Sydenham Society; 1872. pp. 129-156]

[8] Modi P, Ayajuddin M, Phom L, et al. Understanding pathophysiology of sporadic Parkinson's disease in *Drosophila* model: Potential opportunities and notable limitations. In: Jolanta Dorszewska and Wojciech Kozubski, editors. Challenges in Parkinson's Disease. European Union: In-tech Croatia; 2016. pp. 217-244

[9] Postuma RB, Berg D, Stern M, et al. MDS clinical diagnostic criteria for Parkinson's disease. Movement Disorders. 2015;**30**(12):1591-1601

[10] Jellinger KA. Formation and development of Lewy pathology: A critical update. Journal of Neurology. 2009;**256**(3):270-279

[11] Hornykiewicz O, Kish SJ. Biochemical pathophysiology of Parkinson's disease. Advances in Neurology. 1987;**45**:19-34

[12] Pfeiffer RF. Non-motor symptoms in Parkinson's disease. Parkinsonism & Related Disorders. 2016;**22**:S119-S122

[13] Levy G, Schupf N, Tang M, et al. Combined effect of age and severity on the risk of dementia in Parkinson's disease. Annals of Neurology. 2002;**51**(6):722-729

[14] Di Monte DA, Lavasani M, Manning Bog AB. Environmental factors in Parkinson's disease. Neurotoxicology. 2002;**23**:487-502

[15] Polymeropoulos MH, Lavedan C, Leroy E, et al. Mutation in the alpha-synuclein gene identified in families with Parkinson's disease. Science. 1997;**276**:2045-2047

[16] Simon-Sanchez J, Schulte C, Bras J. Genome-wide association study reveals genetic risk underlying Parkinson's disease. Nature Genetics. 2009;**41**:1308-1312

[17] Elbaz A, Clavel J, Rathouz PJ, et al. Professional exposure to pesticides and Parkinson disease. Annals of Neurology. 2009;**66**:494-504

[18] Ritz BR, Manthripragada AD, Costello S, et al. Dopamine transporter genetic variants and pesticides in Parkinson's disease. Environmental Health Perspectives 2009;**117**:964-969

[19] Coulom H, Birman S. Chronic exposure to rotenone models sporadic Parkinson's disease in *Drosophila melanogaster*. The Journal of Neuroscience. 2004;**24**(48):10993-10998

[20] Liou HH, Tsai MC, Chen CJ, et al. Environmental risk factors and Parkinson's disease – A case-control study in Taiwan. Neurology. 1997;**48**(6):1583-1588

[21] Ballard PA, Tetrud JW, Langston JW. Permanent human parkinsonism due to 1-methyl-4-phenyl-1,2,3,6-tetrahydropyridine (MPTP): Seven cases. Neurology. 1985;**35**(7):949-956

[22] Langston JW, Ballard P, Tetrud JW, Irwin I. Chronic parkinsonism in humans due to a product of meperidine-analog synthesis. Science. 1983;**219**:979-980

[23] Priyadarshi A, Khuder SA, Schaub EA, et al. A meta-analysis of Parkinson's disease and exposures to pesticides. Neurotoxicology. 2000;**21**:435-440

[24] Steenland K, Hein MJ, Cassinelli RT, et al. Polychlorinated biphenyls and neurodegenerative disease mortality in an occupational cohort. Epidemiology. 2006;**17**(1):8-13

[25] Goldman SM, Quinlan PJ, Ross GW, et al. Solvent exposures and Parkinson disease risk in twins. Annals of Neurology. 2012;**71**(6):776-784

[26] Gorell JM, Johnson CC, Rybicki BA, et al. Occupational exposures to metals as risk factors for Parkinson's disease. Neurology. 1997;**48**(3):650-658

[27] Finkelstein MM, Jerrett M. A study of the relationships between Parkinson's disease and markers of traffic-derived and environmental manganese air pollution in two Canadian cities. Environmental Research. 2007;**104**(3):420-432

[28] Vitalis T, Fouquet C, Alvarez C, et al. Developmental expression of monoamine oxidases A and B in the central and peripheral nervous systems of the mouse. The Journal of Comparative Neurology. 2002;**442**(4):331-347

[29] Singer TP, Ramsay RR, McKeown K, et al. Mechanism of the neurotoxicity of 1-methyl-4-phenylpyridinium (MPP+), the toxic bioactivation product of 1-methyl-4-phenyl-1,2,3,6-tetrahydropyridine (MPTP). Toxicology. 1988;**49**(1):17-23

[30] Ramsay RR, Singer TP. Energy-dependent uptake of N-methyl-4-phenylpyridinium, the neurotoxic metabolite of 1-methyl-4-phenyl-1,2,3,6-tetrahydropyridine, by mitochondria. The Journal of Biological Chemistry. 1986;**261**(17):7585-7587

[31] Grant H, Lantos PL, Parkinson C. Cerebral damage in paraquat poisoning. Histopathology. 1980;**4**:185-195

[32] Bové J, Prou D, Perier C, et al. Toxin-induced models of Parkinson's disease. NeuroRx. 2005;**2**(3):484-494

[33] Mohammadi-Bardbori A,, Ghazi-Khansari M. Alternative electron acceptors: Proposed mechanism of paraquat mitochondrial toxicity. Environmental Toxicology and Pharmacology. 2008;**26**(1):1-5

[34] Peng J, Ou Mao X, Stevenson FF, et al. The herbicide paraquat induces dopaminergic nigral apoptosis through sustained activation of the JNK pathway. Journal of Biological Chemistry. 2004;**279**(31):32626-32632

[35] Sherer TB, Betarbet R, Testa CM, et al. Mechanism of toxicity in rotenone models of Parkinson's disease. The Journal of Neuroscience. 2003;**23**(34):10756-10764

[36] Schapira AH, Cooper JM, Dexter D, et al. Mitochondrial complex I deficiency in Parkinson's disease. Journal of Neurochemistry. 1990;**54**(3):823-827

[37] Di Monte DA: The environment and Parkinson's disease:is the nigrostriatal system preferentially targeted by neurotoxins? Lancet Neurol. 2003;**2**(9):531-538

[38] Ryu EJ, Harding HP, Angelastro JM, et al. Endoplasmic reticulum stress and the unfolded protein response in cellular models of Parkinson's disease. Journal of Neuroscience. 2002;**22**:10690-10698

[39] Sherer TB, Betarbet R, Stout AK, et al. An in vitro model of Parkinson's disease: Linking mitochondrial impairment to altered a-synuclein metabolism and oxidative damage. Journal of Neuroscience. 2002;**22**:7006-7015

[40] Schapira A, Jenner P. Etiology and pathogenesis of Parkinson's disease. Movement Disorders. 2011;**26**:1049-1055

[41] Polymeropoulos MH, Higgins JJ, Golbe LI, et al. Mapping of a gene for Parkinson's disease to chromosome 4q21–q23. Science. 1996;**274**:1197-1199

[42] Spillantini MG, Crowther RA, Jakes R, et al. Alpha-Synuclein in filamentous inclusions of Lewy bodies from Parkinson's disease and dementia with lewy bodies. Proceedings of the National Academy of Sciences of the United States of America. 1998;**95**:6469-6473

[43] Sheerin UM, Houlden H, Wood NW. Advances in the genetics of Parkinson's disease: A guide for the clinician. Movement Disorders Clinical Practice. 2014;**1**:3-13

[44] Funayama M, Hasegawa K, Kowa H, et al. A new locus for Parkinson's disease (PARK8) maps to chromosome 12p11.2–q13.1. Annals of Neurology. 2002;**51**:296-301

[45] Zimprich A, Biskup S, Leitner P, et al. Mutations in LRRK2 cause autosomal-dominant parkinsonism with pleomorphic pathology. Neuron. 2004;**44**:601-607

[46] Healy DG, Falchi M, O'Sullivan SS, et al. Phenotype, genotype, and worldwide genetic penetrance of LRRK2-associated Parkinson's disease: A case-control study. Lancet Neurology. 2008;**7**:583-590

[47] Di Fonzo A, Tassorelli C, De Mari M, et al. Comprehensive analysis of the LRRK2 gene in sixty families with Parkinson's disease. European Journal of Human Genetics. 2006;**14**:322-331

[48] Khan NL, Brooks DJ, Pavese N, et al. Progression of nigrostriatal dysfunction in a parkin kindred: An [18F]dopa PET and clinical study. Brain. 2002;**125**:2248-2256

[49] Kitada T, Asakawa S, Hattori N, et al. Mutations in the parkin gene cause autosomal recessive juvenile parkinsonism. Nature. 1998;**392**:605-608

[50] Matsumine H, Saito M, Shimoda-Matsubayashi S, et al. Localization of a gene for an autosomal recessive form of juvenile Parkinsonism to chromosome 6q25.2-27. American Journal of Human Genetics. 1997;**60**(3):588-596

[51] Matsumine H, Yamamura Y, Hattori N, et al. A microdeletion of D6S305 in a family of autosomal recessive juvenile parkinsonism (PARK2). Genomics. 1998;**49**(1):143-146

[52] Lohmann E, Periquet M, Bonifati V, et al. How much phenotypic variation can be attributed to parkin genotype? Annals of Neurology. 2003;**54**:176-185

[53] Bonifati V, Rizzu P, Van Baren MJ, et al. Mutations in the DJ-1 Gene associated with autosomal recessive early-onset parkinsonism. Science. 2003;**299**:256-259

[54] Abou-Sleiman PM, Healy DG, Quinn N, et al. The role of pathogenic DJ-1 mutations in Parkinson' s disease. Annals of Neurology 2003;**54**:283-286

[55] Massano J, and Bhatia KP. Clinical approach to Parkinson's disease: Features, diagnosis, and principles of management. Cold Spring Harbor Perspectives in Medicine 2012;**2**(6):a008870. DOI: 10.1101/cshperspect.a008870

[56] Valente EM, Abou-Sleiman PM, Caputo V, et al. Hereditary early-onset Parkinson's disease caused by mutations in PINK1. Science. 2004;**304**:1158-1160

[57] Samaranch L, Lorenzo-Betancor O, Arbelo JM, et al. PINK1-linked parkinsonism is associated with Lewy body pathology. Brain. 2010;**133**:1128-1142

[58] Nuytemans K, Theuns J, Cruts M, et al. Genetic etiology of Parkinson disease associated with mutations in the SNCA, PARK2, PINK1, PARK7, and LRRK2 genes: A mutation update. Human Mutation. 2010;**31**:763-780

[59] Vilariño-Güell C, Wider C, Ross OA, et al. VPS35 mutations in Parkinson disease. American Journal of Human Genetics. 2011;**89**(1):162-167

[60] Zimprich A, Benet-Pagès A, Struhal W, et al. A mutation in VPS35, encoding a subunit of the retromer complex, causes late-onset Parkinson disease. American Journal of Human Genetics. 2011;**89**:168-175

[61] Najim al-Din AS, Wriekat A, Mubaidin A, et al. Pallido-pyramidal degeneration, supranuclear upgaze paresis and dementia: Kufor-Rakeb syndrome. Acta Neurologica Scandinavica. 1994;**89**:347-352

[62] Hampshire DJ, Roberts E, Crow Y, et al. Kufor-Rakeb syndrome, pallido-pyramidal degeneration with supranuclear up gaze paresis and dementia, maps to 1p36. Journal of Medical Genetics. 2001;**38**:680-682

[63] Ning YP, Kanai K, Tomiyama H, et al. PARK9-linked parkinsonism in eastern Asia: Mutation detection in ATP13A2 and clinical phenotype. Neurology. 2008;**70**:1491-1493

[64] Schneider SA, Coro Paisan-Ruiz, Niall P et al.: ATP13A2 mutations (PARK9) cause neurodegeneration with brain iron accumulation. Movement Disorders, The Official Journal of MDS. 2010;**25**:979-984

[65] Paisan-Ruiz C, Bhatia KP, Li A, et al. Characterization of PLA2G6 as a locus for dystonia-parkinsonism. Annals of Neurology. 2009;**65**:19-23

[66] Yoshino H, Tomiyama H, Tachibana N, et al. Phenotypic spectrum of patients with PLA2G6 mutation and PARK14-linked parkinsonism. Neurology. 2010;**75**:1356-1361

[67] Baburina I, Jackowski S. Cellular responses to excess phospholipid. The Journal of Biological Chemistry. 1999;**274**:9400-9408

[68] Leroy E, Boyer R, Auburger G, et al. The ubiquitin pathway in Parkinson's disease. Nature. 1998;**395**:451-452

[69] Ragland M, Hutter C, Zabetian C, et al. Association between the ubiquitin carboxyl-terminal esterase L1 gene (UCHL1) S18Y variant and Parkinson's disease: A HuGE review and meta analysis. American Journal of Epidemiology. 2009;**170**:1344-1357

[70] Ardley HC, Scott GB, Rose SA, et al. UCH-L1 aggresome formation in response to proteasome impairment indicates a role in inclusion formation in Parkinson's disease. Journal of Neurochemistry. 2004;**90**:379-391

[71] Strauss KM, Martins LM, Plun-Favreau H, et al. Loss of function mutations in the gene encoding Omi/HtrA2 in Parkinson's disease. Human Molecular Genetics. 2005;**14**: 2099-2111

[72] Bogaerts V, Nuytemans K, Reumers J, et al. Genetic variability in the mitochondrial serine protease HTRA2 contributes to risk for Parkinson disease. Human Mutation. 2008; **29**:832-840

[73] Suzuki Y, Imai Y, Nakayama H, et al. A serine protease, HtrA2, is released from the mitochondria and interacts with XIAP, inducing cell death. Molecular Cell. 2001;**8**:613-621

[74] Whitworth AJ, Lee JR, Ho VM, et al. Rhomboid-7 and HtrA2/Omi act in a common pathway with the Parkinson's disease factors Pink1 and parkin. Disease Models & Mechanisms. 2008;**1**:168-174

[75] Yun J, Cao JH, Dodson MW, et al. Loss of function analysis suggests that Omi/HtrA2 is not an essential component of the PINK1/PARKIN pathway in vivo. The Journal of Neuroscience. 2008;**283**:14500-14510

[76] Chartier-Harlin MC, Dachsel J, Hulihan M, et al. EIF4G1 mutations in familial parkinsonism. Parkinsonism & Related Disorders. 2009;**15**:145-146

[77] Nuytemans K, Bademci G, Inchausti V, et al. Whole exome sequencing of rare variants in EIF4G1 and VPS35 in Parkinson disease. Neurology. 2013;**80**:982-989

[78] Wooten GF. Functional anatomical and behavioral consequences of dopamine receptor stimulation. Annals of the New York Academy of Sciences. 1997;**835**:153-156

[79] Jellinger KA. The pathology of Parkinson's disease. Advances in Neurology. 2001;**86**:55-72

[80] Lewy FH. Paralysis agitans. Pathologische anatomie. In: Lewandowski M, editor. Handbuch der Neurologie. Berlin: Springer-Verlag; 1912. pp. 920-933

[81] Pienaar IS, Gotz J, Feany MB. Parkinson's disease: Insights from non-traditional model organisms. Progress in Neurobiology. 2010;**92**:558-571

[82] Bilen J, Bonini NM. Drosophila as a model for human neurodegenerative disease. Annual Review of Genetics. 2005;**39**:153-171

[83] Venken KJ, Bellen HJ. Emerging technologies for gene manipulation in *Drosophila melanogaster*. Nature Reviews. Genetics. 2005;**6**:167-178

[84] White KE, Humphrey DM, Hirth F. The dopaminergic system in the aging brain of *Drosophila* Frontier in Neuroscience. 2010;**4**(205):1-12

[85] Lessing D, Bonini NM. Maintaining the brain: Insight into human Neurodegeneration from *Drosophila* mutants. Nature Reviews. Genetics. 2009;**10**:359-370

[86] Lloyd TE, Taylor JP. Flightless flies: Drosophila models of neuromuscular disease. Annals of the New York Academy of Sciences. 2010;**1184**:1-20

[87] Pandey UB, Nichols CD. Human disease models in *Drosophila melanogaster* and the role of the fly in therapeutic drug discovery. Pharmacological Reviews. 2011;**63**:411-436

[88] Prüßing K, Voigt A, Schulz JB. *Drosophila melanogaster* as a model organism for Alzheimer's disease. Molecular Neurodegeneration. 2013;**8**:35-45

[89] Mayer F, Mayer N, Chinn L, Pinsonneault RL et al. Evolutionary conservation of vertebrate blood-brain barrier chemoprotective mechanisms in Drosophila. The Journal of Neuroscience. 2009;**29**:3538-3550

[90] Feany MB, Bender WW. A *Drosophila* model of Parkinson's disease. Nature. 2000;**404**: 394-398

[91] Botella JA, Bayersdorfer F, Gmeiner F et al. Modelling Parkinson's disease in Drosophila. NeuroMolecular Medicine. 2009;**11**:268-280

[92] Hirth F. *Drosophila melanogaster* in the study of human neurodegeneration. CNS & Neurological Disorders – Drug Targets. 2010;**9**:504-523

[93] Varga SJ, Qi C, Podolsky E, et al. A new Drosophila model to study the interaction between genetic and environmental factors in Parkinson's disease. Brain Research 2014;**1583**: 277-286

[94] Meulener M, Whitworth AJ, Armstrong-Gold CE, et al. Drosophila DJ-1 mutants are selectively sensitive to environmental toxins associated with Parkinson's disease. Current Biology. 2005;**15**:1572-1577

[95] Menzies FM, Yenisetti SC, Min KT. Roles of Drosophila DJ-1 in survival of dopaminergic neurons and oxidative stress. Current Biology. 2005;**15**:1578-1582

[96] Venderova K, Kabbach G, Abdel-Messih E, et al. Leucine-rich repeat kinase 2 interacts with parkin, DJ-1 and PINK-1 in a Drosophila Melanogaster model of Parkinson's disease. Human Molecular Genetics. 2009;**18**:4390-4404

[97] Wang C, Lu R, Ouyang X, et al. Drosophila overexpressing parkin R275W mutant exhibits dopaminergic neuron degeneration and mitochondrial abnormalities. The Journal of Neuroscience. 2007;**27**(32):8563-8570

[98] Clark IE, Dodson MW, Jiang C, et al. *Drosophila* pink1 is required for mitochondrial function and interacts genetically with parkin. Nature. 2006;**441**(7097):1162-1166

[99] Chaudhuri A, Bowling K, Funderburk C, et al. Interaction of genetic and environmental factors in a Drosophila parkinsonism model. The Journal of Neuroscience. 2007;**27**(10):2457-2467

[100] Park J, Kim SY, Cha GH, et al. Drosophila DJ-1 mutants show oxidative stress-sensitive locomotive dysfunction. Gene. 2005;**361**:133-139

[101] Hosamani R, Muralidhara. Acute exposure of *Drosophila melanogaster* to Paraquat causes oxidative stress and mitochondrial dysfunction. Archives of Insect Biochemistry and Physiology. 2013;**83**(1):25-40

[102] Ebrahimi-Fakhari D, Wahlster L, McLean P. Protein degradation pathways in Parkinson's disease: Curse or blessing. Acta Neuropathologica. 2012;**124**:153-172

[103] Wang A, Costello S, Cockburn M, et al. Parkinson's disease risk from ambient exposure to pesticides. European Journal of Epidemiology. 2011;**26**(7):547-555

[104] Martin CA, Barajas A, Lawless G, et al. Synergistic effects on dopamine cell death in a Drosophila model of chronic toxic exposure. Neurotoxicology. 2014;**44**:344-351

[105] Sun F, Kanthasamy A, Anantharam V, et al. Envionmental neurotoxic chemicals-induced ubiquitin proteosome system dysfunction in the pathogenesis and progression of parkinson's disease. Pharmacology & Therapeutics. 2007;**114**:327-344

[106] Wang C, Ko HS, Thomas B, et al. Stress-induced alterations in parkin solubility promote parkin aggregation and compromise parkin's protective function. Human Molecular Genetics. 2005;**14**(24):3885-3897

[107] Yao D, Gu Z, Nakamura T, et al. Nitrosative stress linked to sporadic Parkinson's disease: S-nitrosylation of parkin regulates its E3 ubiquitin ligase activity. Proceedings of the National Academy of Sciences of the United States of America. 2004;**101**(29):10810-10814

[108] Casarejos MJ, Menendez J, Solano RM, et al. Susceptibility to rotenone isincreased in neurons from parkin null mice and is reduced by minocycline. Journal of Neurochemistry. 2006;**97**(4):934-946

[109] Liao J, Morin LW, Ahmad ST. Methods to characterize spontaneous and startle induced locomotion in a rotenone induced parkinson's disease model of Drosophila. Journal of Visualized Experiments. 2014;**90**:1-6

[110] Shu YU, Meng J, Ying X, et al. Rotenone induces more serious learning and memory impairment that $\alpha$-synuclein A30P does in Drosophila. Journal of Shanghai Jiaotong University. 2011;**15**(3):229-234

[111] Khatri DK, Juvekar AR. Abrogation of locator impairment in a rotenone-induced Drosophila melanogaster and zebrafish model of parkinson's disease by ellagic acid and curcumin. International Journal of Nutrition, Pharmacology, Neurological Diseases. 2016;**6**(2):90-96

[112] Subramanian P, Prasanna V, Jayapalan JJ, et al. Role of *Bacopa monnieri* in the temporal regulation of oxidative stress in clock mutant (cryb) of *Drosophila melanogaster*. Journal of Insect Physiology. 2014;**65**:37-44

[113] Vanhauwaert R, Verstreken P: Flies with Parkinson's disease. Experimental Neurology. 2015;**274**(Pt A):42-51

[114] Greene JC, Whitworth AJ, Kuo I, et al. Mitochondrial pathology and apoptotic muscle degeneration in *Drosophila* parkin mutants. Proceedings of the National Academy of Sciences of the United States of America. 2003;**100**(7):4078-4083

[115] Park J, Lee SB, Lee S, et al. Mitochondrial dysfunction in *Drosophila* PINK1 mutants is complemented by parkin. Nature. 2006;**441**(7097):1157-1161

[116] Dodson MW, Guo M. Pink1, Parkin, DJ-1 and mitochondrial dysfunction in Parkinson's disease. Current Opinion in Neurobiology. 2007;**17**:331-337

[117] Wood-Kaczmar A, Gandhi S, Wood NW. Understanding the molecular causes of Parkinson's disease. Trends in Molecular Medicine. 2006;**12**:521-528

[118] Bonifati V, Rohé CF, Breedveld GJ, et al. Early-onset parkinsonism associated with PINK1 mutations: Frequency, genotypes, and phenotypes. Neurology. 2005;**65**(1):87-95

[119] Yang Y, Gehrke S, Imai Y, et al. Mitochondrial pathology and muscle and dopaminergic neuron degeneration caused by inactivation of *Drosophila* Pink1 is rescued by Parkin. Proceedings of the National Academy of Sciences of USA. 2006;**103**:10793-10798

[120] Vincow ES, Merrihew G, Thomas RE, et al. The PINK1-parkin pathway promotes both mitophagy and selective respiratory chain turnover in vivo. Proceedings of the National Academy of Sciences of the United States of America. 2013;**110**(16):6400-6405

[121] Holt IJ, Harding AE, Petty RK, Morgan-Hughes JA. A new mitochondrial disease associated with mitochondrial DNA heteroplasmy. American Journal of Human Genetics. 1990;**46**:428-433

[122] Zeviani M, Servidei S, Gellera C, et al. An autosomal dominant disorder with multiple deletions of mitochondrial DNA starting at the D-loop region. Nature. 1989;**339**:309-311

[123] Moraes CT, Shanske S, Tritschler HJ, et al. mtDNA depletion with variable tissue expression: A novel genetic abnormality in mitochondrial diseases. American Journal of Human Genetics. 1991;**48**:492-501

[124] Zeviani M. Nucleus-driven mutations of human mitochondrial DNA. Journal of Inherited Metabolic Disease. 1992;**15**:456-471

[125] Lightowlers RN, Taylor RW, Turnbull DM. Mutations causing mitochondrial disease: What is new and what challenges remain? Science. 2015;**349**:1494-1499

[126] Payne BA, Chinnery PF. Mitochondrial dysfunction in aging: Much progress but many unresolved questions. Biochimica et Biophysica Acta. 1847;**2015**:1347-1353

[127] Lane RK, Hilsabeck T, Rea SL. The role of mitochondrial dysfunction in age-related diseases. Biochimica et Biophysica Acta. 1847;**2015**:1387-1400

[128] Grunewald A, Rygiel KA, Hepplewhite PD, et al. Mitochondrial DNA depletion in respiratory chain-deficient Parkinson disease neurons. Annals of Neurology. 2016;**79**:366-378

[129] Bender A, Krishnan KJ, Morris CM, et al. High levels of mitochondrial DNA deletions in substantia nigra neurons in aging and Parkinson disease. Nature Genetics. 2006;**38**:515-517

[130] Kandul NP, Zhang T, Hay BA, Guo M. Selective removal of deletion-bearing mitochondrial DNA in heteroplasmic *Drosophila*. Nature Communications. 2016;**7**:13100

[131] Detmer SA, Chan DC. Functions and dysfunctions of mitochondrial dynamics. Nature Reviews. Molecular Cell Biology. 2007;**8**:870-879

[132] Chen H, Chomyn A, Chan DC. Disruption of fusion results in mitochondrial heterogeneity and dysfunction. The Journal of Biological Chemistry. 2005;**280**:26185-26192

[133] Hales KG, Fuller MT. Developmentally regulated mitochondrial fusion mediated by a conserved, novel, predicted GTPase. Cell. 1997;**90**:121-129

[134] Skulachev VP. Mitochondrial filaments and clusters as intracellular power-transmitting cables. Trends in Biochemical Sciences. 2001;**26**:23-29

[135] Hoppins S, Lackner L, Nunnari J. The machines that divide and fuse mitochondria. Annual Review of Biochemistry. 2007;**76**:751-780

[136] Hwa JJ, Hiller MA, Fuller MT, Santel A. Differential expression of the *Drosophila* mitofusin genes fuzzy onions (fzo) and dmfn. Mechanisms of Development. 2002;**116**: 213-216

[137] Yarosh W, Monserrate J, Tong JJ, et al. The molecular mechanisms of OPA1-mediated optic atrophy in *Drosophila* model and prospects for antioxidant treatment. PLoS Genet. 2008;**4**:e6

[138] Verstreken P, Ly CV, Venken KJ, et al. Synaptic mitochondria are critical for mobilization of reserve pool vesicles at *Drosophila* neuromuscular junctions. Neuron. 2005; **47**(3):365-378

[139] Maraganore DM, de Andrade M, Lesnick TG, et al. High-resolution whole-genome association study of Parkinson disease. American Journal of Human Genetics. 2005;**77**: 685-693

[140] Hardy J, Singleton A. Genomewide association studies and humandisease. The New England Journal of Medicine. 2009;**360**:1759-1768

[141] Pankratz N, Foroud T. Genetics of Parkinson disease. Genetics in Medicine. 2007;**9**: 801-811

[142] Satake W, Nakabayashi Y, Mizuta I, et al. Genome-wide association study identifies common variants at four loci as genetic risk factors for Parkinson's disease. Nature Genetics. 2009;**41**:1303-1307

[143] Pankratz N, Wilk JB, Latourelle JC, et al. Genome wide association study for susceptibility genes contributing to familial Parkinson disease. Human Genetics. 2009;**124**:593-605

[144] Lill CM, Roehr JT, McQueen MB, et al. Comprehensive research synopsis and systematic meta-analyses in Parkinson's disease genetics: The PDgene database. PLoS Genetics. 2012;**8**:4-13

[145] Nalls MA, Pankratz N, Lill CM, et al. Large-scale meta-analysis of genome-wide association data identifies six new risk loci for Parkinson's disease. Nature Genetics. 2014;**056**:1-7

[146] Escott-Price V, Nalls MA, Morris HR, et al. Polygenic risk of Parkinson disease is correlated with disease age at onset. Annals of Neurology. 2015;**77**:582-591

[147] Hamza TH, Zabetian CP, Tenesa A, et al. Common genetic variation in the HLA region is associated with late-onset sporadic Parkinson's disease. Nature Genetics. 2010; **42**:781-785

[148] Do CB et al. Web-based genome-wide association study identifies two novel loci and a substantial genetic component for Parkinson's disease. PLoS Genetics. 2011;**7**(6): e1002141

[149] Ivatt RM, Sanchez-Martinez A, Godena VK, et al. Genome-wide RNAi screen identifies the Parkinson disease GWAS risk locus SREBF1 as a regulator of mitophagy. Proceedings of the National Academy of Sciences of USA. 2014;**111**:8494-8499

[150] Bodmer W, Bonilla C. Common and rare variants in multifactorial susceptibility to common diseases. Nature Genetics. 2008;**40**:695-701

[151] Keller MF, Saad M, Bras J et al. Using genome-wide complex trait analysis to quantify "missing heritability" in Parkinson's disease. Human Molecular Genetics 2012; **21**:4996-5009

[152] Brand AH, Perrimon N. Raf acts downstream of the EGF receptor to determine dorsoventral polarity during Drosophila oogenesis. Genes & Development. 1994;**8**:629-639

[153] Singleton AB, Farrer M, Johnson J, et al. Alpha-Synuclein locus triplication causes Parkinson's disease. Science 2003;**302**:841

[154] Fire A, Xu S, Montgomery MK, et al. Potent and specific genetic interference by double-stranded RNA in *Caenorhabditis elegans*. Nature. 1998;**391**:806-811

[155] Meister G, Tuschl T. Mechanisms of gene silencing by double-stranded RNA. Nature. 2004;**431**:343-349

[156] Silva J, Chang K, Hannon GJ, Rivas FV. RNA-interference-based functional genomics in mammalian cells: Reverse genetics coming of age. Oncogene. 2004;**23**:8401-8409

[157] Ghildiyal M, Zamore PD. Small silencing RNAs: An expanding universe. Nature Reviews. Genetics. 2009;**10**:94-108

[158] Cronin SJ, Nehme NT, Limmer S, et al. Genome-wide RNAi screen identifies genes involved in intestinal pathogenic bacterial infection. Science. 2009;**325**:340-343

[159] Yamamoto-Hino M, Kanie Y, Awano W, et al. Identification of genes required for neural-specific glycosylation using functional genomics. PLoS Genetics. 2010;**6**:e1001254

[160] Valakh V, Naylor SA, Berns DS, DiAntonio A. A large-scale RNAi screen identifies functional classes of genes shaping synaptic development and maintenance. Developmental Biology. 2012;**366**:163-171

[161] Samuel AH, Lesley AK, Koji Y, et al. High-content genome-wide RNAi screens identify-regulators of parkin upstream of mitophagy. Nature. 2013;**504**:291-295

[162] Cong L, Ran FA, Cox D, et al. Multiplex genome engineering using CRISPR/Cas systems. Science. 2013;**339**:819-823

[163] Chen BH, Gilbert LA, Cimini BA, et al. Dynamic imaging of genomic loci in living human cells by an optimized CRISPR/cas system. Cell. 2014;**155**:1479-1491

[164] Wang X, Cao C, Huang J, et al. One-generation of triple gene-targeted pigs using CRISPR/Cas9system. Scientific Reports. 2016;**6**:20620

[165] Heidenreich M, Zhang F. Applications of CRISPR-Cas systems in neuroscience. Nature Reviews. Neuroscience. 2016;**17**:36-44

[166] Walters BJ, Azam AB, Gillon CJ, et al. Advanced in vivo use of CRISPR/Cas9 and anti-sense DNA inhibition for gene manipulation in the brain. Frontiers in Genetics. 2016;**6**:362

[167] Yang W, Tu Z, Sun Q, Li XJ. CRISPR/Cas9: Implications for Modeling and therapy of neurodegenerative diseases. Frontiers in Molecular Neuroscience. 2016;**9**:30

[168] Pendleton RG, Parvez F, Sayed M, et al: Effects of pharmacological agents upon a transgenic model of Parkinson's disease in *Drosophila melanogaster*. Journal of Pharmacology and Experimental Therapeutics. 2002;**300**(1):91-96

[169] Auluck PK, Chan HYE, Trojanowski JQ, et al. Chaperone suppression of $\alpha$- synuclein toxicity in a Drosophila model for Parkinson's disease. Science. 2002;**295**(5556):865-868

[170] Auluck PK, Meulener MC, Bonini NM. Mechanisms of suppression of $\alpha$-synuclein neurotoxicity by geldanamycin in Drosophila. Journal of Biological Chemistry. 2005; **280**(4):2873-2878

[171] Auluck PK, Bonini NM. Pharmacological prevention of Parkinson disease in Drosophila. Nature Medicine. 2002;**8**(11):1185-1186

[172] Outeiro TF, Kontopoulos E, Altmann SM, et al. Sirtuin 2 inhibitors rescue $\alpha$-synuclein-mediated toxicity in models of Parkinson's disease. Science. 2007;**317**(5837):516-519

[173] Trinh K, Moore K, Wes PD, et al. Induction of the phase II detoxification pathway suppresses neuron loss in Drosophila models of Parkinson's disease. Journal of Neuroscience. 2008;**28**(2):465-472

[174] Wassef R, Haenold R, Hansel A, et al. Methionine sulfoxide reductase A and a dietary supplement S-methyl-L-cysteine prevent Parkinson's-like symptoms. Journal of Neuroscience. 2007;**27**(47):12808-12816

[175] Long J, Gao H, Sun L, et al. Grape extract protects mitochondria from oxidative damage and improves Locomotor dysfunction and extends lifespan in a Drosophila Parkinson's disease model. Rejuvenation Research. 2009;**12**(5):321-331

[176] Faust K, Gehrke S, Yang Y, et al. Neuroprotective effects of compounds with antioxidant and anti-inflammatory properties in a Drosophila model of Parkinson's disease. BMC Neuroscience. 2009;**10**:109

[177] Lavara-Culebras E, Muñoz-Soriano V, Gómez-Pastor R, et al. Effects of pharmacological agents on the lifespan phenotype of Drosophila DJ-1β mutants. Gene. 2010;**462**(1-2):26-33

[178] Brunn GJ, Hudson CC, Sekuli'c A, et al. Phosphorylation of the translational repressor PHAS-I by the mammalian target of rapamycin. Science. 1997;**277**(5322):99-101

[179] Burnett PE, Barrow RK, Cohen NA, et al. RAFT1 phosphorylation of the translational regulators p70 S6 kinase and 4E-BP1. Proceedings of the National Academy of Sciences of the United States of America. 1998;**95**(4):14321437

[180] Tain LS, Mortiboys H, Tao RN, et al. Rapamycin activation of 4E-BP prevents parkinsonian dopaminergic neuron loss. Nature Neuroscience. 2009;**12**(9):1129-1135

[181] Saini N, Schaffner W. Zinc supplement greatly improves the condition of parkin mutant Drosophila. Biological Chemistry. 2010;**391**(5):513-518

[182] Jimenez-Del-Rio M, Guzman-Martinez C, Velez-Pardo C. The effects of polyphenols on survival and locomotor activity in *Drosophila melanogaster* exposed to iron and paraquat. Neurochemical Research. 2010;**35**(2):227-238

[183] Liu Z, Hamamichi S, Lee BD et al: Inhibitors of LRRK2 kinase attenuate neurodegeneration and Parkinson-like phenotypes in *Caenorhabditis elegans* and Drosophila Parkinson's disease models. Human Molecular Genetics. 2011;**20**:3933-3942

[184] Neef DW, Jaeger AM, Thiele DJ. Heat shock transcription factor 1 as a therapeutic target in neurodegenerative diseases. Nature Reviews Drug Discovery. 2011;**10**(12):930-944

[185] Thakur P, Nehru B. Long-termheat shock proteins (HSPs) induction by carbenoxolone improves hallmark features of Parkinson's disease in a rotenone-based model. Neuropharmacology. 2014;**79**:190-200

[186] Kilpatrick K, Novoa JA, Hancock T, et al. Chemical induction of Hsp70 reduces $\alpha$-synuclein aggregation in neuroglioma cells. ACS Chemical Biology. 2013;**8**(7):1460-1468

[187] Chiosis G, Tao H. Purine-scaffold Hsp90 inhibitors. IDrugs: The Investigational Drugs Journal. 2006;**9**(11):778-782

[188] McLean PJ, Klucken J, Shin Y, et al. Geldanamycin induces Hsp70 and prevents $\alpha$-synuclein aggregation and toxicity in vitro. Biochemical and Biophysical Research Communications. 2004;**321**(3):665-669

[189] Shen HY, He JC, Wang Y, et al. Geldanamycin induces heat shock protein 70 and protects against MPTP-induced dopaminergic neurotoxicity in mice. Journal of Biological Chemistry. 2005;**280**(48):39962-39969

[190] Putcha P, Danzer KM, Kranich LR et al. Brain-permeable small-molecule inhibitors of Hsp90 prevent $\alpha$-synuclein oligomer formation and rescue $\alpha$-synuclein-induced toxicity. Journal of Pharmacology and Experimental Therapeutics. 2010;**332**(3):849-857

[191] Wang D, Qian L, Xiong H, et al. Antioxidants protect PINK1-dependent dopaminergic neurons in Drosophila. Proceedings of the National Academy of Sciences of the United States of America. 2006;**103**(36):13520-13525

[192] Cooper AA, Gitler AD, Cashikar A, et al. $\alpha$-synuclein blocks ER-Golgi traffic and Rab1 rescues neuron loss in Parkinson's models. Science. 2006;**313**(5785):324-328

[193] Ng CH, Guan MS, Koh C, et al. AMP kinase activation mitigates dopaminergic dysfunction and mitochondrial abnormalities in Drosophila models of Parkinson's disease. The Journal of Neuroscience. 2012;**32**(143):11-17

[194] Buttner S, Broeskamp F, Sommer C, et al. Spermidine protects against alpha-synuclein neurotoxicity. Cell Cycle. 2014;**13**(3903):8. DOI: 10.4161/15384101.2014.973309

[195] Kontopoulos E, Parvin JD, Feany MB. Alpha-synuclein acts in the nucleus to inhibit histone acetylation and promote neurotoxicity. Human Molecular Genetics. 2006; **15**(20):3012-3023

[196] Pinna A. Adenosine A2A receptor antagonists in Parkinson's disease: Progress in clinical trials from the newly approved istradefylline to drugs in early development and those already discontinued. CNS Drugs. 2014;**28**(5):455-475

[197] Stocchi F, Rascol O, Destee A, et al. AFQ056 in Parkinson patients with levodopa-induced dyskinesia: 13-week, randomized, dose-finding study. Movement Disorders. 2013;**28**(13):1838-1846

[198] Borgohain R, Szasz J, Stanzione P, et al. For the study 016 investigators. Randomized trial of safinamide add-on to levodopa in Parkinson's disease with motor fluctuations. Movement Disorders 2014;**29**(2):229-237

[199] Sureda FX, Junyent F, Verdaguer E, et al. Antiapoptotic drugs: A therapautic strategy for the prevention of neurodegenerative diseases. Current Pharmaceutical Design. 2011;**17**(3):230-245

[200] Holmay MJ, Terpstra M, Coles LD, et al. N-Acetylcysteine boosts brain and blood glutathione in Gaucher and Parkinson diseases. Clinical Neuropharmacology. 2013; **36**(4):103-106

[201] Palfi S, Gurruchaga JM, Ralph GS, et al. Long-term safety and tolerability of ProSavin, a lentiviral vector-based gene therapy for Parkinson's disease: A dose escalation, open-label, phase 1/2 trial. Lancet. 2014;**383**(9923):1138-1146

[202] Nishimura K, Takahashi J. Therapeutic application of stem cell technology toward the treatment of Parkinson's disease. Biological & Pharmaceutical Bulletin. 2013; **36**(2):171-175

[203] Harraz MM, Dawson TM, Dawson VL. MicroRNAs in Parkinson's disease. Journal of Chemical Neuroanatomy. 2011;**42**(2):127-130

Chapter 9

# *Drosophila* Model in the Study Role of UCH-L1

Dang Thi Phuong Thao

Additional information is available at the end of the chapter

http://dx.doi.org/10.5772/intechopen.73578

**Abstract**

UCH-L1 (ubiquitin carboxyl-terminal hydrolase L1) is a protein, which plays an important role in ubiquitin-proteasome system. Many previous reports showed the relation between UCH-L1 and neurodegenerative diseases, diabetes, as well as cancer. However, the mechanism still remains unclear. In the aim to investigate the functions and regulatory mechanism of UCH-L1 in living organism, *Drosophila melanogaster* model was utilized to examine the role of UCH-L1. This chapter provides a summary on recent findings related to the roles of UCH-L1 based on the model. First, abnormal expression of *Drosophila* ubiquitin carboxyl-terminal hydrolase (dUCH) leads to the defects on fly tissue development and function. Gain function of dUCH in the eye imaginal discs induced a rough eye phenotype in the adult, partly resulting from induction of caspase-dependent apoptosis, upset of photoreceptor cell distribution and ommatidium apical mispatterning. Interestingly, the dUCH overexpression of induced rough eye phenotype was completely recused by co-expression either Sevenless or Draf of the mitogen-activated protein kinase pathway. Besides, knockdown dUCH in dopaminergic neurons resulted in some Parkinson's disease—like phenotypes in fly. Taken together, those findings in *Drosophila* model contributed a significant dUCH in tissue development and function.

**Keywords:** *Drosophila melanogaster*, UCH-L1, human diseases, eye development, anti-dUCH antibody

## 1. Introduction

Ubiquitin carboxyl-terminal hydrolase L1 (UCH-L1), a protein of 223 amino acids (aa), weighs about 24,824 Da, a period lasting for more than 48 half-hour. UCH-L1 is an abundant protein in neurons, accounting for 1–2% of the total protein in the human brain [1]. In addition to the brain, UCH-L1 is also expressed strongly in the peripheral nervous system, including sensory and nervous system activity. UCH-L1 belongs to remove the tagged enzyme (deubiquitinating enzyme (DUB)), an important protein in ubiquitin proteasome system (UPS). UCH-L1 hydrolases the peptide bond between ubiquitins and also plays a function as a ligase when it be in dimer form [2, 3]. UCH-L1 is an enzyme which binds to the polyubiquitin chains and released the single ubiquitin in the ubiquitin proteasome system. However, when UCH-L1 is in binary form, UCH-L1 leads to the formation of a polyubiquitin chain linked through lysine 63 (K63). Although the main activity of UCH-L1 is still unclear, UCH-L1 has been believed to play its role through maintaining a pool of free monomeric ubiquitin which is important for the function of ubiquitin proteasome system [4]. Abnormal function of UCH-L1 leads to the reduction of protein degradation, followed by the accumulation of ubiquitinated protein [5–7]. UCH-L1, therefore, may relate to many biological processes which dependent to ubiquitination including DNA repair, cell signalling, trafficking, endocytosis and degradation.

In 1998, a missense mutation of UCH-L1 (I93M) was first identified in a German family with Parkinson's disease (PD) [8]. By contrast, another variant of UCH-L1 (S18Y) was discovered as a factor in the risk reduction of PD [9]. Other studies also found that UCH-L1 was related to abnormal accumulation and aggregation of $\alpha$-synuclein which leads to formation of Lewy bodies [3]. Furthermore, gracile axonal dystrophy (GAD) mouse which carries a deletion within UCH-L1 gene manifested motor ataxia, axonal degeneration and a reduction in the monoubiquitin level in neurons [10–12].

On the other hand, many studies indicated that UCH-L1 involved too many types of human cancer [4]. High expression of UCH-L1 was found in many types of cancers such as breast cancer, non-small cell lung cancer [13, 14]. UCH-L1 expression can be self-upregulated via oncogenic $\beta$-catenin/TCF activation. The UCH-L1 upregulates oncogenic $\beta$-catenin by which feedback regulates the expression of *uch-l1* gene [15]. UCH-L1 may also promote cancer metastasis via $\beta$-catenin-induced epithelial-to-mesenchymal transition [16, 17]. High levels of UCH-L1 may promote oncogenic transformation, invasion and metastasis, and the function of UCH-L1 might due to the enhancement of Akt signalling in vitro and in vivo [16, 18, 19].

By contrast, UCH-L1 had been also reported as a tumor suppressor in many other studies. The downregulation of UCH-L1 was observed in various types of cancer such as esophageal cancer, breast cancer, prostate cancer and pancreatic cancer [20–24]. Reduction in UCH-L1 expression leads to cell proliferation arrest and p53-mediated apoptosis [22, 25].

In humans, the gene coding for UCH-L1 is located in the short arm of chromosome 4 at position 14, from base pair 40,953,685 to 40,965,202, 11,518 base pairs long [26]. In *Drosophila melanogaster*, ubiquitin carboxyl-terminal hydrolase (dUCH) encoded by CG4265 gene is a homolog of human UCH-L1 (hUCH-L1). The identity and similarity between dUCH and hUCH-L1 are 44.5 and 75.7%, respectively. In this chapter, we provide a summary on recent findings related to the roles of UCH-L1 in living organisms by *Drosophila* models. Those findings indicated that dUCH (ortholog of human UCH-L1 in *Drosophila*) plays an important role tissue development and involves in Parkinson's disease.

http://dx.doi.org/10.5772/intechopen.73578

## 2. *Drosophila* model in the study role of UCH-L1

### 2.1. Homolog of human ubiquitin carboxyl-terminal hydrolase L1 (UCH-L1) in *Drosophila melanogaster*

The survey of the *Drosophila* genome database allowed an identification of the CG4265 as a homolog of the human UCH-L1. The CG4265 gene, named as dUCH (*Drosophila* ubiquitin carboxyl-terminal hydrolase), encodes a 224-amino-acid protein that shows 44.5% identity and 75.7% similarity with human UCH-L1. The Cys residue at amino acid (aa) position 90 and the His residue at aa 161, both of which are essential for hydrolase activity of human UCH-L1 [27–29], are conserved in *Drosophila melanogaster* along with several other species including *Mus musculus* and *Caenorhabditis elegans* (**Figure 1**).

### 2.2. Generation of anti-dUCH antibody

Since *Drosophila melanogaster* has been shown to be a compatible model for studying human diseases, the UCH-L1 homologous protein in *Drosophila melanogaster* (dUCH) is utilized for analyzing the role of UCH-L1 in living system. Thereby, anti-dUCH antibody is essential for research and needs to be generated. The produced anti-dUCH antibody was shown to have high specificity and sensitivity to the dUCH protein. The affinity of the antibody is 1:320,000 at 7.81 ng/μl antigen concentration. The 1:40,000 dilution-produced antibodies can detect antigen at a low concentration 0.98 ng/μl [30]. Besides, the antibody showed a high specificity for

```
hUCH-L1    1 --MQLKPMEINPEMLNKVLSRLGVAGQWRFVDVLGLEEESLGSVPAPACALLLLFPLTAQ 58
mUCH-L1    1 --MQLKPMEINPEMLNKVLAKLGVAGQWRFADVLGLEEETLGSVPSPACALLLLFPLTAQ 58
dUCH       1 -MLTWTPLESNPEVLTKYIHKLGVSPAWSVTDVIGLEDDTLEWIPRPVKAFILLFPCSET 59
cUBH-1     1 MAAPWTPLESNPSVINPMIEKMGVSGVK-TVDVLFFDDESIGK---PQHAVILCFPEYKK 56

hUCH-L1   59 HENFRKKQIEELKGQEVSPKVYFMKQTIGNSCGTIGLIHAVANNQDKLGFEDGSVLKQFL 118
mUCH-L1   59 HENFRKKQIEELKGQEVSPKVYFMKQTIGNSCGTIGLIHAVANNQDKLEFEDGSVLKQFL 118
dUCH      60 HRAEEHDRIKEVEEQ-HPEDLFYMRQFTHNACGTVALIHSVANNKEV--DIDRGVLKDFL 116
cUBH-1    57 VDEIMKPIYEQ--AKAADDSVFFMKQKISNACGTFALFHSLANLEDRINLGDGSFA-KWL 113

hUCH-L1  119 SETEKMSPEDRAKCFEKNEAIQAAHDAVAQEGQCRV--DDKVNFHFILFNNVDGHLYELD 176
mUCH-L1  119 SETEKLSPEDRAKCFEKNEAIQAAHDSVAQEGQCRV--DDKVNFHFILFNNVDGHLYELD 176
dUCH     117 EKTASLSPEERGRALEKDEKFTADHEALAQEGQTNAANHEKVIHHFIALVNKEGTLYELD 176
cUBH-1   114 AEAKKVGIEERSDFLANNAELAGIHAAAATDGQTAP--SGDVEHHFICFVGKNGILYEID 171

hUCH-L1  177 GRMPFPVNHGASSEDTLLKDAAKVCREFTEREQGEVRFSAVALCKAA- 223
mUCH-L1  177 GRMPFPVNHGASSEDSLLQDAAKVCREFTEREQGEVRFSAVALCKAA- 223
dUCH     177 GRKSFPIKHGPTSEETFVKDAAKVCKEFMARDPNEVRFTVLALTAAQQ 224
cUBH-1   172 SRRPFAREIGPTSDATLVKDAGAACQHLIEKLD-NVSFSAIAVVNQ-- 216
```

**Figure 1.** Amino acid sequences of UCH-L1 protein between human (hUCH-L1), mouse (mUCH-L1), *Drosophila* (dUCH-L1) and *C. elegans* (cUBH-L1). The identity and similarity between human and *Drosophila* were 44.5 and 75.7%, respectively. Identical amino acids are shaded in dark grey, and similar amino acids are shaded in light grey. The red letters indicate the identical amino acids at active sites. Clustal Omega (1.2.4) multiple sequence alignment was applied.

*Drosophila* either in Western blot or in immunostaining. When the dUCH was overexpressed in fly eye imaginal discs using the GAL4/UAS system, the dUCH protein level was specifically recognized by the anti-dUCH antibody, and the antibody sensitivity showed different levels of the dUCH target protein in *Drosophila* tissues either in Western blot or in immunostaining (**Figure 2**). Success in producing dUCH antibody provides a good material for further experiments in the study role of UCH-L1 by *Drosophila* model.

### 2.3. *Drosophila* model for studying the UCH-L1 role in tissue development

Being a member of ubiquitin proteasome system (UPS), UCH-L1 is thought to be involved in many different processes in living organisms, such as cell proliferation and differentiation. In *Drosophila* model, tissue-specific knockdown of dUCH resulted in abnormal phenotype in adult flies. When dUCH was knocked down in posterior area of eye imaginal discs by the combination of GMR-Gal4 driver and UAS-duchIR cassette (GMR-Gal4 > UAS-duchIR), the *duch* knocked-down adult compound eye exhibited a rough eye phenotype, and ommatidium was bulged and sticked together, while the control fly showed a normal phenotype. Knockdown dUCH in the thorax by Pnr-Gal4 driver gave hair-deformed defection. The wing of the knocked-down dUCH flies also showed some extraordinary phenotype as the vein in the wing disappeared or deformed. When TH-Gal4 drives the synthesis of *duch* dsRNA

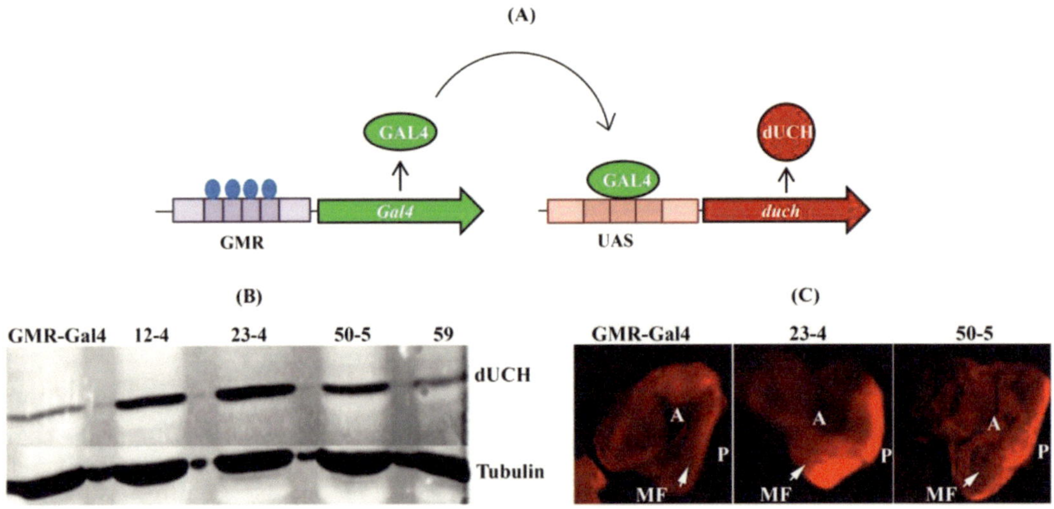


**Figure 2.** Generation of polyclonal anti-dUCH antibody for studying UCH-L1 function in *Drosophila melanogaster* model. (A) GAL4/UAS system is used for overexpressing dUCH protein in transgenic flies. Gal4 protein was expressed under GMR driving promoter in *Drosophila* posterior eye imaginal discs. Then, the expressed Gal4 bound to UAS element on the upstream of duch gene in transgenic *Drosophila* and caused the *duch* gene expression. (B) Western blot analysis of total protein from eye imaginal discs with polyclonal anti-dUCH antibody (above) and monoclonal anti-alpha tubulin antibody (below). GMR-Gal4: total protein from eye imaginal discs of transgenic fly, which showed dUCH endogenous protein. 12-4, 23-4, 50-5, 59: total protein from four different transgenic fly lines, which overexpresses dUCH protein under GMR-Gal4 driver. (C) Immunohistochemistry analysis on eye imaginal discs from the third instar *Drosophila melanogaster* larvae with polyclonal anti-dUCH antibody. GMR-Gal4: eye imaginal discs of transgenic fly, which showed dUCH endogenous protein. 23-4, 50-5: eye imaginal discs from two different transgenic fly lines, which overexpresses dUCH protein under GMR-Gal4 driver.

in *Drosophila* brain tissue, the third larval crawling ability was strongly defected (**Figure 3**). Emphatically, knockdown dUCH in whole bodies of the flies by Act5C-Gal4 resulted in pupal lethal effects. These observations strongly suggested that the dUCH plays an important role in maintaining normal *Drosophila* tissue development.

On the other hand, overexpression of dUCH in *Drosophila melanogaster* showed an apoptosis induction in eye imaginal discs and resulted in rough eye phenotype in adult flies. The apoptosis induction was vanished by co-expression of P35, a vacuolar viral protein that inhibits downstream effecter caspases. The apoptosis induction is followed by compensatory proliferation (**Figure 4**) [31].

Furthermore, dUCH overexpression also caused the upset in distribution of photoreceptor clusters in fly pupal retina (**Figure 5**).

In *Drosophila* pupal retinae, the ommatidia were arranged precisely. Different cell types appeared in typical shape and position. However, overexpression of dUCH in pupal retinae increased apical mispatterning. In many regions of dUCH-overexpressing retinae, ommatidia showed defects in alignment and orientation. Cone cell clusters are in different sizes and distorted. In addition, the morphology of pigment cells was aberrant. Defects in the shape and the number of primary pigment cells were detected. The shape of secondary and tertiary pigment cells (interommatidial pigment cells) was altered. In addition to the morphological

| GAL4 line | Expression of GAL4 | Phenotype | | |
|---|---|---|---|---|
| | | Control | Knockdown of dUCH | |
| Act5C | All tissues | Normal | Pupal lethal | |
| GMR | Cells behind the morphogenetic furrow of eye dics |  | 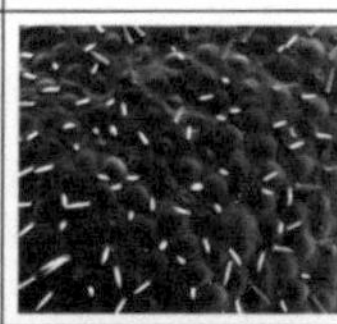 | Rough eye |
| Pnr | Dorsal wing dics | 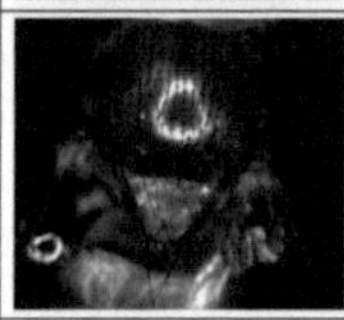 | 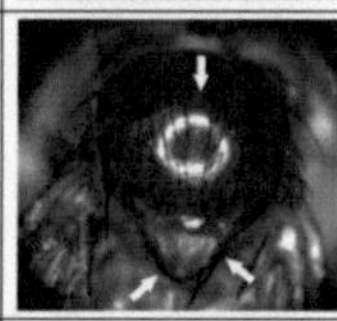 | Disappeared/ deformed some hairs in thorax |
| MS1096 | Dorsal wing dics | 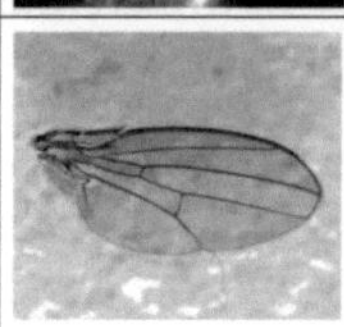 | 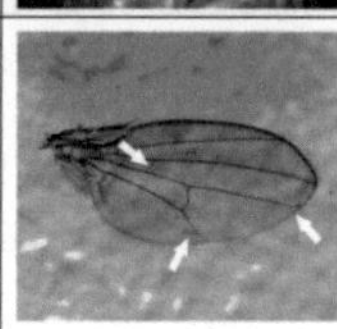 | Disappeared/ deformed some veins in wings |
| TH | Dopaminergic cells |  |  | Locomotor dysfunction in the third larvae |

**Figure 3.** Tissue-specific knockdown of dUCH resulted in defects in adult flies.

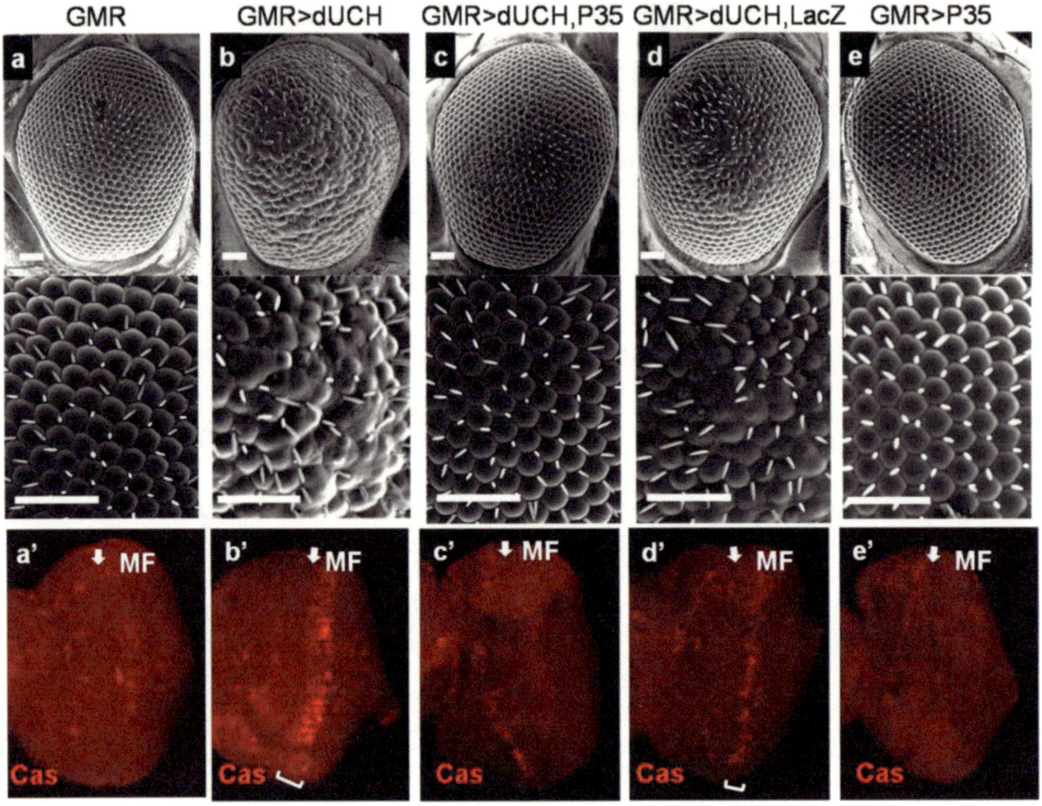


**Figure 4.** Overexpression of dUCH induces caspase-dependent apoptosis in eye imaginal discs. (a–e) Scanning electron micrographs of adult compound eyes. (a'-e') Immunostaining of the eye imaginal discs with anti-active caspase-3 antibody. (a,a') GMR-GAL4; (b,b') GMR-GAL4;UAS-dUCH/+; (c,c') GMR-GAL4;UAS-dUCH/+;UAS-P35/+; (d,d')GMR-GAL4;UAS-dUCH/+;UAS-LacZ/+; (e,e') GMR-GAL4; UAS-P35/+. Note the increased number of caspase-3 positive cells (brackets) behind the morphogenetic furrow of eye discs overexpressing dUCH (b') and the lack of signals detected in eye discs co-expressing both dUCH and P35 (c'). The arrow indicates the morphogenetic furrow (MF). The bars are for 50 μm.

changes, the alignment of these cells was confused. In many regions, adjacent ommatidia were separated by more than one layer of interommatidial pigment cells. As a consequence, ommatidia in abnormal region did not maintain hexagonal shape. Bristles were misplaced, possibly due to the aberrance of pigment cells (**Figure 6**).

Interestingly, co-expressing dUCH with Sevenless or Draf in eye imaginal discs could suppress the rough eye phenotype induced by overexpressing dUCH. It is therefore likely that overexpression of dUCH downregulates the MAPK pathway, resulting in impairment of eye development (**Figure 7**) [31].

### 2.4. *Drosophila* model for studying the UCH-L1 role in Parkinson's disease

UCH-L1 was first linked to PD when mutation UCH-L1I93M was found in two siblings from a family with autosomal dominant PD [8]. Transgenic mice that overexpression of UCH-L1I93M showed an accumulation of $\alpha$-synuclein with ubiquitin in the brain [3]. UCH-L1-deficient

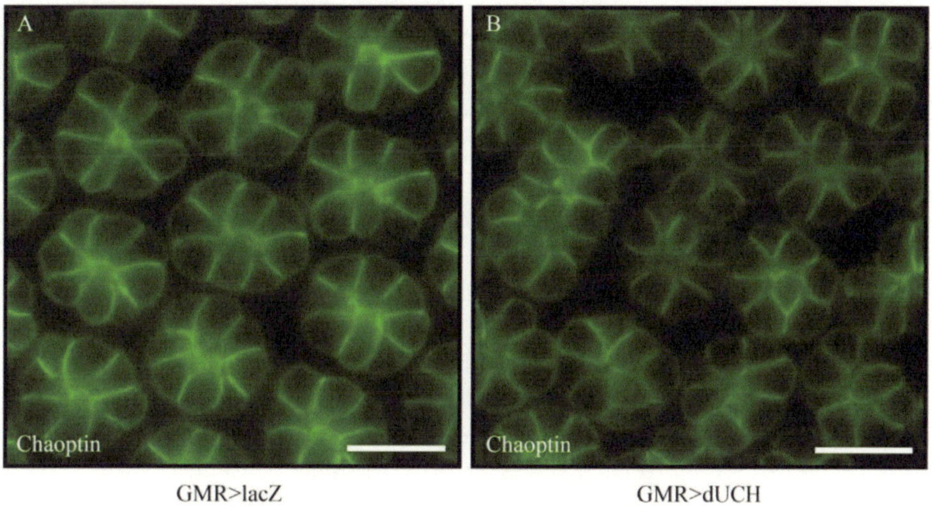


**Figure 5.** Immunostaining of retinae at 42 h after puparium formation (APF) with anti-chaoptin antibody. (A) Control retina and (B) dUCH-overexpressing retina. The bars indicate 10 μm.

mice showed neuronal loss in the spinal gracile tract and exhibit early development sensory and progressive motor ataxia [7]. However, another mutation UCH-L1S18Y is dedicated that decreased rick in PD by antioxidant and neuron-protective function [32]. Therefore, the mechanism of UCH-L1 still remains unclear. In *Drosophila* model, specific knockdown dUCH in dopaminergic neuron caused a degeneration of DA neurons and resulted in locomotor dysfunctions (**Figures 8** and **9**).

## 2.5. Materials and methods

### *2.5.1. Fly stocks*

Fly stocks were maintained at 25°C on standard food containing 0.7% agar, 5% glucose and 7% dry yeast. Wild-type strain Canton-S was obtained from the Bloomington *Drosophila* Stock Center (BDSC). RNAi lines carrying UAS-dUCH-IR fusion (GD#26468) for knockdown *Drosophila* ubiquitin carboxyl-terminal hydrolase (dUCH, CG4265) were received from the Vienna *Drosophila* Resource Center (VDRC). GAL4 drivers were used to perform the targeted knockdown of dUCH in various tissues of *D. melanogaster*: Act5C-GAL4 (BDSC#3954), GMR-GAL4 (line #16), MS1096-GAL4 (BDSC#8860), pnr-GAL4 (BDSC#3039) and TH-GAL4 (BDSC#8848).

### *2.5.2. Western immunoblot analysis*

Wild-type and transgenic adult flies carrying GMR-GAL4 > UAS-dUCH were frozen in liquid nitrogen and homogenized in a solution containing 50 mM Tris-HCl (pH 7.5); 5 mM $MgCl_2$; 150 mM NaCl; 10% glycerol; 0.1% Triton X-100; 0.1% NP-40; 1 mM phenylmethylsulfonyl fluoride; 5 mM β-mercaptoethanol; 10 g/ml each of aprotinin, leupeptin and pepstatin A; and 1 g/ml each of antipain, chymostatin and phosphoramidon. Homogenates were centrifuged, and extracts (200 g of protein) were electrophoretically separated on SDS-polyacrylamide gels containing 10%

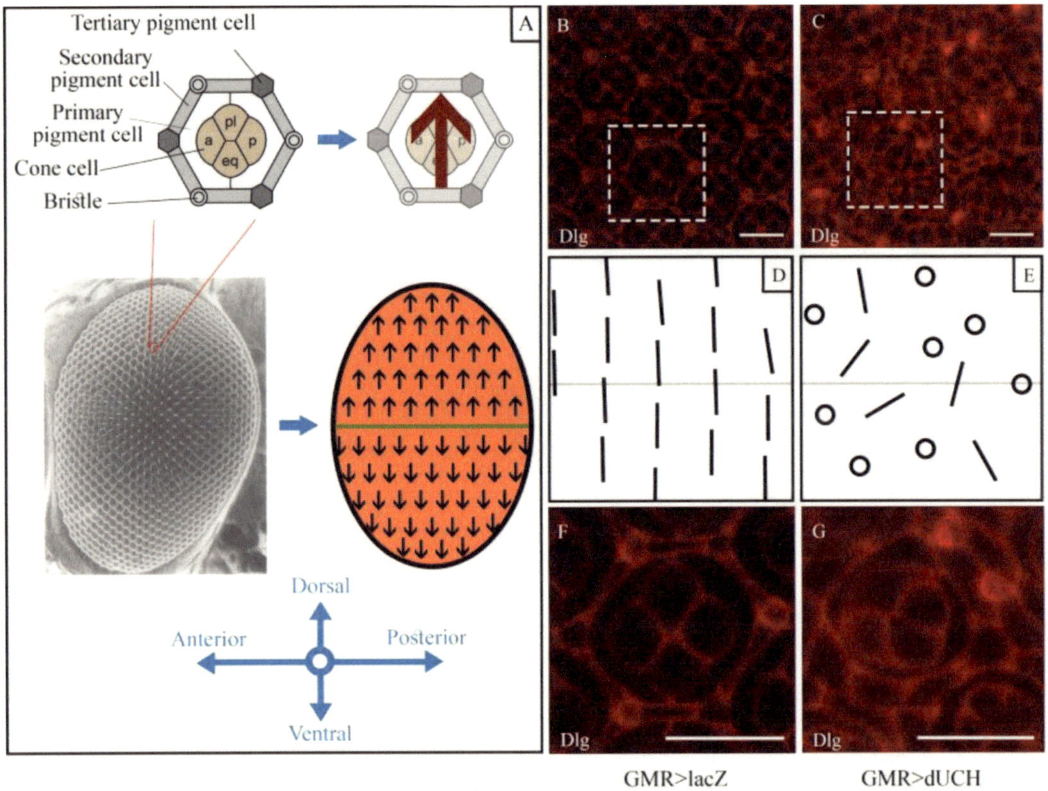


**Figure 6.** Overexpression of dUCH-induced apical mispatterning of 42 h APF retinae. (A) Normal adult *Drosophila* eye schematically representing orientation of the ommatidia with the green line representing the equator and schematically representing cross-sectional structure of a pupal ommatidium at the apical level with a, anterior cone cell; p, posterior cone cell; pl., polar cone cell; eq, equatorial cone cell. Red arrow marks equatorial-polar axis. (B-C) Immunostaining of retinae at 42 h APF with anti-Dlg antibody, (B) control retina and (C) dUCH-overexpressing retina. (D-E) Diagrams show orientation of the ommatidia in control fly (D) and dUCH-overexpressing fly (E). Black segments represent apical orientation of the ommatidia, black circles represent unclear cases and grey lines represent the anterior-posterior axis of the retinae. (F-G) The magnification of the ommatidia in control fly (F) and dUCH-overexpressing fly (G). Bars in all figures indicate 10 μm.

acrylamide and then transferred to polyvinylidene difluoride membranes (Bio-Rad). The blotted membranes were blocked with TBS/0.05% Tween-20 containing 5% skim milk for 1 h at 25°C, followed by incubation with rabbit polyclonal anti-dUCH at 1:1000 dilution or mouse monoclonal anti-$\alpha$ tubulin (Developmental Studies Hybridoma Bank (DSHB)) at 1:5000 dilution for 16 h at 4°C. After washing, the membranes were incubated with HRP-conjugated secondary antibodies (GE Healthcare Bioscience) at 1:10,000 dilution for 1 h at 25°C. Detection was performed with ECL Western blotting detection reagents (GE Healthcare Bioscience), and images were analyzed with a Lumivision Pro HSII image analyzer (Aisin Seiki).

### 2.5.3. *Immunostaining*

Larval and adult brains were dissected in cold phosphate-buffered saline (PBS) and fixed in 4% paraformaldehyde at 25°C for 15 min. After washing with 0.3% PBS-T (PBS containing 0.3% Triton-X100) twice, the samples were blocked in blocking solution (0.15% PBS-T containing

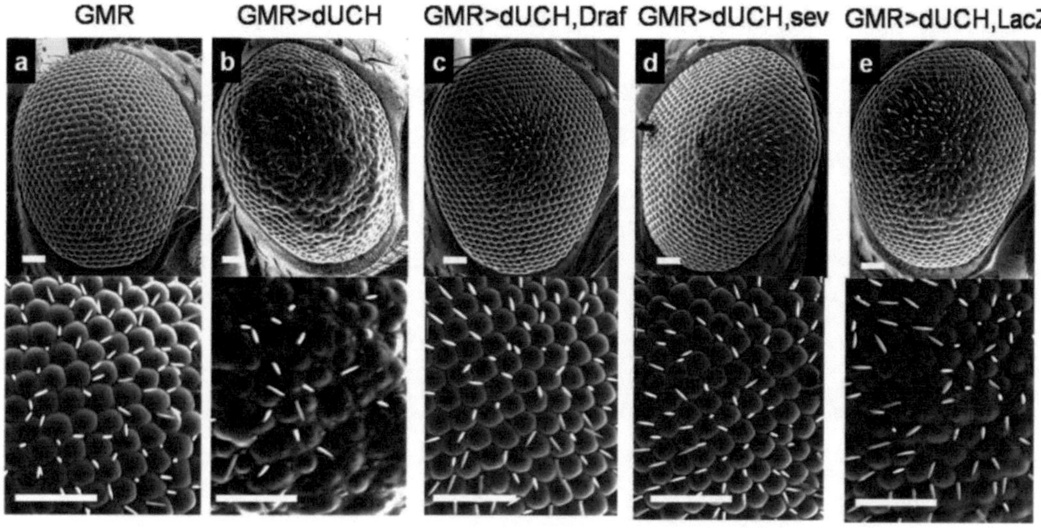


**Figure 7.** Suppression of the dUCH-induced rough eye phenotype by co-expression of sev or Draf. (a) GMR-GAL4;+; (b) GMR-GAL4;UAS-d;CH/+; (c) GMR-GAL4;UAS-dUCH/+;hsp-Draf/+; (d) GMR-GAL4/hsp-sev;UAS-dUCH/+; (e) GMR-GAL4;+;UAS-LacZ/+. Magnifications are 200× for the upper and 700× for the lower panels. Flies were reared at 28°C. The bars indicate 50 μm.

10% normal goat serum) at 25°C for 20 min. Samples were then incubated with the following primary antibodies diluted in blocking solution: rabbit anti-*Drosophila* ubiquitin carboxyl-terminal hydrolase (anti-dUCH; 1:500) at 4°C for 16 h or rabbit anti-tyrosine hydroxylase (anti-TH; 1:250; Millipore, AB152) at 4°C for 20 h. After washing with 0.3% PBS-T, samples were incubated with secondary antibodies conjugated with Alexa 488 or FITC (1500, Invitrogen) at 25°C for 2 h and then washed and mounted in VECTASHILED Antifade Mounting Medium (Vector Laboratories, Japan). Finally, the samples were inspected by a confocal laser scanning microscope (Olympus FluoView FV10i or Olympus BX41 Microscope).

### 2.5.4. *Crawling assay*

Male larvae in the early third instar stage were collected randomly and washed with PBS to discard food traces. After that, larvae were transferred to agar plates containing 2% agar with a density of 2–4 larvae per plate. The movement of larvae was recorded by a digital camera for 60 s. The recorded videos were then converted into AVI type by MOV to AVI converter (Pazera Jacek, Poland) and then analyzed by ImageJ (NIH, USA) with wrMTrck plugin (developed by Dr. Jesper Søndergaard Pedersen) to track larval movement and draw motion paths.

### 2.5.5. *Climbing assay*

Newly eclosed adult male flies were collected and transferred to conical tubes which have heights of 15 cm and diameters of 2 cm. After that, the tubes were tapped to collect the flies to the bottom, and the length of time to record the movement of flies was 30 s. The procedures were repeated five times and recorded by a digital camera. For all of the climbing experiments, the height which each fly climbed to was scored as follows: 0 (less than 2 cm),

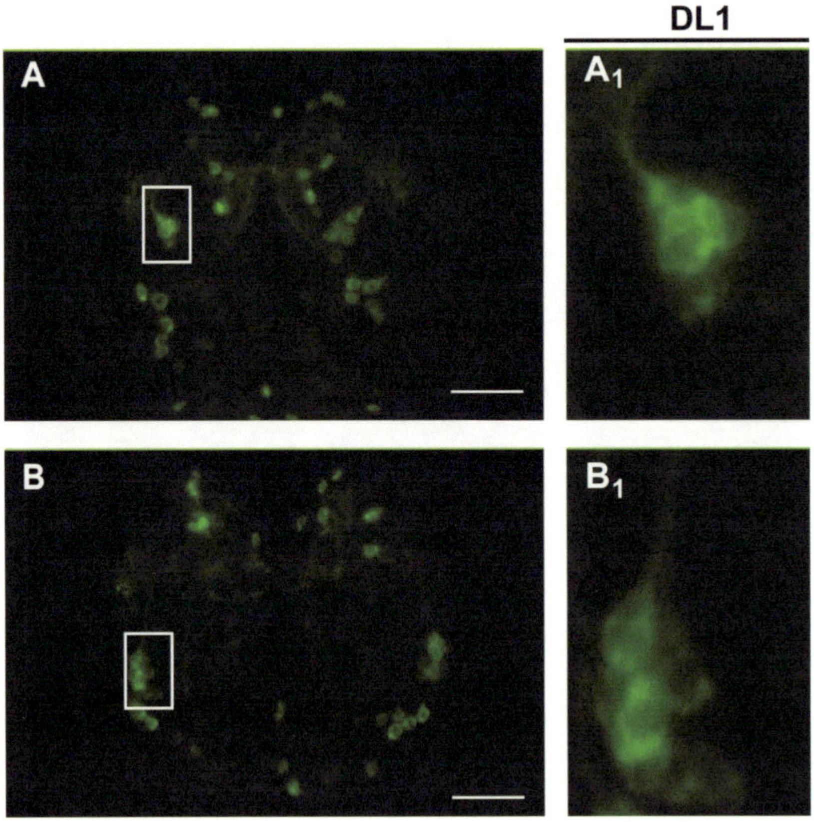


**Figure 8.** Loss of DL1 dopaminergic (DA) neurons in dUCH knockdown brain lobe. DA neuron clusters in the third instar larval were stained by anti-tyrosine hydroxylase antibody (anti-TH (green)). (A) Whole brain lobe with DA clusters in dUCH knockdown fly: TH-GAL4/UAS-dUCH-IR (TH > dUCH-IR). (A1) The magnification of DL1 DA cluster in knockdown fly brain lobe. (B) Whole brain lobe with DA clusters in control fly: TH-GAL4/+. (B1) The magnification of DL1 DA cluster in control fly brain lobe.

1 (between 2 and 4 cm), 2 (between 4 and 6 cm), 3 (between 6 and 8 cm), 4 (between 8 and 10 cm) and 5 (more than 10 cm). The climbing assay was performed every 5 days until all flies lose their locomotor abilities.

### *2.5.6. Conclusion and perspective*

UCH-L1 was known as a complex and unclear function protein. It has several irrelevant activities as hydrolase and ligase, which are also related to ubiquitin. Previous reports showed that abnormal UCH-L1 functioning, caused by mutations or change in levels of protein expression. Those reports also implied that UCH-L1 could have many negative effects, with impacts on cell proliferation, cell cycling and cell death through activation of many genes [33, 34]. In this chapter, some data compatibly demonstrated that overexpression of dUCH, a homolog of human UCH-L1 in *Drosophila melanogaster*-induced apoptosis, interfered eye development by upset distribution of photoreceptor cell distribution and caused

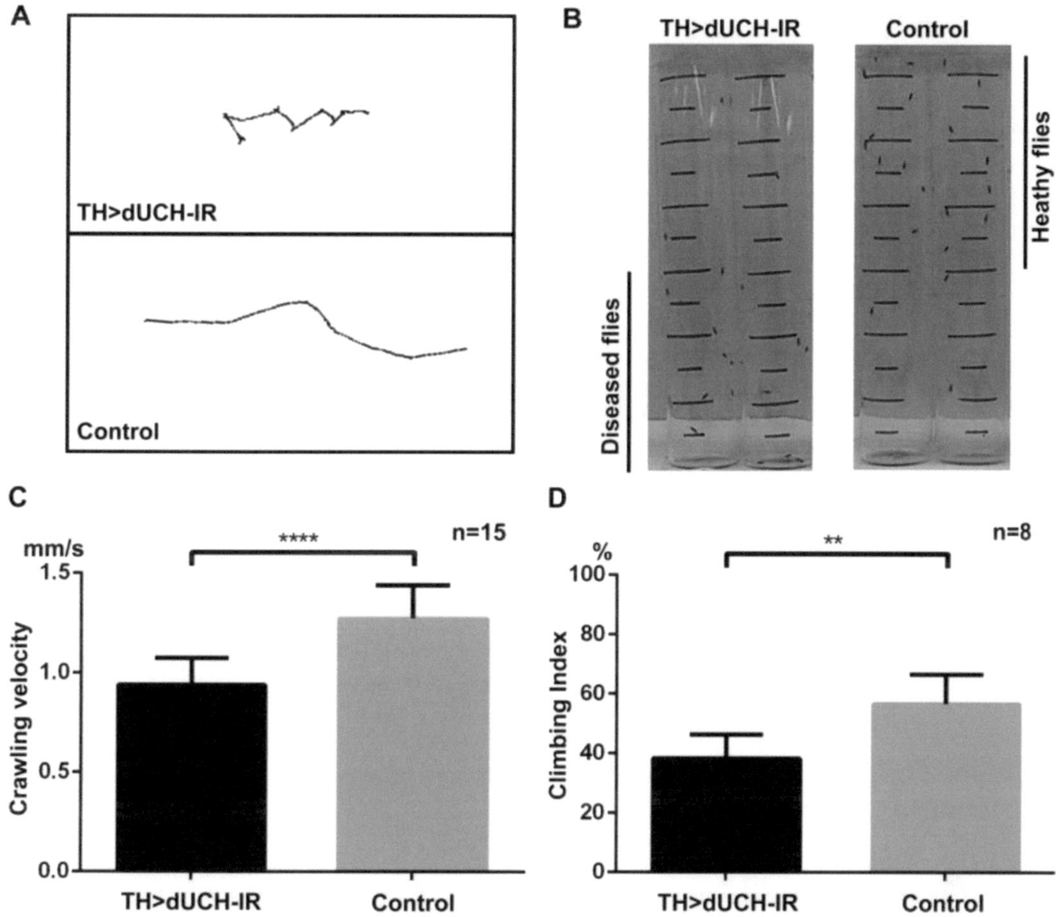


**Figure 9.** The dysfunction in locomotor in dopaminergic neuron-specific dUCH knockdown flies. (A) Motion paths of larvae: control and dUCH knockdown larvae (TH > dUCH-IR). Knockdown larvae exhibit shorter and disorder crawling paths (upper panel) compared to control (below panel). (B) Climbing assay for measurement of adult fly locomotor ability. (C) Crawling velocity of control (TH) and knockdown larvae (TH > dUCH-IR). Knockdown larvae showed the reduction in crawling pace and parametric unpaired t test with Welch's correction, ****$p < 0.0001$; error bars present SD. (D) Climbing ability of control (TH) and dUCH knockdown adult flies (TH > dUCH-IR). Knockdown flies start to exhibit the decline in climbing ability at 5 days after eclosion, repeatedly measuring two way ANOVA with Bonferroni's post hoc test, **$p < 0.01$; error bars present SEM.

apical mispatterning in ommatidium. The effects of dUCH overexpression may involve in mitogen-activated protein kinase pathway. On the other hand, knockdown dUCH resulted in defect of tissue development and function. Particularly, knockdown dUCH in dopaminergic neuron impaired fly locomotion and degenerated dopaminergic neurons. Besides the *Drosophila* model's benefits, as well as the correlation between *Drosophila* UCH (dUCH) and human UCH (UCH-L1), these data strongly demonstrated that *Drosophila melanogaster* is an advantage model to investigate the functions and regulatory mechanism of UCH-L1 in living organism.

## Acknowledgements

I am grateful to Professor Yamaguchi Masamitsu and Professor Tran Linh Thuoc for their great supports to our research. I would also like to extend my warm thanks to my students Cao Thi Thuy Trang, Huynh Man Anh and Vuu My Dung for their great contribution on the manuscript.

## Author details

Dang Thi Phuong Thao

Address all correspondence to: dtpthao@hcmus.edu.vn

University of Science, VNU-HCMC, Ho Chi Minh City, Vietnam

## References

[1] Wilkinson KD, Lee KM, Deshpande S, Duerksen-Hughes P, Boss JM, et al. The neuron-specific protein PGP 9.5 is a ubiquitin carboxyl-terminal hydrolase. Science. 1989; **246**:670-673

[2] Larsen CN, Krantz BA, Wilkinson KD. Substrate specificity of deubiquitinating enzymes: ubiquitin C-terminal hydrolases. Biochemistry. 1998;**37**(10):3358-3368

[3] Liu Y, Fallon L, Lashuel HA, Liu Z, Lansbury PT Jr. The UCH-L1 gene encodes two opposing enzymatic activities that affect alpha-synuclein degradation and Parkinson's disease susceptibility. Cell. 2002;**111**(2):209-218

[4] Hurst-Kennedy J, Chin LS, Li L. Ubiquitin C-terminal hydrolase L1 in tumorigenesis. Biochemistry Reaserch International. 2012;**2012**:123706

[5] Tan YY, Zhou HY, Wang ZQ, Chen SD. Endoplasmic reticulum stress contributes to the cell death induced by UCH-L1 inhibitor. Molecular and Cellular Biochemistry. 2008; **318**(1-2):109-115

[6] Bifsha P, Landry K, Ashmarina L, Durand S, Seyrantepe V, Trudel S, Quiniou C, Chemtob S, Xu Y, Gravel RA, Sladek R, Pshezhetsky AV. Altered gene expression in cells from patients with lysosomal storage disorders suggests impairment of the ubiquitin pathway. Cell Death and Differentiation. 2007;**14**(3):511-523

[7] Saigoh K, Wang YL, Suh JG, Yamanishi T, Sakai Y, Kiyosawa H, Harada T, Ichihara N, Wakana S, Kikuchi T, Wada K. Intragenic deletion in the gene encoding ubiquitin carboxy-terminal hydrolase in gad mice. Nature Genetics. 1999;**23**(1):47-51

[8] Leroy E et al. The ubiquitin pathway in Parkinson's disease. Nature. 1998;**395**:451-452. DOI: 10.1038/26652

[9] Lincoln S et al. Low frequency of pathogenic mutations in the ubiquitin carboxy-terminal hydrolase gene in familial Parkinson's disease. Neuroreport. 1999;**10**:427-429

[10] Yamazaki K, Wakasugi N, Tomita T, Kikuchi T, Mukoyama M, et al. Gracile axonal dystrophy (GAD), a new neurological mutant in the mouse. Proceedings of the Society for Experimental Biology and Medicine. 1988;**187**:209-215

[11] Ichihara N, Wu J, Chui DH, Yamazaki K, Wakabayashi T, et al. Axonal degeneration promotes abnormal accumulation of amyloid beta-protein in ascending gracile tract of gracile axonal dystrophy (GAD) mouse. Brain Research. 1995;**695**:173-178

[12] Osaka H, Wang YL, Takada K, Takizawa S, Setsuie R, et al. Ubiquitin carboxy-terminal hydrolase L1 binds to and stabilizes monoubiquitin in neuron. Human Molecular Genetics. 2003;**12**:1945-1958

[13] Miyoshi Y, Nakayama S, Torikoshi Y, Tanaka S, Ishihara H, Taguchi T, Tamaki Y, Noguchi S. High expression of ubiquitin carboxy-terminal hydrolase-L1 and -L3 mRNA predicts early recurrence in patients with invasive breast cancer. Cancer Science. 2006; **97**(6):523-529

[14] Hibi K, Westra WH, Borges M, Goodman S, Sidransky D, Jen J. PGP9.5 as a candidate tumor marker for non-small-cell lung cancer. The American Journal of Pathology. 1999; **155**(3):711-715

[15] Bheda A, Shackelford J, Pagano JS. Expression and functional studies of ubiquitin C-terminal hydrolase L1 regulated genes. PLoS One. 2009;**4**(8):e6764

[16] Jang MJ, Baek SH, Kim JH. UCH-L1 promotes cancer metastasis in prostate cancer cells through EMT induction. Cancer Letters. 2011;**302**(2):128-135

[17] Kim K, Lu Z, Hay ED. Direct evidence for a role of beta-catenin/LEF-1 signaling pathway in induction of EMT. Cell Biology International. 2002;**26**(5):463-476

[18] Ma Y, Zhao M, Zhong J, Shi L, Luo Q, Liu J, Wang J, Yuan X, Huang C. Proteomic profiling of proteins associated with lymph node metastasis in colorectal cancer. Journal of Cellular Biochemistry. 2010;**110**(6):1512-1519

[19] Hussain S, Foreman O, Perkins SL, Witzig TE, Miles RR, van Deursen J, Galardy PJ. The de-ubiquitinase UCH-L1 is an oncogene that drives the development of lymphoma in vivo by deregulating PHLPP1 and Akt signaling. Leukemia. Official Journal of the Leukemia Society of America, Leukemia Research Fund, U.K. 2010;**24**(9):1641-1655

[20] Tokumaru Y, Yamashita K, Kim MS, Park HL, Osada M, Mori M, Sidransky D. The role of PGP9.5 as a tumor suppressor gene in human cancer. International Journal of Cancer. Journal International du Cancer. 2008;**123**(4):753-759

[21] Takase T, Hibi K, Yamazaki T, Nakayama H, Taguchi M, Kasai Y, Ito K, Akiyama S, Nagasaka T, Nakao A. PGP9.5 overexpression in esophageal squamous cell carcinoma. Hepato-Gastroenterology. 2003;**50**(53):1278-1280

[22] Xiang T, Li L, Yin X, Yuan C, Tan C, Su X, Xiong L, Putti TC, Oberst M, Kelly K, Ren G, Tao Q. The ubiquitin peptidase UCHL1 induces G0/G1 cell cycle arrest and apoptosis through stabilizing p53 and is frequently silenced in breast cancer. PLoS One. 2012; **7**(1):e29783

[23] Ummanni R, Jost E, Braig M, Lohmann F, Mundt F, Barett C, Schlomm T, Sauter G, Senff T, Bokemeyer C, Sultmann H, Meyer-Schwesinger C, Brummendorf TH, Balabanov S. Ubiquitin carboxyl-terminal hydrolase 1 (UCHL1) is a potential tumour suppressor in prostate cancer and is frequently silenced by promoter methylation. Molecular Cancer. 2011;**10**:129

[24] Sato N, Fukushima N, Maitra A, Matsubayashi H, Yeo CJ, Cameron JL, Hruban RH, Goggins M. Discovery of novel targets for aberrant methylation in pancreatic carcinoma using high-throughput microarrays. Cancer Research. 2003;**63**(13):3735-3742

[25] Caballero OL, Resto V, Patturajan M, Meerzaman D, Guo MZ, Engles J, Yochem R, Ratovitski E, Sidransky D, Jen J. Interaction and colocalization of PGP9.5 with JAB1 and p27(Kip1). Oncogene. 2002;**21**(19):3003-3010

[26] Meray RK, Lansbury PT Jr. Reversible monoubiquitination regulates the Parkinson disease-associated ubiquitin hydrolase UCH-L1. The Journal of Biological Chemistry. 2007;**282**(14):10567-10575

[27] Larsen CN, Price JS, Wilkinson KD. Substrate binding and catalysis by ubiquitin C-terminal hydrolases: Identification of two active site residues. Biochemistry. 1996; **35**(21):6735-6744

[28] Case A, Stein RL. Mechanistic studies of ubiquitin C-terminal hydrolase L1. Biochemistry. 2006;**45**(7):2443-2452

[29] Boudreaux DA, Maiti TK, Davies CW, Das C. Ubiquitin vinyl methyl ester binding orients the misaligned active site of the ubiquitin hydrolase UCHL1 into productive conformation. Proceedings of the National Academy of Sciences of the United States of America. 2010;**107**:9117-9122. DOI: 10.1073/pnas.0910870107

[30] Tram NTQ, Trang NTT, Thao DTP, Thuoc TL. Production of polyclonal anti-dUCH (drosophila ubiquitin carboxyl-terminal hydrolase) antibodies. Monoclonal Antibodies in Immunodiagnosis and Immunotherapy. 2013;**32**(2):105-112

[31] Thao DT, An PN, Yamaguchi M, LinhThuoc T. Overexpression of ubiquitin carboxyl terminal hydrolase impairs multiple pathways during eye development in Drosophila melanogaster. Cell and Tissue Research. 2012;**348**(3):453-463

[32] Ragland M, Hutter C, Zabetian C, Edwards K. Association between the ubiquitin carboxyl-terminal esterase L1 gene (UCHL1) S18Y variant and Parkinson's Disease: A HuGE review and meta-analysis. American Journal of Epidemiology. 2009;**170**:1344-1357. DOI: 10.1093/aje/kwp288

[33] Chu KY, Li H, Wada K, Johnson JD. Ubiquitin C-terminal hydrolase L1 is required for pancreatic beta cell survival and function in lipotoxic conditions. Diabetologia. 2012; **55**:128-140

[34] Kabuta T, Mitsui T, Takahashi M, Fujiwara Y, Kabuta C, et al. Ubiquitin C-terminal hydrolase L1 (UCH-L1) acts as a novel potentiator of cyclin-dependent kinases to enhance cell proliferation independently of its hydrolase activity. The Journal of Biological Chemistry. 2013;**288**:12615-12626

Chapter 10

# *Swiss Cheese, Drosophila* Ortholog of Hereditary Spastic Paraplegia Gene *NTE*, Maintains Neuromuscular Junction Development and Microtubule Network

Elena Ryabova, Nataliya Matiytsiv, Olena Trush,
Iryna Mohylyak, Galina Kislik, Pavel Melentev and
Svetlana Sarantseva

Additional information is available at the end of the chapter

http://dx.doi.org/10.5772/intechopen.73077

**Abstract**

Neuropathy target esterase (NTE) is a molecular target for the organophosphorus compound-induced delayed neuropathy (OPIDN) and also one of the genetic factors responsible for the development of the hereditary spastic paraplegia (HSP), characterized by axon degeneration of motoneurons causing progressive lower-limb spastic paralysis. Both HSP and OPIDN are characterized by the distal axonopathy. The molecular mechanisms underlying the axonopathy involved in HSP and OPIDN are poorly understood. In order to have a better understanding of the mechanisms that NTE is involved in, we used one of the homologs, human NTE. *Swiss cheese* (*sws*) is a *Drosophila melanogaster* ortholog of *NTE* with 39% homology. Mutations in *sws* as it was shown before lead to age-dependent neurodegeneration, structure alteration of glia cells, and reduced insect life span. To study SWS functions, we used the system of the third-instar larval neuromuscular junctions of *D. melanogaster*. In this study, we show that mutations in *sws* ($sws^1$ and $sws^{76\text{–}1}$) and SWS knockdown alter neuromuscular junction's morphology and synaptic microtubules organization.

**Keywords:** axon degeneration, *Drosophila melanogaster*, neuromuscular junctions, neuropathy target esterase (NTE), *Swiss cheese* (*sws*)

## 1. Introduction

It has recently been established that mutations in the particular region of the neuropathy target esterase (*NTE*) gene coding for the catalytic domain of the NTE protein cause an autosomal-recessive form of HSP (SPG39) Gordon-Holmes syndrome, Boucher-Neuhäuser syndrome, Laurence-Moon syndrome, Oliver-McFarlane syndrome, and Leber's congenital amarosis [1–4]. Initially, NTE was found in human brain homogenates as an enzyme, the activity of which could be inhibited by organophosphates, leading to the development of organophosphorus compound-induced delayed neuropathy (OPIDN) [5]. HSP and OPIDN are both characterized by the distal degeneration of motor and sensory axons [6, 7].

NTE is a highly conserved protein with homology among many organisms (from yeast to humans), particularly in its esterase catalytic domain [8]. The *NTE* ortholog in *Drosophila melanogaster* (*D. melanogaster*, Meigen, 1830) is called *Swiss cheese* (*sws*). All the currently known *sws* mutants were obtained and described by Kretzschmar et al. [9]. These mutants developed axonal and glial pathology in the brain and neuronal apoptosis [9]. The level of phosphatidylcholine was increased in the mutants [10]. SWS and NTE share high structural (39%) and functional homology [10]. These proteins are widely expressed in the nervous system [8, 10], localized on the endoplasmic reticulum (ER), and also considered to have esterase activity [11, 10]. *NTE* knockout mice (mouse and human *NTE* genes have 95% homology) die on the ninth day of the embryonic development, unlike mice without *NTE* expression only in neurons. Mice without *NTE* neuronal expression develop a phenotype similar to *sws* mutants, including vacuolization, myelin production, and neuronal death [11]. *NTE* knockdown in zebra fish leads to development defects, axon shortening, and reduction of axonal arborization [12]. There have been many studies dedicated to NTE/sws; however, we still do not know much about their exact roles in the development and functioning of the nervous system.

In our study, we identified a new allele in *sws* using genome screening. We also carried out functional studies of *sws in vivo*, using larval neuromuscular junctions (NMJs) of *D. melanogaster* as a good system of HSP modeling. We showed that *sws* is widely expressed in the larval nervous system, especially in glial cells. We also established that mutations in the *sws* gene alter NMJ morphology, the distribution of synaptic markers, microtubule (MT) network, and synaptic microtubules organization.

## 2. The new *sws* allele

During the massive screen tests searching for X-linked mutants with age-dependent neurodegeneration, we analyzed paraffin-embedded histological sections of the brain tissue in mutant flies with life span reduction [13]. Sections from a number of mutants showed strong brain vacuolization similar to *sws* phenotype described by Kretzschmar et al. [9]. However, in the $sws^{76-15}$ line, we revealed a different phenotype with small vacuoles in all brain regions, which is not a hallmark of *sws* [13] (**Figure 1C** and **H**). Therefore, we believe that it is a new allele of *sws*, and we used deficiency mapping on 7D1–D5 band (Df (1) C128) which uncovered a *sws* phenotype [9]. To confirm localization in this region, we rescued a neurodegenerative phenotype with 7D1 band duplication (Dp(1;3)$sn^{13a1}$). We were successful in mapping the $sws^{76-15}$ mutant using deficiency. In addition, heterozygous individuals used for complementation analysis with $sws^{1}$ ($sws^{1}/sws^{76-15}$) and $sws^{4}$ ($sws^{4}/sws^{76-15}$, data not shown) showed a mutant phenotype, which corresponds to $sws^{76-15}$-like phenotype (**Figure 1E** and **J**). We suggested that this allele is neomorphic and dominant to other *sws* alleles but recessive to the wild type.

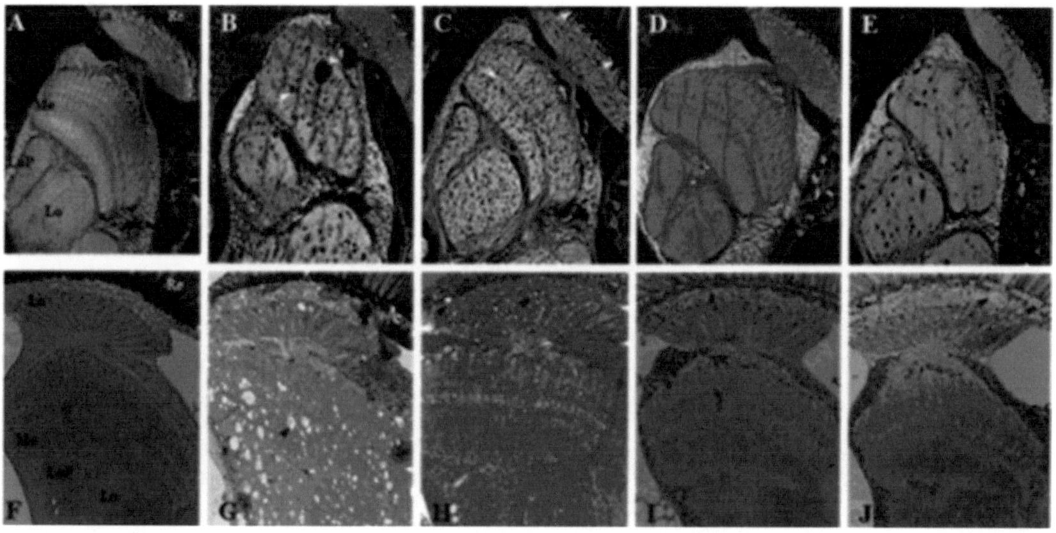

**Figure 1.** Histological sections of a brain's optic lobe in a 20-day-old *Drosophila melanogaster*. (A–E) Horizontal 7-μm paraffin-embedded sections. (F–J) Horizontal 1-μm plastic semi-thin sections. (A and F) Homozygous wild-type tissue of *Oregon* males used as controls (Bloomington Stock Centre). Neuropile and glia have very regular structure in all layers of the lobe. (B and G) Homozygous *sws*[1]-mutant males, degeneration tissue is clearly seen as vacuoles in all neuropile and highly stained glia cells hyperwrapping in the lamina cortex. (C and H) Homozygous *sws*[76–15] males with small vacuoles in neuropile. (D and I) Heterozygous *sws*[1]/*Oregon* old females used as controls in complementation test, glia and neuropile show wild-type phenotype. (E and J) Heterozygous *sws*[1]/*sws*[76–15] old females with mutant phenotypes in neuropile and lamina cortex glia, suggesting mutations in the same gene. Re, retina; La, lamina; Me, medulla; Lo, lobula; LoP, lobula plata. Paraffin-embedded sections were analyzed using a Nikon LSM A1 Clem confocal microscope; plastic-embedded semi-thin sections were analyzed using Karl Zeiss light microscope. Microscopic pictures were taken at the same level of the brain.

**Figure 2.** Conceptual sequence of SWS protein with identified location of *sws* mutations.

Previously, two transcripts of *sws* gene were described [14, 9]. The larger transcript SWS-RA of 5.4 kb is expressed at all developmental stages, prominently in young embryos and adult heads and bodies, whereas the smaller SWS-RB 1.7 kb transcript is only detectable in adult heads. Later, the third transcript SWS-RC of 5.4 kb was discovered (FlyBase).

The products are alternatively spliced transcripts from the same transcription unit: SWS-B is the first four exons of SWS-RA; SWS-RC differs from SWS-RA only in the fourth exon. Previously identified $sws^1$ and $sws^4$ mutations are specific to a larger SWS-RA [9]: in $sws^1$, a C to A nucleotide exchange at 1616 position produces a stop codon in place of a codon for serine, while in $sws^4$, nucleotide 3357 is changed from G to A, substituting asparagine for glycine. We also have examined the ORF of SWS-RA transcript in $sws^{76-15}$ line and determined a new, significant single nucleotide substitution. In $sws^{76-15}$, the G at position 4233 was replaced by a C, causing the substitution of glutamic acid to aspartic acid (**Figure 2**).

## 3. Expression of SWS protein in the larval neuromusculature

To analyze the SWS expression pattern, we used an immunohistochemical method with anti-sws and anti-horseradish peroxidase (HRP) (a widely used marker of neuronal membrane)

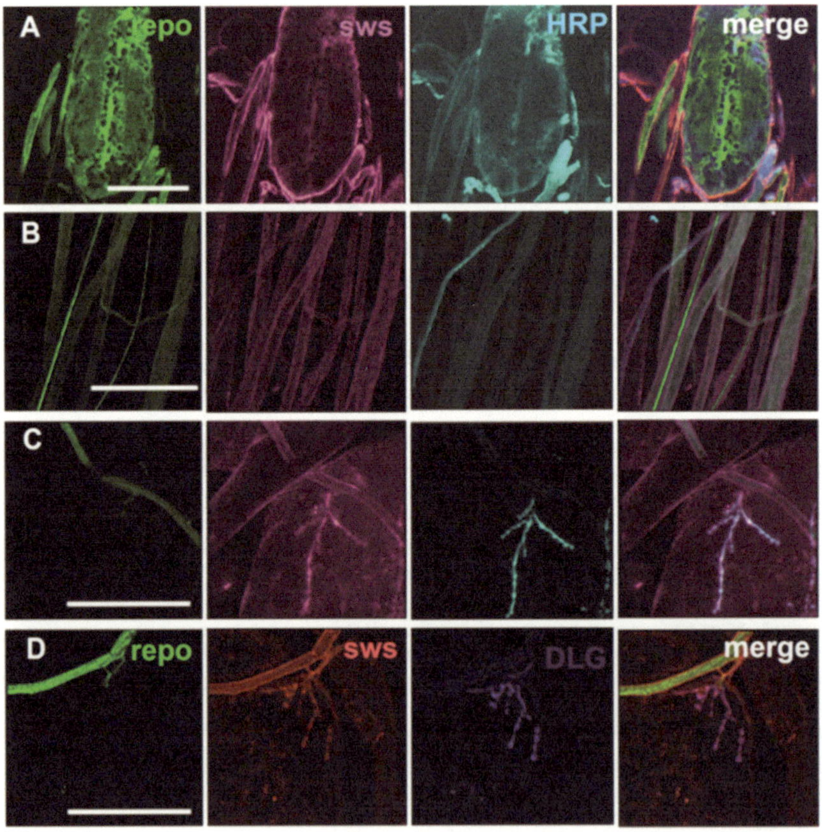


**Figure 3.** Distribution of SWS protein in the larval neuromusculature. SWS is localized at the presynaptic sides of larval NMJs. Figures (A–C) represent confocal imaging analysis of control *Canton S* larval brain (A), axons (B), NMJs on muscle 4 at abdominal segments A3–A4 (C) immunohistochemically stained with repo antibody (green), SWS antibody (purple), presynaptic marker HRP (blue), and repo, SWS, and HRP merged (green, purple, and blue). Figure D represents confocal imaging analysis of SWS in larval NMJs on muscle 4 at abdominal segments A3–A4 immunohistochemically stained with repo antibody (green), SWS antibody (red), postsynaptic marker Dlg (indigo), and repo, SWS, and Dlg merged. Scale bar: 50 μm.

antibodies [15]. Results showed that in wild-type larvae, SWS was expressed in cells of the ventral nerve cord (VNC) and localized in axons; here, a smaller amount of sws was found in NMJ in the same area as the HRP signal confirming its presynaptic localization (**Figure 3A–C**). To detect SWS localization in the postsynaptic region of NMJ synapses, we analyzed larvae expressing the postsynaptic density marker protein Discs Large (Dlg) [15]. Double labeling with an antibody against Dlg revealed that SWS immunoreactivity does not extend into the postsynaptic area (**Figure 3D**). To determine whether *sws* is expressed in glial cells, we used anti-glia marker—repo [16]. The analysis showed that *sws* is expressed in glial cells, but only those located around the brain and axons (**Figure 3A–C**). A similar pattern of *sws* expression was observed in adult *D. melanogaster* brain [9, 10].

## 4. *sws* mutants and SWS knockdown display a changed number of satellite boutons at the NMJ

The localization of SWS in synaptic boutons assumes its participation in NMJ development. In order to study this possible role of SWS, we used a transgenic line *CD8;D42* expressing green fluorescent protein (GFP) in the neuronal membrane. The NMJs (abdominal segment 3 and muscle 4) of the offspring from *CD8;D42* and *SWS-RNAi* and *sws* mutants (*sws*$^{1}$ and *sws*$^{76-15}$) were analyzed using confocal microscopy [17] (**Figure 4**). Morphology analysis was done by the estimation of synaptic bouton number and NMJ area. The analysis showed that the NMJ

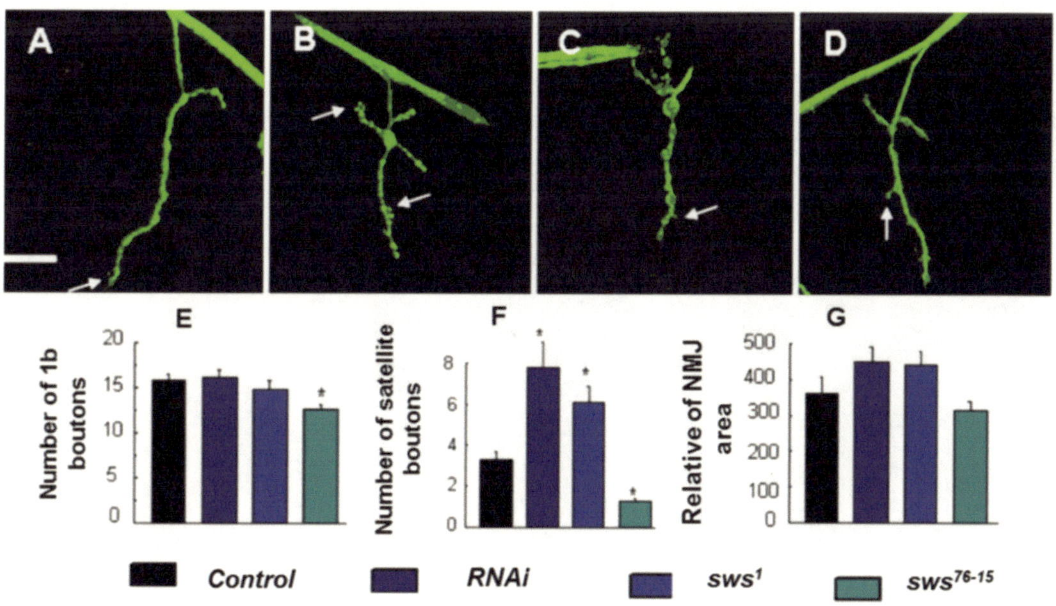


**Figure 4.** NMJ morphology is altered in *sws* mutants and SWS knockdown in motor neurons. Representative confocal images of larval NMJs (muscle 4, hemi-segments 3–4) with illuminated neuronal membrane in controls (*CD8/+;D42/+*) (A), *SWS-RNAi//CD8/+;D42/+* (B), *sws*$^{1}$*;CD8/+;D42/+* (C), *sws*$^{76-15}$*; CD8/+;D42/+* (D). Arrowheads point to satellite boutons. Bar = 25 μm. (E–G) Quantitative analysis of NMJ morphology in every genotype: total number of 1b boutons (F) and number of satellite boutons (G), relative area of NMJs (H) . *P < 0.05.

area (Control—362.5 ± 47.1; *SWS-RNAi*—448.6 ± 39.2; *sws*$^{4}$—436.7 ± 37.1; *sws*$^{76-15}$—314.4 ± 25.2, p > 0.05) and the total number of 1b boutons were not significantly different in the mutants compared to the control, except for *sws*$^{76-15}$, with a slightly decreased total number of synaptic boutons (**Figure 4F** and **G**). Interestingly, different mutations altered a number of satellite boutons (boutons budding off the parental large synaptic boutons or interboutonal space) differently (**Figure 4A–D**). We observed an increased number in satellite boutons in *SWS-RNAi* and *sws*$^{1}$, but a decreased number of them in *sws*$^{76-1}$ (**Figure 4D** and **G**).

## 5. The abundance of synaptic marker Bruchpilot and Dlg is altered in *sws* mutants and SWS knockdown

We further examined potential defects in synaptic organization of *sws* mutant NMJs using synaptic markers. We analyzed presynaptic marker Bruchpilot and the postsynaptic marker Dlg in *sws* mutants to check whether the satellite bouton change is associated with alterations in the subcellular localization of synaptic proteins. For active zone analysis, we performed immunohistochemical staining with nc82 antibodies against a *D. melanogaster*-active zone protein—Bruchpilot [18]. Active zone number in *SWS-RNAi* and *sws*$^{1}$ was not different from the control but was reduced in *sws*$^{76-15}$ (**Figure 5**), which is, in our view, with a decrease in the number of satellite boutons. Next, we determined the number of active zones per one bouton. As seen in **Figure 5**, the average number of active zone contained in one bouton was significantly lower compared to that of the control (**Figure 5E**).

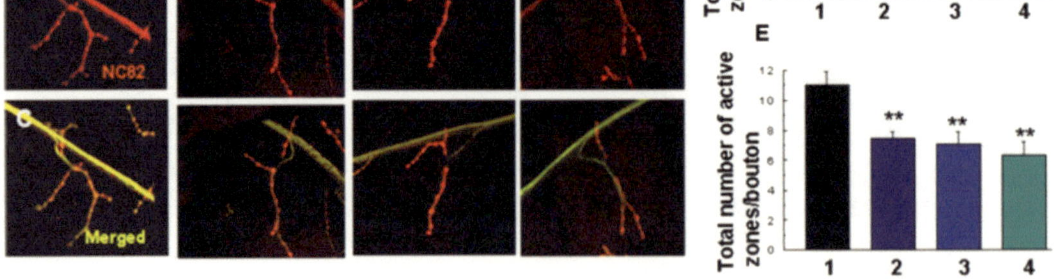


**Figure 5.** Distribution of active zones in SWS knockdown and *sws* mutants. (A–C) Representative confocal images of larval NMJs (muscle 4, hemi-segments 3–4) stained for active zones (nc82) in controls. Bar = 10 µm. (D–E) Quantification of the total number of active zones in a single NMJ and in a single bouton (F) in every genotype. 1, *Control*; 2, *RNAi*; 3, *sws*$^{1}$; 4, *sws*$^{76-15}$. $^{**}P < 0.01$, $^{*}P < 0.05$.

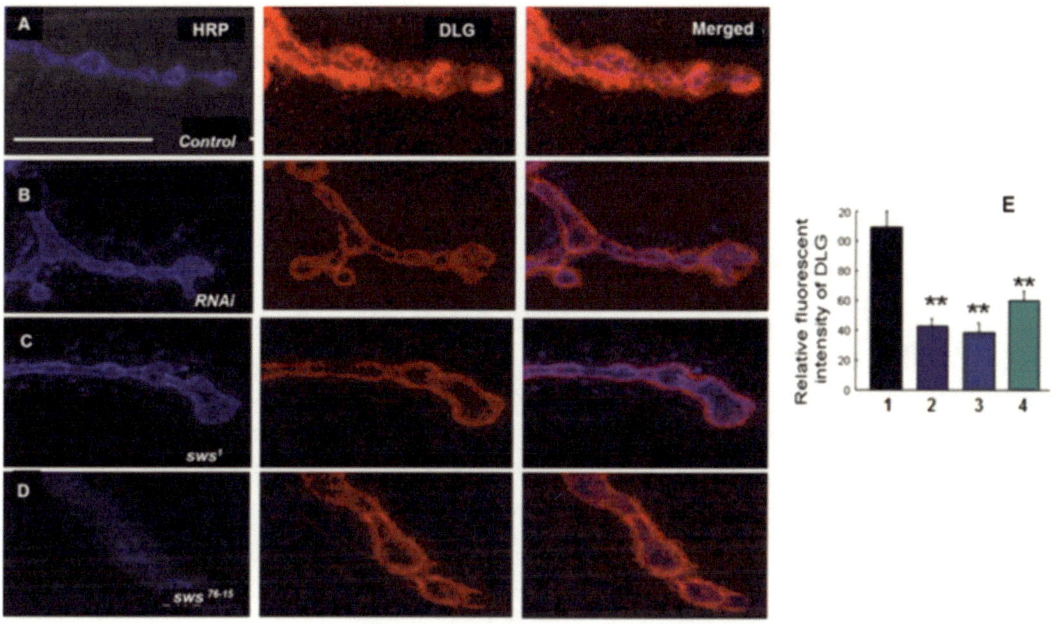


**Figure 6.** The synaptic levels of Dlg are reduced in SWS knockdown and *sws* mutants. (A–E) Representative confocal images of larval NMJs (muscle 4, hemi-segments 3–4) stained for HRP and DLG in controls (A), *RNAi* (B), $sws^1$ (C), $sws^{76-15}$ (D). Bar = 20 μm. (E) Quantification of relative fluorescent intensity of Dlg in NMJs. 1, *Control*; 2, *RNAi*; 3, $sws^1$; 4, $sws^{76-15}$. $^{**}P < 0.01$.

For postsynaptic density analysis, we used antibodies against Dlg protein, the PSD 95 homolog in mammals [19]. It has been previously shown that Dlg localizes in the subsynaptic reticulum (SSR), located around Ib NMJ boutons, and regulates its development [20]. Anti-Dlg staining in the control line was very intensive, with a halo-like pattern around Ib boutons (**Figure 6A**), whereas the Dlg intensity level in SWS knockdown and *sws* mutants was significantly reduced (**Figure 6B–E** and **F**).

## 6. *sws* mutants and SWS knockdown display aberrant microtubule organization

Many synaptic bouton NMJ overgrowth phenotypes, particularly those with excess satellite boutons as well as disrupted distribution and reduced level of synaptic proteins, often cause an alteration of microtubule (MT) organization [15, 21, 22]. MT structural and dynamical regulation is one of the most important elements of synapse formation control. MTs are also necessary for synaptic terminal stabilization in synaptic development. To analyze presynaptic MTs of NMJs, we used mAb 22C10 antibodies against MT-associated neuronal protein Futsch [23]. During normal synaptic growth of the *D. melanogaster* NMJ, Futsch is found in association with loops of bundled microtubules typically observed within stable boutons. Loops are usually present at wild-type *D. melanogaster* NMJ, though in small amounts, and especially

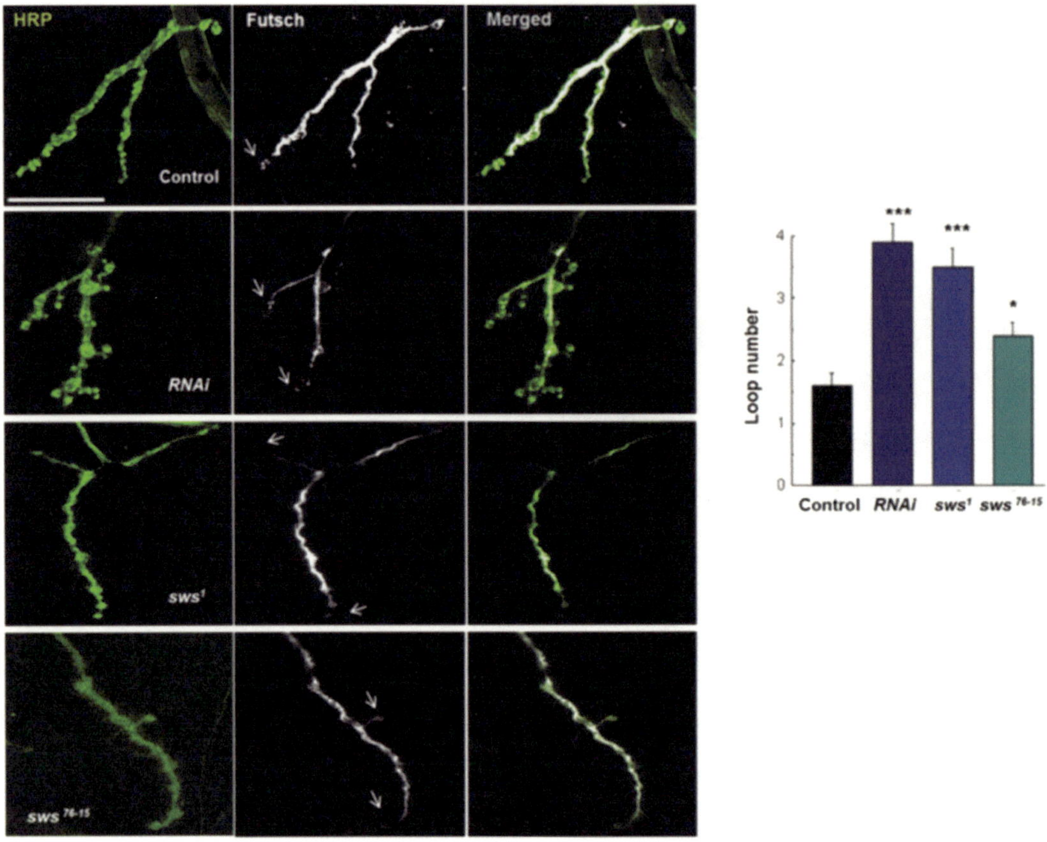


**Figure 7.** The number of presynaptic Futsch-positive loops is increased at synapses in SWS knockdown and *sws*-mutant larvae. (A–E) NMJ terminals of wild type (A), SWS-RNAi (B), $sws^1$ (C), $sws^{76-15}$ (D), co-stained with anti-HRP (green) and anti-Futsch (white). Futsch-positive loops are indicated by arrowheads. Scale bar: 25 μm. (E) Statistical comparison of the number of Futsch-positive loops at the NMJ terminals in different genotypes. The number of NMJs analyzed for each genotype is given on the bars. ***P < 0.001, *P < 0.05; error bars indicate s.e.m.

at the branch points and within terminal boutons [21]. An increase in looped MTs in the presynaptic terminal suggests MT abnormalities. Quantification of the number of Futsch-positive loops on the NMJ innervating muscle 4 showed that SWS knockdown and $sws^1$-mutant NMJs contained significantly more number of microtubule loops compared with controls (P < 0.001; **Figure 7A–C** and **F**). Instead, within the many of mutant boutons, $sws^{76-15}$ larvae exhibit only a slightly elevated number of MT loops at the NMJ (**Figure 7D** and **F**).

## 7. *sws* mutations and SWS knockdown cause disrupted mitochondrial organization in NMJs

Mitochondria play an important role in energy-dependent processes of synaptogenesis [24, 25], and, as shown earlier, mitochondrial transport depends on the MT maintenance [25]. In order to analyze mitochondria at NMJ synapses, we analyzed the distribution of mitochondria labeled by

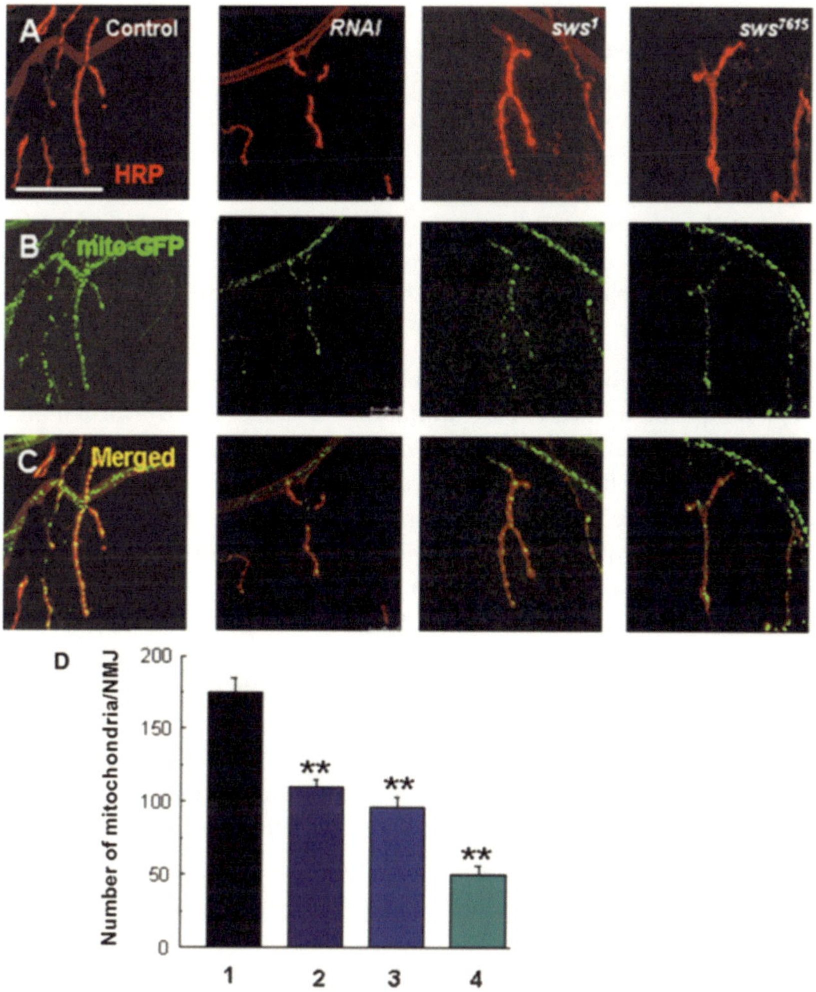


**Figure 8.** NMJ distribution of mitochondria in *sws* mutants. (A–C) Representative confocal images of larval NMJs (muscle 4, hemi-segments 3–4) visualized with mitochondria expressing GFP in controls (*mito-GFP/+;D42/+*). Bar = 10 μm. (D) Quantification of mitochondria in NMJs in every genotype. 1, *Control*; 2, *RNAi*; 3, *sws*$^1$; 4, *sws*$^{76-15}$. $^{**}$P < 0.01.

a green fluorescent protein (GFP) tag (mito-GFP) in control strain and *sws* mutants. The mito-GFP tag was constructed by fusing the N-terminal 31 amino acid (mitochondria-targeting) sequence from human cytochrome C oxidase subunit VIII with N-terminus of GFP [26]. Mitochondria expressing GFP are very easy to observe in NMJs of control motor neurons (**Figure 8A**). However, the number of mitochondria was significantly reduced in mutant lines compared to that in control (**Figure 8A–D** and **E**).

## 8. Summary and concluding remarks

In this chapter, we investigated the role of *sws* in the development and functioning of the nervous system using an excellent well-developed model of *D. melanogaster* larval NMJs. We showed that *sws* is widely expressed in the larval nervous system and localized in the brain, axons, and NMJs. However, SWS is predominantly found in glial cells of the brain and axons. These data correlate with the sws distribution in the brain of adult flies [10, 27]. We also described a new mutation in the *sws* gene ($sws^{76-15}$) and compared its phenotypic manifestations in the nervous system of the larvae to the previously described $sws^{1}$ mutation and SWS knockdown. We show the presynaptic depletion of SWS in the motor neurons using GAL4-inducible fly line *SWS-RNAi*, and null $sws^{1}$ mutations result in a significant expansion of synaptic bouton number. Interestingly, the number of 1b and satellite boutons in the $sws^{76-15}$ mutant was decreased in the mutant as compared with that in the control. These results indicate that SWS controls synaptic bouton formation and differentiation of the NMJ during the third-instar larval stage.

Satellite bouton production has been observed for certain *D. melanogaster* proteins Spastin [28] and Atlastin [29] implicated in HSP. How mutations in *sws* lead to the formation of excess or deficiency satellite boutons remains to be elucidated. However, as previously shown, mutants characterized by extra satellite boutons can display changing the number of microtubule loops [15]. We found that SWS knockdown $sws^{1}$ and $sws^{76-15}$ mutations resulted in more stable MT loops in the NMJ. One of our discoveries is that *sws* mutations cause mitochondrial dysfunction in NMJs, which is a common effect in other HSP forms [30–32]. Mitochondrial transport to synapse is tightly regulated to provide sufficient energy for synaptic transmission [33, 34]. Decreased synaptic transmission has been reported to be associated with a reduced number of functional mitochondria [35, 36]. Indeed, glial SWS knockdown induces defects in neuronal transmission; however, levels of Bruchpilot, an active-site marker, were unchanged [26]. Here, we show that although there was an increase in bouton number at NMJs of *sws* mutants, the boutons did not show a reduction in the abundance in the active zone synaptic function marker Bruchpilot, while their number was significantly lower in the single bouton, suggesting a decrease in their neurotransmitter function.

Six alleles of the *D. melanogaster sws* gene were described, and two of these, $sws^{1}$ and $sws^{5}$, mutations and SWS knockdown were characterized at the biochemical level [9, 37, 26]. $sws^{1}$ is the most completely studied *sws* mutation and results in a truncated protein of about a fourth of the original length [9]. Previous reports on *sws* mutants have described abnormal morphology only in the brain and eye of the adult fly, but have correlated these changes with the loss of SWS catalytic activity [9, 10]. Novel point mutant, $sws^{76-15}$, is located in the region of SWS, which has no homology with NTE. On this basis, we believe that the mutant has not broken esterase function. In our study, SWS knockdown $sws^{1}$ and $sws^{76-15}$ acted practically identically, which shows, for our point of view, that we revealed the functions of SWS, which are not connected with esterase function. To confirm this, it is necessary to investigate the esterase activity of $sws^{76-15}$ mutant.

When discussing about SWS functions in the nervous system, we should remember the fact that the most prominent expression *sws* was found in glial cells on the surface of the brain and axons. Indeed, a recent study showed that the loss of SWS in glia impairs neuronal function, strongly suggesting that the loss of glial SWS plays an important role in the phenotypes observed in the *sws* mutant [26]. We did not observe significant and gross changes in glia morphology, presumably because of the short developmental larval stage.

In summary, our research showed the role of *sws* in the regulation of NMJ functioning. Further studies about processes regulated by *sws* will help us to better understand the molecular mechanisms underlying the pathogenesis of diseases.

## 9. Materials and methods

### 9.1. *D. melanogaster* lines

Ethylmethane sulfonate (EMS)-induced mutant alleles of *sws* were used as a research material and wild-type *Oregon R* line as a control. All mutants were isolated in screens for structural brain defects using the histology brain assay [38]: $sws^{1}$ is previously described by Kretzschmar et al. [9], $sws^{76\text{–}15}$ (also referred as 76-15 line) isolated by Shcherbata et al. [13]. Another strain was obtained from the Bloomington Drosophila Stock Centre (Indiana University, USA). All strains were kept on standard a medium at 25°C.

### 9.2. Tissue sections

#### *9.2.1. Paraffin sections*

The mass histology procedure by Heisenberg and Bohl [38] was used for general neurodegenerative phenotype analysis in adult flies. Flies were placed into collars and fixed in Carnoy solution (ethanol-chloroform-acetic acid, 6:3:1) at 4°C, which was followed by their dehydration in ethanol (30 min), methyl benzoate (30 min), and twice in paraffin (60 min). Paraffin blocks were used to prepare 7-μm thick sections. Paraffin slices were washed with xylene and covered with DPX ("Fluka," USA). The preparations were examined on a Carl Zeiss Jena microscope at 12 × 40 magnifications in UV light for eye auto-fluorescence (no staining). In the experiment, we tested at least 25 flies (20–22 days old) of each genotype.

#### *9.2.2. Semi-thin plastic sections*

Fly heads were dissected on ice and fixed in 2% glutaraldehyde and 2% osmium solutions for 6 h. After removing osmium, the heads were dehydrated in ethanol solutions of increasing concentrations (30, 50, 70, 90, and 100%) and subsequently incubated two times for 2 min in propylene oxide solution at room temperature. Afterwards, the propylene oxide was replaced with a propylene oxide-resin mixture and left to incubate overnight. The mixture was replaced with pure resin, and the heads were incubated for three more hours. Molds for block preparation were filled with resin, and the heads were separately placed in the molds with proper orientation. Molds with oriented heads were left overnight at 70°C for the resin in the blocks to solidify. Then, blocks were cracked from molds and placed into signed tubes, with the subsequent preparation of 1-μm thick sections using a semiautomatic Historange Microtome using a diamond knife. Sections were stained with the Toluidine Blue solution, washed with distilled water, and covered with glass using DPX-mountant for histology [38].

### 9.3. Genetic analysis of *sws*$^{76\text{–}15}$

#### *9.3.1. Mapping*

Complementation analysis includes obtaining of trans-heterozygotes to control recessive mutations of the same trait. Mapping was performed by crossing mutants with a deficiency line (Df (1) C128/FM6) and a line with duplication (Df(1)ct-J4, In(1)dl-49, f(1)/C(1)DX, y(1) w(1) f(1); Dp(1;3)sn(13a1)/+) in the same band of 7D1 X-chromosome. Both lines were kindly provided by the Bloomington Stock Centre.

#### *9.3.2. Molecular identification*

Total RNA was extracted from 22-day-old fly heads using TRIZOL LS ("Life Technologies," Switzerland) according to a standard technique [37]. The extracted RNA was dissolved in $MQH_2O$ with 0.1% DEPC (diethyl pyrocarbonate) and stored at −80°C. Using the Primer Select software, we developed eight pairs of primers to the ORF SWS-RA transcript (4274 bp) sequence. Expected fragments were from 620 to 829 bp and overlapped with each other. The cDNA was synthesized using RNaseOUT Recombinant Ribonuclease Inhibitor and Super Script II RNase $H^-$Reverse Transcriptase ("Invitrogen," USA). cDNA was used as a template in polymerase chain reaction (PCR) with Expand High-Fidelity PCR System ("Roche Diagnostics," Germany). After polymerization, 0.5 ml Taq-polymerase ("Roche Diagnostics," Germany) and 0.5 ml dATF were added to the mixture and incubated for 30 min at 72°C to form the polyadenine tail for subsequent T/A cloning in pGEM-T-Easy Vector. The size of the fragments was estimated by their electrophoretic movement in 1% agarose gel compared to 1 kb Plus DNA Ladder. DNA fragments were eluted from gel using GFP PCR DNA and Gel Band Purification Kit ("Amersham Biosciences," GB). DNA sequencing was performed by Macrogen Inc. (Korea); each fragment from PCR of different clones was tested three times in both directions. Using the DNA Star software, we performed comparative analysis between the cDNA fragments of *sws* gene nucleotide sequence from the Gene Bank database (NCBI) in wild-type strain *Oregon-R* and *sws*$^{76\text{–}15}$ mutant.

### 9.4. Sample preparation and assay of neuromuscular junction morphology

The third-stage larvae were dissected in freshly prepared HL3 solution (110 mM NaCl, 5 mM KCl, 10 mM $NaHCO_3$, 5 mM HEPES, 30 mM sucrose, 5 mM trehalose, 10 mM $MgCl_2$, pH 7.2) (Budnik and Ruiz-Canada, 2006). Then, they were fixed with 4% formaldehyde (Sigma-Aldrich, United States) for 15 min, washed with phosphate buffer saline (PBS) and mounted in VectaShield mounting medium (Vector Laboratories, USA). Samples were analyzed with Leica TCS-SP5 laser confocal microscope (Leica, Germany). Bouton number, axon branch number, and neuromuscular junction length were estimated with the ImageJ software (http://imagej.nih.gov/ij/) and LAS AF Lite software (Leica, Germany). Then, 6–8 larvae of each genotype were analyzed. Each experiment was done in triplicate. For comparison between genotypes, all samples were processed simultaneously and imaged using identical microscopic acquisition parameters. All images were also corrected for any background before any intensity measurements.

### 9.5. Mitochondria assay

Larvae were dissected in HL3, fixed with 4% paraformaldehyde for 15 min, washed with PBS (3 × 15 min), and mounted in VectaShield mounting medium (Vector Laboratories, USA). Samples were analyzed using a Leica TCS-SP5 laser confocal microscope (Leica, Germany) at 488 nm. Relative fluorescence was estimated using the ImageJ software. In total, 6–8 larvae of each genotype were analyzed. Each experiment was performed in triplicate.

### 9.6. Immunohistochemistry

Third-instar larvae were dissected in PBS, fixed in 4% paraformaldehyde for 20 min, and washed with PBS (3 × 15 min). Then, larvae were blocked in blocking buffer BlockPRO (Visual Protein Biotechnology Corporation, USA) for 1 h at room temperature, followed by overnight incubation at 4°C in primary antibodies (diluted in BlockPRO) and washing in PBS (3 × 15 min). Afterward, larvae were incubated in secondary antibodies (diluted in BlockPRO) for 2 h, followed by washing in PBS (3 × 15 min) and mounting in VectaShield medium (Vector Laboratories, USA). We used the following primary antibodies: rabbit anti-sws (1:100; Almabion, Russia), mouse anti-Brp (Bruchpilot) (1:200; mAb NC82; Developmental Studies Hybridoma Bank (DSHB), USA), mouse anti-Dlg ((1:200; mAb 4F3; Developmental Studies Hybridoma Bank (DSHB), USA), mouse anti-Futsch ((1:200; Developmental Studies Hybridoma Bank (DSHB), USA), and secondary antibodies: goat anti-mouse Cy3-conjugated (1:400, Jackson ImmunoResearch, USA). Antibodies were obtained from the Developmental Studies Hybridoma Bank developed under the auspices of the NICHD and maintained by The University of Iowa, Department of Biology, Iowa City, IA.

### 9.7. Statistics

Statistical analyses were performed using the KyPlot software (KyensLab Inc.). A one-way analysis of variance (ANOVA) was followed by planned multiple comparisons between relevant groups with Tukey-Kramer test.

## Acknowledgements

We thank Dr. Doris Kretzschmar for providing anti-sws antibody vector. We also thank Prof. Christos Samakovlis and Prof. Karl-Friedrich Fischbach for insight, discussion, advice, and support.

## Conflict of interest statement

The authors have declared that no competing interests exist.

## Funding

This work was supported by the Russian Fund for Basic Research (15-04-09041 to S.S.), INTAS (04-83-3248 to N.M.), and DAAD fellowship for young scientists to I.M.

## Author details

Elena Ryabova[1], Nataliya Matiytsiv[2], Olena Trush[2], Iryna Mohylyak[2], Galina Kislik[1], Pavel Melentev[1] and Svetlana Sarantseva[1*]

*Address all correspondence to: svesar1@yandex.ru

1 B.P. Konstantinov Petersburg Nuclear Physics Institute, National Research Centre "Kurchatov Institute", Gatchina, Russia

2 Department of Genetics and Biotechnology, Ivan Franko National University of Lviv, Lviv, Ukraine

## References

[1] Rainier S, Bui M, Mark E, Thomas D, Tokarz D, Ming L, Delaney C, Richardson RJ, Albers JW, Matsunami N, Stevens J, Coon H, Leppert M, Fink JK. Neuropathy target esterase gene mutations cause motor neuron disease. American Journal of Human Genetics. 2008;**82**:780-785. DOI: 10.1016/j.ajhg.2007.12.018

[2] Synofzik M, Gonzalez MA, Lourenco CM, Coutelier M, Haack TB, Rebelo A, Hannequin D, Strom TM, Prokisch H, Kernstock C, Durr A, Schöls L, Lima-Martínez MM, Farooq A, Schüle R, Stevanin G, Marques W Jr, Züchner S. PNPLA6 mutations cause Boucher-Neuhäuser and Gordon Holmes syndromes as part of a broad neurodegenerative spectrum. Brain. 2014;**137**(Pt 1):69-77. DOI: 10.1093/brain/awt326

[3] Hufnagel RB, Arno G, Hein ND, Hersheson J, Prasad M, Anderson Y, Krueger LA, Gregory LC, Stoetzel C, Jaworek TJ, Hull S, Li A, Plagnol V, Willen CM, Morgan TM, Prows CA, Hegde RS, Riazuddin S, Grabowski GA, Richardson RJ, Dieterich K, Huang T, Revesz T, Martinez-Barbera JP, Sisk RA, Jefferies C, Houlden H, Dattani MT, Fink JK, Dollfus H, Moore AT, Ahmed ZM. Neuropathy target esterase impairments cause Oliver-McFarlane and Laurence-Moon syndromes. Journal of Medical Genetics. 2015;**52**:85-94. DOI: 10.1136/jmedgenet-2014-102856

[4] Kmoch S, Majewski J, Ramamurthy V, Cao S, Fahiminiya S, Ren H, MacDonald IM, Lopez I, Sun V, Keser V, Khan A, Stránecký V, Hartmannová H, Přistoupilová A, Hodaňová K, Piherová L, Kuchař L, Baxová A, Chen R, Barsottini OG, Pyle A, Griffin H, Splitt M, Sallum J, Tolmie JL, Sampson JR, Chinnery P, Care4Rare Canada, Banin E, Sharon D, Dutta S, Grebler R, Helfrich-Foerster C, Pedroso JL, Kretzschmar D, Cayouette M, Koenekoop RK. Mutations in PNPLA6 are linked to photoreceptor degeneration and various forms of childhood blindness. Nature Communications. 2015;**6**:5614. DOI: 10. 1038/ncomms6614

[5] Johnson MK. The delayed neurotoxic action of some organophosphorus compounds. Identification of the phosphorylation site as an esterase. The Biochemical Journal. 1969; **114**:11-17

[6] Glynn P. A mechanism for organophosphate-induced delayed neuropathy. Toxicology Letters. 2006;**162**:94-97. DOI: 10.1016/j.toxlet.2005.10.012

[7] Fink JK. Hereditary spastic paraplegia. Current Neurology and Neuroscience Reports. 2006;**6**:65-76. DOI: 10.1055/s-0034-1386767

[8] Moser M, Stempfl T, Li Y, Glynn P, Büttner R, Kretzschmar D. Cloning and expression of the murine sws/NTE gene. Mechanisms of Development. 2000;**90**:279-282

[9] Kretzschmar D, Hasan G, Sharma S, Heisenberg M, Benzer S. The *Swiss cheese* mutant cause glial hyperwrapping and brain degeneration in *Drosophila*. The Journal of Neuroscience. 1997;**17**:7425-7432

[10] Mühlig-Versen M, da Cruz AB, Tschäpe JA, Moser M, Büttner R, Athenstaedt K, Glynn P, Kretzschmar D. Loss of Swiss cheese/neuropathy target esterase activity causes disruption of phosphatidylcholine homeostasis and neuronal and glial death in adult Drosophila. The Journal of Neuroscience. 2005;**25**:2865-2873. DOI: 10.1523/JNEUROSCI.5097-04.2005

[11] Akassoglou K, Malester B, Xu J, Tessarollo L, Rosenbluth J, Chao MV. Brain-specific deletion of neuropathy target esterase/Swiss cheese results in neurodegeneration. Proceedings of the National Academy of Sciences of the United States of America. 2004;**101**:5075-5080. DOI: 10.1073/pnas.0401030101

[12] Song Y, Wang M, Mao F, Shao M, Zhao B, Song Z, et al. Knockdown of Pnpla6 protein results in motor neuron defects in zebra fish. Disease Models & Mechanisms. 2013;**6**: 404-413. DOI: 10.1242/dmm.009688

[13] Shcherbata GR, Matiĭtsiv NP, Chernik IAI, Iatsenko AS, Radysh VV, Kucherenko MM, Maksimiv DV. Genetic analysis of X-chromosome *Drosophila melanogaster* neurodegenerative mutants induced by ethylmethanesulphonate and nitrosoethylurea. Genetika. 2004;**40**:1286-1292

[14] Hasan G. Molecular cloning of an olfactory gene from *Drosophila melanogaster*. Proceedings of the National Academy of Sciences of the United States of America. 1990;**87**:9037-9041

[15] Franco B, Bogdanik L, Bobinnec Y, Debec A, Bockaert J, Parmentier ML, Grau Y. Shaggy, the homolog of glycogen synthase kinase 3, controls neuromuscular junction growth in Drosophila. The Journal of Neuroscience. 2004;**24**:6573-6577. DOI: 10.1523/JNEUROSCI.1580-04.2004

[16] Xiong WC, Okano H, Patel NH, Blendy JA, Montell C. Repo encodes a glial-specific homeo domain protein required in the Drosophila nervous system. Genes & Development. 1994; **8**:981-994

[17] Budnik V, Ruiz-Canada C. The Fly Neuromuscular Junction: Structure and Function. Int. Rev. Neurobiol. 2nd ed. San Diego: Academic Press; 2006

[18] Wagh DA, Rasse TM, Asan E, Hofbauer A, Schwenkert I, Dürrbeck H, Buchner S, Dabauvalle MC, Schmidt M, Qin G, Wichmann C, Kittel R, Sigrist SJ, Buchner E. Bruchpilot, a protein with homology to ELKS/CAST, is required for structural integrity and function of synaptic active zones in *Drosophila*. Neuron. 2006;**49**:833-844. DOI: 10.1016/j.neuron.2006.02.008

[19] Cohen RS, Blomberg F, Berzins K, Siekevitz P. The structure of postsynaptic densities isolated from dog cerebral cortex. I. Overall morphology and protein composition. The Journal of Cell Biology. 1997;**74**:181-203

[20] Budnik V, Koh YH, Guan B, Hartmann B, Hough C, Woods D, Gorczyca M, Budnik V, Koh YH, Guan B, Hartmann B, Hough C, Woods D, et al. Regulation of synapse structure and function by the *Drosophila* tumor suppressor gene dlg. Neuron. 1996;**17**:627-640

[21] Roos J, Hummel T, Ng N, Klämbt C, Davis GW. *Drosophila* Futsch regulates synaptic microtubule organization and is necessary for synaptic growth. Neuron. 2000;**26**:371-382

[22] Viquez NM, Li CR, Wairkar YP, DiAntonio A. The B' protein phosphatase 2A regulatory subunit well-rounded regulates synaptic growth and cytoskeletal stability at the *Drosophila* neuromuscular junction. The Journal of Neuroscience. 2006;**26**:9293-9303. DOI: 10.1523/JNEUROSCI.1740-06.2006

[23] Hummel T, Krukkert K, Roos J, Davis G, Klambt C. *Drosophila* Futsch/22C10 is a MAP1B-like protein required for dendritic and axonal development. Neuron. 2000;**26**:357-370

[24] Zenisek D, Matthews G. The role of mitochondria in presynaptic calcium handling at a ribbon synapse. Neuron. 2000;**25**:229-237

[25] Saxton WM, Hollenbeck PJ. The axonal transport of mitochondria. Journal of Cell Science. 2012;**125**(Pt 9):2095-2104. DOI: 10.1242/jcs.053850

[26] Pilling AD, Horiuchi D, Lively CM, Saxton WM. Kinesin-1 and dynein are the primary motors for fast transport of mitochondria in *Drosophila* motor axons. Molecular Biology of the Cell. 2006;**17**:2057-2068. DOI: 10.1091/mbc.E05-06-0526

[27] Dutta S, Rieche F, Eckl N, Duch C, Kretzschmar D. Glial expression of Swiss cheese (SWS), the *Drosophila* orthologue of neuropathy target esterase (NTE), is required for neuronal ensheathment and function. Disease Models & Mechanisms. 2016;**9**:283-294. DOI: 10.1242/dmm.022236

[28] Sherwood NT, Sun Q, Xue M, Zhang B, Zinn K. *Drosophila* spastin regulates synaptic microtubule networks and is required for normal motor function. PLoS Biology. 2004;**2**:e429. DOI: 10.1371/journal.pbio.0020429

[29] Lee M, Paik SK, Lee MJ, Kim YJ, Kim S, Nahm M, SJ O, Kim HM, Yim J, Lee CJ, Bae YC, Lee S. *Drosophila* Atlastin regulates the stability of muscle microtubules and is required for synapse development. Developmental Biology. 2009;**330**:250-262. DOI: 10.1016/j.ydbio.2009.03.019

[30] Lu J, Rashid F, Byrne PC. The hereditary spastic paraplegia protein spartin localises to mitochondria. Journal of Neurochemistry. 2006;**98**:1908-1919. DOI: 10.1111/j.1471-4159.2006.04008.x

[31] Goizet C, Depienne C, Benard G, Boukhris A, Mundwiller E, Solé G, Coupry I, Pilliod J, Martin-Négrier ML, Fedirko E, Forlani S, Cazeneuve C, Hannequin D, Charles P, Feki I, Pinel JF, Ouvrard-Hernandez AM, Lyonnet S, Ollagnon-Roman E, Yaouanq J, Toutain A, Dussert C, Fontaine B, Leguern E, Lacombe D, Durr A, Rossignol R, Brice A, Stevanin G. REEP1 mutations in SPG31: Frequency, mutational spectrum, and potential association with mitochondrial morpho-functional dysfunction. Human Mutation. 2011;**32**:1118-1127. DOI: 10.1002/humu.21542

[32] Verny C, Guegen N, Desquiret V, Chevrollier A, Prundean A, Dubas F, Cassereau J, Ferre M, Amati-Bonneau P, Bonneau D, Reynier P, Procaccio V. Hereditary spastic paraplegia-like disorder due to a mitochondrial ATP6 gene point mutation. Mitochondrion. 2011;**11**:70-75. DOI: 10.1016/j.mito.2010.07.006

[33] Ly CV, Verstreken P. Mitochondria at the synapse. The Neuroscientist. 2006;**12**:291-299. DOI: 10.1177/1073858406287661

[34] Wang X, Schwarz TL. Imaging axonal transport of mitochondria. Methods in Enzymology. 2009;**457**:319-333. DOI: 10.1016/S0076-6879(09)05018-6

[35] Goldstein AY, Wang X, Schwarz TL. Axonal transport and the delivery of pre-synaptic components. Current Opinion in Neurobiology. 2008;**18**:495-503. DOI: 10.1016/j.conb.2008.10.003

[36] Verstreken P, Ly CV, Venken KJ, Koh TW, Zhou Y, Bellen HJ. Synaptic mitochondria are critical for mobilization of reserve pool vesicles at drosophila neuromuscular junctions. Neuron. 2005;**47**:365-378. DOI: 10.1016/j.neuron.2005.06.018

[37] Sujkowski A, Rainier S, Fink JK, Wessells RJ. Delayed induction of human NTE (PNPLA6) rescues neurodegeneration and mobility defects of *Drosophila swiss cheese* (*sws*) mutants. PLoS One. 2015;**10**:e0145356. DOI: 10.1371/journal.pone.0145356

[38] Heisenberg M, Bohl K. Isolation of anatomical brain mutants of *Drosophila* by histological means. Naturforsch. 1979;**34**:143-147

Chapter 11

# Substrate Specificities and Kinetic Parameters of Recombinant *Drosophila melanogaster* Glutathione *S*-Transferases E6 and E7

Vennobaashini Venu and Zazali Alias

Additional information is available at the end of the chapter

http://dx.doi.org/10.5772/intechopen.72970

**Abstract**

*D. melanogaster* glutathione transferases E6 and E7 (DmGSTE6 and DmGSTE7) were successfully cloned, purified, and biochemically characterized. The recombinant proteins were readily purified using the combination of both anionic and BSP/GSH-agarose affinity chromatography. Although both GSTs have significant identity in their amino acid sequence, each enzyme displayed unique biochemical characteristics. Both recombinant proteins were only active toward 1-chloro-2,4-dinitrobenzene (CDNB), 1,2-dichloro-4-nitrobenzene (DCNB), and *p*-nitrobenzyl chloride (*p*-NBC) with significant difference in catalytic activities. The findings have shown that neither GSTE6 nor GSTE7 was able to counter oxidative stress. Comparatively, GSTE7 was a more efficient enzyme at turning over DCNB and *p*-NBC, based on its kcat/Km values which were of 0.183 and 2.25 $min^{-1}$ $mM^{-1}$, respectively. Thin-layer chromatography analysis showed that both isoforms were not able to conjugate several tested insecticides. The inhibition kinetics of natural products and dyes toward GSTs in vitro revealed that phenol red possessed inhibition effects only on GSTE6 while rose bengal and cardiogreen inhibit significantly on both GSTE6 and GSTE7. In contrast, methylene blue dye and *trans*-chalcone have been shown to stimulate GSTE7 activity toward CDNB.

**Keywords:** detoxification, *D. melanogaster*, glutathione transferases, insecticide resistance, kinetic parameters

## 1. Introduction

One of the most popular classes of detoxification enzymes that constitute in almost all living organisms is the glutathione transferases (GSTs). GSTs conjugate the thiol groups of reduced glutathione (GSH) toward the negative charge center of lipid soluble compounds (xenobiotics)

to make it water soluble and easy to excrete out. These enzymes have wide distribution in nature and are found ubiquitously in almost all living things including plants, animals, and even bacteria. GSTs are expressed in sex, age, tissue, organ, species, and tumor-specific patterns of expression, and their composition differs significantly [1]. Insect GSTs can be categorized into six classes including Delta, Epsilon, Theta, Omega, Sigma, and Zeta, but it is the Delta and Epsilon classes that are most commonly associated with resistance [2, 3]. Several studies also reported that Epsilon classes in Dipteran organisms are to confer insecticide resistance, and their catalytic diversity would likely promote their role in detoxification [4–7]. Another study [8] also suggested that the expression of the epsilon class GSTs, SlGSTE2, and SlGSTE3 genes in *Spodoptera litura,* a Lepidoptera detoxifies carbaryl, DDT, RH5992, malathion, and deltamethrin, which is a synthetic chemical insecticide.

A work in housefly isozymes suggested that MdGST6A and MdGST6B which belong to the epsilon class function as key enzymes in the detoxification of insecticides such as methyl parathion and lindane [9]. It was demonstrated that the expression of GSTE6 and GSTE7 in *D. melanogaster* significantly increased by more than 50% upon exposure to paraquat (1,1-dimethyl-4,4`-bipyridylium) and phenobarbital (PhB) [10]. Besides that, acute insecticide exposure of methyl parathion results in significant increase in expression of GSTE6 (100%) and GSTE7 (72%) [11]. This suggested their immediate involvement in insecticide metabolism.

Therefore, the current study wanted to investigate the behavior of two epsilon class *D. melanogaster* GSTs, namely, GSTE6 and GSTE7, against varieties of xenobiotics and therefore characterizes their biochemical contribution. This may give insight the potential role the GSTs could have played during xenobiotic metabolism. This could help anticipate the behavior of insecticides related GSTs in other dipterans of economic importance.

## 2. Materials and methods

### 2.1. Chemicals and source of insect

Unless otherwise stipulated, chemicals employed were of the highest grade obtainable. Buffer components, pesticides, dyes, and GST substrates used were purchased from Sigma-Aldrich. *D. melanogaster* obtained from the Genetic Department of the University of Malaya was reared on oats and glucose-based diet at room temperature. Only 5 days post emerged, flies were used for the experiments.

### 2.2. DNA and protein analysis

Nucleotide and deduced amino acid sequences of *DmGSTE6* and *DmGSTE7* obtained from sequencing were compared to existing sequences in Gene Bank by BLAST searching (http://www.ncbi.nlm.nih.gov). The software used for sequence analysis, matrix table of percentage amino acid identities sequence and alignment with CLUSTAL W, was BioEdit software version 7.2.0.

### 2.3. Cloning, expression, and purification of recombinant proteins

Total DNA was isolated from adult tissues using DNeasy Blood & Tissue Kit (Qiagen) according to the manufacturer's instruction. The primers for amplifying GSTE6 were forward primer containing Nde1 site: 5′-GGAATTC CATATG gtgaaattgactttatac-3′ and reverse primer containing EcoR1 site: 5′-CG GAATTC tcatgcttcgaatgtgaa-3′; the primers for amplifying GSTE7 were forward primer containing Nde1 site: 5′-GGAATTC CATATGcccaaattgatactgtac-3′ and reverse primer containing Xho1 site: 5′-CCG CTCGAGttaattcgatgcgaaagt-3′. Genomic DNA was denatured at 95°C for 3 min, followed by 32-cycle amplification (95°C for 30 seconds, 60°C for 30 seconds, 72°C for 1 min) with final extension at 72°C for 7 min. PCR products were analyzed using 1.0% agarose gel and stained with ethidium bromide. These two PCR products were cloned into pET 30a(+) expression vector (Novagen). The recombinant plasmids were analyzed by sequencing and transformed into *E. coli* BL21 (DE3) pLysS (Invitrogen) independently and grown at 37°C in 400 mL Luria-Bertani (LB) media containing 30 μg/mL kanamycin. Protein expression was induced by the addition of isopropyl β-D-thiogalactoside (IPTG) at final concentration of 1 mM. Incubation was continued for further 4 h at 37°C, after which the cells were harvested by centrifugation at 4°C (5000 rpm for 15 min) and resuspended in 5 mL binding buffer (25 mM sodium phosphate, pH 7.4). 100 μL of lysozyme (10 mg/mL) (Sigma-Aldrich) was added and incubated at room temperature for 1 h. The cell suspension was homogenized; the cell debris was palleted by centrifugation at 100,000× *g* for 1 h (4°C). The clear obtained supernatant was used as source for recombinant purification.

Chromatography was carried out using an ÄKTA Purifier FPLC equipped with UNICORN software Version 5.1 and a fraction collector (Frac900) for greater automation of the purification process. The recombinant proteins were purified using Hi-Trap™ Q HP column (5 ml, GE Healthcare) followed by BSP-GSH-Sepharose column (packed in TriconTM, 1 ml) (BSP-GSH-Sepharose matrix was a gift from Dr. AG Clark, Victoria University of Wellington). The BSP (2 mM) used to elute the proteins from affinity matrix was removed from eluate using Hi-Trap™ Desalting column (GE Healthcare).

The Hi-Trap™ Q HP column was equilibrated with 25 mM phosphate buffer and pH 7.4, and 5 mL lysate was applied to the column. Both GSTE6 and GSTE7 enzymes did not bind to the Hi-Trap™ Q HP column, so the flow through was collected and loaded to a BSP-GSH-Sepharose column which was pre-equilibrated with 25 mM phosphate buffer and pH 7.4. The column was washed with 1 M NaCl, and the proteins were eluted with 2 mM BSP in 25 mM phosphate buffer and pH 7.4. Purified enzymes were desalted using Hi-Trap™ Desalting column (15 ml) (GE Healthcare). Proteins were concentrated using Vivaspin 20 (10,000 MWCO, Sartorius). The pooled purified enzymes from subsequent purification were freeze-dried and stored in −20°C for further analysis.

### 2.4. Protein quantification (Bradford assay)

Protein concentration was determined using Coomassie Brilliant Blue G-250, and Bovine serum albumin was used as standard [12].

### 2.5. Molecular weight estimation

Sodium dodecyl sulfate-polyacrylamide gel electrophoresis (SDS-PAGE) was carried out in a 12% polyacrylamide gel [13] using BioRad Mini-PROTEAN system. Mark 12™ unstained standard (Invitrogen, USA) was used as the protein marker, and the gels were stained with Colloidal Coomassie Blue G-250 [14]. Stained gels were scanned with Image Scanner III (GE Healthcare) and visualized and analyzed with Image Master Software.

### 2.6. Enzymatic assay and kinetic evaluation

All assays were performed using a Jasco V-630 spectrophotometer equipped with temperature controller. Enzymatic assays with 1-chloro-2,4-dinitrobenzene (CDNB), ethacrynic acid (EA), sulfobromophthalein (BSP), *p*-nitrobenzyl chloride (NBC), and *trans*-4-phenyl-3-buten-2-one (PBO) [15] and 1,2-dichloro-4-nitrobenzene (DCNB) [16] were determined accordingly. The ability to conjugate trans-2-octenal, hexa-2,4-dienal, *trans*-hex-2-enal, and *trans,trans*-2,4-heptadienal were performed accordingly [17]. The peroxidase activities were determined using cumene hydroperoxide and hydrogen peroxide as substrates as previously described [18].

For the determination of kinetic parameters, initial velocity data was obtained by varying the concentration of hydrophobic substrates (CDNB, NBC, and DCNB) at concentration range of 0.01 to 2.5 mM at fixed GSH concentration (1 mM). Kinetic data were evaluated by non-linear regression analysis with Michaelis Menten equation ($\upsilon = Vmax \times [S]/Km + [S]$), using SigmaPlot 12.0 graph and analysis software. The catalytic constant, *Kcat*, was calculated using the equation Kcat = Vmax/[E], where [E] is the total enzyme concentration. The effect of inhibitors on the catalytic activity was studied by analyzing the reaction rate (nmol/min) in the presence (concentration varied) and the absence of inhibitors. Experiments were repeated at least three times. The IC50 and EC50 were determined by fitting sigmoid concentration-response curves using GraphPad Prism 6.00 software.

### 2.7. Pesticide conjugation

The capability of both recombinant proteins to conjugate selected pesticides (Clodinafop-propargyl, Fenoxaprop-ethyl, Propoxur, isoproturon, and methyl parathion) was evaluated using thin-layer chromatography. An aliquot of 8 μL of reaction products was loaded on a 0.2 mm thick, 8 cm × 10 cm thin-layer chromatography silica gel 60 F2 s4 plate (Merck) and developed using butan-1-ol/acetic acid/water (12,3,5, by vol.) for 1 h. The air-dried plate was stained with ninhydrin (0.25%, w/v, in acetone) [19]. Control reactions using CDNB substrate were performed in a similar manner using CDNB and GSH.

## 3. Results

In the study both Dm*GSTE6* and *DmGSTE7* were amplified using PCR where genomic DNA was used as the template. Their intron-free coding sequences [20] were used directly for

http://dx.doi.org/10.5772/intechopen.72970

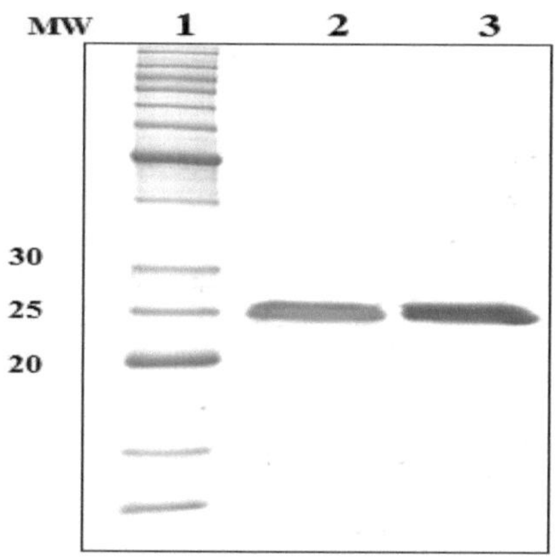


**Figure 1.** SDS-PAGE of purification of GSTE6 and GSTE7 using BSP/GSH-agarose matrix. The bound recombinant proteins were eluted using 2 mM BSP solution in 25 mM phosphate buffer, pH 7.4. Lane 1, BenchMark™ protein ladder; lane 2, purified recombinant GSTE6; and lane 3, purified recombinant GSTE7.

| Substrates | Substrate specificity (nmol/min/mg) | |
|---|---|---|
| | GSTE6 | GSTE7 |
| 1-Chloro-2,4-dinitrobenzene | 80.67 ± 4.42 | 740.33 ± 15.04 |
| 1,2-Dichloro-4-nitrobenzene | 18.11 ± 1.04 | 37.04 ± 2.11 |
| *trans*-Hex-2-enal | ND | ND |
| Hexa-2,4-dienal | ND | ND |
| *trans*-Oct-2-enal | ND | ND |
| *trans*-4-Phenyl-butene-2-one | ND | ND |
| *trans,trans*-Hepta-2,4-dienal | ND | ND |
| Ethacrynic acid | ND | ND |
| *p*-Nitrobenzyl chloride | 3.66 ± 0.58 | 249.67 ± 9.61 |
| Bromosulfophthalein | ND | ND |
| Cumene hydroperoxide | ND | ND |
| Hydrogen peroxide | ND | ND |

Means ± SD of three experiments, each with triplicate determinations. *ND denotes not detected

**Table 1.** Substrate activities of GSTE6 and GSTE7.

expression of the proteins in *E. coli* under induction of IPTG. Proper orientation of the GSTE6 and GSTE7 expression constructs was confirmed by PCR analysis using the constructs as a template, DNA sequencing, and CDNB conjugation activity assay of the extracted bacteria lysate. The proteins were highly expressed and isolated using combination of an anion exchange and BSP/GSH-agarose matrix which has been shown to capture a number of Epsilon class GSTs from *D. melanogaster* [10, 11]. The subunit size of GSTE6 and GSTE7 is predicted to be 25.0146 kDa and 25.5101 kDa, respectively, based on their amino acid compositions. The SDS-PAGE demonstrated that both recombinants migrated at 25 kDa (**Figure 1**).

The work later proceeded into looking at the activities of the recombinants toward different common GST substrates as seen in **Table 1**. The comparison indicated that both recombinants reacted toward the same types of substrates but of different degree of specific activities. Our kinetic analysis as shown in **Table 2** indicates the variation of kinetic parameters of both isoforms on DCNB and NBC. Comparatively GSTE6 has shown to have higher affinity toward DCNB than GSTE7.

To further investigate both isoforms' functional differences, several natural products and dyes were acted upon the enzymes as seen in **Table 3**. Each dye appeared to have different inhibitory effects on each GST isoform. Comparatively, cardiogreen and rose bengal had inhibited the activities of GSTE6 and GSTE7, respectively, at IC50 less than 10 nM.

| **Enzyme** | **Substrate** | **$V_{max}$ (nmol/min)** | **$K_m$ (mM)** | **$K_{cat}$ ($min^{-1}$)** | **$K_{cat}/K_m$ ($min^{-1}mM^{-1}$)** |
|---|---|---|---|---|---|
| GSTE6 | DCNB | 0.029 ± 0.008 | 0.167 ± 0.001 | 0.007 | 0.042 |
| | *p*-NBC | 0.209 ± 0.013 | 0.278 ± 0.005 | 0.051 | 0.183 |
| GSTE7 | DCNB | 0.296 ± 0.033 | 0.415 ± 0.002 | 0.043 | 0.104 |
| | *p*-NBC | 1.311 ± 0.051 | 0.060 ± 0.002 | 0.135 | 2.25 |

Means ± SD of three experiments, each with triplicate determinations.

**Table 2.** Kinetic parameters of GSTE6 and GSTE7 when DCNB and *p*-NBC were used as substrates.

| **Compound** | **Compound concentration (mM)** | **GSTE6 $IC_{50}$ (nM)** | **GSTE7 $IC_{50}$ (nM)** | **$EC_{50}$ (nM)** |
|---|---|---|---|---|
| Sebacic acid | 0–100 | NE | NE | |
| *trans*-chalcone | 0–100 | 86.79 | | 2.958 x $10^5$ |
| Cardiogreen | 0–3 | 4.21 | 9.22 | |
| Crystal violet | 0–10 | 32.24 | 50.59 | |
| Methylene blue | 0–100 | 76.66 | | 1.747 x $10^5$ |
| Rose bengal | 0–3 | 3.68 | 1.07 | |
| Phenol red | 0–10 | 7.29 | 30.36 | |
| Cibacron blue | 0–10 | 82.64 | 210.56 | |

The data are mean value of at least three independent experiments. *NE denotes no effect.

**Table 3.** Effect of selected dyes and natural products on the activities of GSTE6 and GSTE7.

## 4. Discussions

The molecular weights of purified enzymes were approximately 25 kDA, respectively, which corresponded to the calculated molecular masses (**Figure 1**) and were in agreement with data previously reported [6].

It was observable (**Table 1**) that both recombinants have the same affinity toward certain substrates notably of compound containing aromatic ring. However GSTE7 seemed to have higher specific activities toward each conjugated substrate as compared to GSTE6. GSTE7 was reacted 9-, 4-, and 68-fold higher toward CDNB, DCNB, and NBC, respectively, as compared to GSTE6. Both were unable to conjugate lipid peroxidation products, and neither could it have shown peroxidase activities. This implied that neither GSTE6 nor GSTE7 could have directly participated in countering oxidative stress in fruit flies. In another instance, substrate specificity variation was observed when other substrates are used. It was previously reported that GSTE6 reacted toward 4-hydroxynonenal, adrenochrome, phenethyl isothiocyanate, 5-hydroperoxyeicosatetraenoic acid, and hydroxyethyldisulfide, while GSTE7 reacted only to 4-hydroxynonenal, 5-hydroperoxyeicosatetraenoic acid, and 2-hydroxyethyldisulfide [6].

In comparison to *Musca domestica*, DmGSTE6 is 79 and 77% similar and also 62 and 59% identical to MdGST6A and MdGST6B, respectively. Both MdGST6A and MdGST6B showed more or less 40% identity with other *Drosophila* Epsilon class proteins. MdGST6A and MdGST6B were both known involved in insecticide metabolism in houseflies [9]. Further assumption on both isoforms (DmGSTE6 and DmGSTE7) that could have been participating in insecticide conjugation was much supported by the induction study performed previously [10]. Through proteomic analysis, the expression of GSTE6 and GSTE7 was shown to increase in the methyl parathion-challenged adult flies. The authors however cautiously suggested any direct relationship of the isoforms increase expression to metabolism of the tested insecticide. There were reports suggested that only Epsilon class GSTs were able to react with DCNB [21] and the detoxification ability of GSTs against insecticides is correlated to its ability to react with DCNB [9].There was also instances of which GSTE1-1z of a reportedly high specific activity toward DCNB and yet does not confer insecticide resistance in *Anopheles gambiae* [22].

The rate of conjugation reaction, $V_{max}$, for GSTE7 toward DCNB was however ten times faster than GSTE6. The behavior could have implied the differences in the stabilization of the GSH in the hydrophilic pocket that lead to the conjugation of DCNB. This has remarkably affected the Kcat and Kcat/Km values of both isoforms. The turnover and the catalytic efficiency of GSTE7 were, respectively, sixfold and twofold higher than those of GSTE6. GSTE7 has lower $K_m$ value to suggest its higher affinity toward NBC. The behavior of GSH conjugation to NBC could probably differ to what has been with DCNB where the speed of reaction was very much higher in GSTE7 when NBC was the second substrate. The catalytic efficiency, Kcat/Km, of the reaction catalyzed by GSTE7 was shown 12 times higher than of the GSTE6. These observations proposed that GSTE7 was a better isoform for conjugation of the selected substrates than GSTE6.

These differences indicated that GSTE6 and GSTE7 have considerable variations in their secondary structural organization. Such variations in structure may form the basis of differences in their corresponding substrate specificities and in catalytic efficiency [23] although both originated from same cluster and located next to each other on the genomic DNA.

Our preliminary investigative attempt to show that both isoforms could conjugate insecticides was performed using thin-layer chromatography. Pesticides such as temephos, malathion, DDT, fenthion, fenitrothion, permethrin, bromophos, chlorpyrifos, clodinafop-propargyl, fenoxaprop-ethyl, propoxur, isoproturon, and methyl parathion were used in the test. None of the pesticides appeared conjugated by both isoforms. The test suggested both recombinant GSTE6 and GSTE7 did not react and involve in detoxification of insecticides and herbicides directly. Thus so far, the role of GSTE6 and GSTE7 in detoxification of insecticides remains unclear as there is no any promising evidence to prove its involvement in detoxification process.

A comprehensive microarray-based atlas of adult gene expression in multiple *Drosophila* tissues available (http://flyatlas.org) reported that GSTE6 expressed in adult crop, midgut, tubule, hindgut, ovary, and larval hindgut while GSTE7 expressed in adult crop, midgut, tubule, hindgut, virgin spermatheca and larval midgut, and hindgut and fat body. Other reports also indicated that GSTE6 was found abundant in hindgut, while GSTE7 was found abundant in Malpighian tubules [24]. The co-expression of GSTE6 and GSTE5 was suggested to play a role in male reproductive fitness and success [25]. It was identified that a potential DNA transcription factor binding motifs (TFBMs) of cytochrome $P_{450}$, GSTs, and carboxylesterases is expressed in the *D. melanogaster* third instar larval midguts [26].

GSTE6 was reported to have GRE-like, Fox-like, NF-kappaB-like, and E47-like TFBMs while GSTE7 to have GRE-like and E47-like TFBMs. The four mentioned TFBMs are known to have mammalian function and were observed to be linked to the oxidative stress response [26]. The author reported that GSTE6 and GSTE7 responded different levels of dietary hydrogen peroxide. However, the author concluded that there is no solid evidence to prove if some or all of the potential TFBMs are functional or response of the midgut-associated GSTs to the oxidative stressor, dietary H2O2. They may simply be associated with these genes with limited or no role in response to this oxidative stressor. In another study, GSTE7 also appeared to be involved in activation of survival program through immune deficiency (IMD) pathway as it was reported being expressed in a strongly infected airway epithelium of *D. melanogaster* [27].

Exposure of *Drosophila* to toxins evokes coordinated response by the Malpighian tubules, involving both alterations in detoxification pathways as well as enhanced transport through DHR96, the *Drosophila* ortholog of the vertebrate PXR/CAR family of nuclear receptors [21]. In relation with that statement, a study [28] stated that in insects, either two distinct receptors have evolved the ability to regulate a very similar set of genes. More than one receptor pathway exists to regulate similar sets of genes. This suggests the possibilities of induction of GSTE6 and GSTE7 together with other genes. Apart from that, basal expression and induction were detected in the key metabolic tissues, namely, sections of the midgut and the Malpighian tubules.

However, the difference in the expression of both GSTE6 and GSTE7 and its inability to detoxify possibly due to cis-regulatory elements controlling the expression of genes may not be acting independently whereby the substrate models may be acting solely to increase the transcriptional output of the tissue-specific modules [28], and the fact that these two genes are found sequentially on the chromosome may support a model of coordinated regulation [7]. In another instance, reports have indicated that DmGSTE6 was strongly co-expressed with DmGSTE7, DmGSTE8, DmGSTE5, DmGSTE3, DmGSTE9, and DmGSTD1, while DmGSTE7 was strongly co-expressed with DmGSTE6, DmGSTE8, DmGSTE3, and DmGSTE9 [29]. These gave insights of possible role of a selective protein to be the key regulator of sets of genes.

Rose bengal was known to inhibit several drug-metabolizing enzymes such as cytochrome P450 and UDP-glucuronosyltransferase [30]. Phenol red however inhibited GSTE6 much effective than on GSTE7.

Methylene blue dye and *trans*-chalcone had shown significant enhancement of GSTE7 activity toward CDNB with EC50 ranging from 1 to 2 × 105 nM. Chalcones are open chain flavonoids that are widely biosynthesized in plants. A study [31] revealed the pharmacological properties of natural and synthetic chalcones as antioxidant, cytotoxic, anticancer, antimicrobial, antiprotozoal, antiulcer, antihistaminic, and anti-inflammatory activities but mechanism of action of *trans*-chalcone as an inhibitor to GSTE6 while as a stimulator for GSTE7 is remaining unclear. It was reported that certain haloalka(e)nes including ethylene bromide and methylene chloride form a highly reactive episulfonium ion intermediates that catalyze GST activation reactions [32]. The intermediate effect as an inhibitor to GSTE6 while as a stimulator for GSTE7 activity is remaining unclear. Basic triphenylmethane dyes such as crystal violet have been shown to inhibit glutathione *S*-transferases from both insect sources [33] and rat liver [34]. The mode of inhibition of crystal violet appeared to involve competition by the free dye with the electrophilic substrate [35].

## 5. Conclusions

In conclusion, the study has highlighted that both DmGSTDE6 and DmGSTDE7 behave catalytically different toward same substrates, despite their high sequence similarity. DmGSTDE7 appeared to demonstrate high specific activities toward mentioned substrates and catalytically stimulated by *trans*-chalcone and methylene blue. None of the isoforms appeared to be able to conjugate pesticides in vitro and hence contradictory to previous work that may cautiously suggest their involvement in methyl parathion metabolism. Many other related findings may lead to assumption that the increase in expression of both isoforms could have been a result of co-expression with other related genes, and yet they have not involved directly in pesticide metabolism. The work has also indicated that both isoforms were not participating in oxidative stress and its functional role thus far was of the normal detoxification xenobiotic taken by the organism. This may thus generate further interest in their functional roles, and a more suitable approach could be adapted to realize the goal.

## Acknowledgements

The work has been funded by the University of Malaya Postgraduate Research Fund (PPP: PV034/2012A) and the Ministry of Science, Technology, and Innovation (12-02-03-2041).

## Conflict of interest

There is no conflict of interests that could be perceived as prejudicing the impartiality of the research reported.

## Author details

Vennobaashini Venu and Zazali Alias*

*Address all correspondence to: alias@um.edu.my

Institute of Biological Sciences, Faculty of Science, University of Malaya, Kuala Lumpur, Malaysia

## References

[1] Sheehan D, Meade G, Foley VM, Dowd CA. Structure, function and evolution of glutathione transferases: Implications for classification of non-mammalian members of an ancient enzyme superfamily. The Biochemical Journal. 2001;**360**(1):16

[2] Ding Y, Ortelli F, Rossiter LC, Hemingway J, Ranson H. The Anopheles Gambiae glutathione transferase supergene family: Annotation, phylogeny and expression profiles. BMC Genomics. 2003;**4**:35-50

[3] Ranson H, Rossiter L, Ortelli F, Jesen B, Wang X, Roth CW, Collins FH, Hemingway J. Identification of a novel class of insect glutathione *S*-transferases involved in resistance to DDT in the malaria vector *Anopheles gambiae*. The Biochemical Journal. 2001;**359**:295-304

[4] Enayati AA, Ranson H, Hemingway J. Mini review: Insects glutathione transferases and insecticide resistance. J. Insect Molecular Biology. 2005;**14**:3-8

[5] Ketterman AJ, Saisawang C, Wongsantichon J. Insect glutathione transferases. Drug Metabolism Reviews. 2011;**43**:253-265

[6] Saisawang C, Wongsantichon J, Ketterman AJA. Preliminary characterization of the cytosolic glutathione transferase proteome from *Drosophila melanogaster*. The Biochemical Journal. 2011;**442**:181-190

[7] Lumjuan N, Rajatileka S, Changsom D, Wicheer J, Leelapat P, Prapanthadara LA, Somboon P, Lycett G, Ranson H. The role of the *Aedes aegypti* epsilon glutathione transferases in conferring resistance to DDT and pyrethroid insecticides. Insect Biochemistry and Molecular Biology. 2011;**41**:203-209

[8] Deng H, Huang Y, Feng Q, Zheng S. Two epsilon glutathione *S*-transferase cDNAs from the common cutworm, *Spodoptera litura*: Characterization and developmental and induced expression by insecticides. Journal of Insect Physiology. 2009;**55**:1174-1183

[9] Wei SH, Clark AG, Syvanen M. Identification and cloning of a key insecticide metabolizing glutathione *S*-transferase (MdGST-6A) from a hyper insecticide resistant strain of the housefly *Musca domestica*. Insect Biochemistry and Molecular Biology. 2001;**31**:1145-1153

[10] Alias Z, Clark AG. Studies on the glutathione *S*-transferase proteome of adult *Drosophila melanogaster*. Responsiveness to chemical challenge. Proteomics. 2007;**7**:3618-3628

[11] Alias Z, Clark AG. Adult *Drosophila melanogaster* glutathione *S*-transferases: Effects of acute treatment with methyl parathion. Pesticide Biochemistry and Physiology. 2010; **98**:94-98

[12] Bradford MM. A rapid and sensitive method for the quantitation of microgram quantities of protein utilizing the principle of protein-dye binding. Analytical Biochemistry. 1976;**72**:248-254

[13] Laemmli UK. Cleavage of structural proteins during the assembly of the head of bacteriophage T4. Nature. 1970;**227**:680-685

[14] Neuhoff V, Arnold N, Taube D, Ehrhardt W. Improved staining of proteins in polyacrylamide gels including isoelectric focusing gels with clear background at nanogram sensitivity using Coomassie Brilliant Blue G-250 R-250. Electrophoresis. 1988;**9**:255-262

[15] Habig WH, Pabst MJ, Jakoby WB. Glutathione *S*-transferases: The first enzymatic step in mercapturic acid formation. The Journal of Biological Chemistry. 1974;**249**:7130-7139

[16] Motoyama N, Dauterman WC. Purification and properties of housefly glutathione S- transferase. Insect Biochemistry. 1977;**7**:361-369

[17] Brophy PM, Southan C, Barrett J. Glutathione transferases in the tapeworm *Moniezia expansa*. The Biochemical Journal. 1989;**262**:939-946

[18] Paglia DE, Valentine WN. Studies on the quantitative and qualitative characterization of erythrocyte glutathione peroxidase. The Journal of Laboratory and Clinical Medicine. 1967;**70**:158-169

[19] Ben-Arie N, Khen M, Lancet D. Glutathione *S*-transferases in rat olfactory epithelium: Purification, molecular properties and odorant biotransformation. The Biochemical Journal. 1993;**292**:379-384

[20] Sawicki R, Singh SP, Mondal AK, Benes H, Zimniak P. Cloning, expression and biochemical characterization of one epsilon-class (GST-3) and ten Delta-class (GST-1) glutathione *S*-transferase from *Drosophila melanogaster*, and identification of additional nine members of the epsilon class. The Biochemical Journal. 2003;**370**:661-669

[21] Wang Y, Qiu L, Ranson H, Lumjuan N, Hemingway J, Setzer WN, Meehan EJ, Chen L. Structure of an insect epsilon class glutathione *S*-transferase from the malaria vector *Anopheles gambiae* provides an explanation for the high DDT-detoxifying activity. Journal of Structural Biology. 2008;**164**:228-235

[22] Ortelli F, Rossiter J, Vontas H, Ranson J, Hemingway J. Heterologous expression of four glutathione transferase genes genetically linked to a major insecticide-resistance locus from the malaria vector *Anopheles gambiae*. The Biochemical Journal. 2003;**272**:957-963

[23] Tang AH, Biochemical TCP. Characterization of Drosophila glutathione *S*- transferases D1 and D21. The Journal of Biological Chemistry. 1994;**269**:27876-27884

[24] Yang J, McCart C, Woods DJ, Terhzaz S, Greenwood KG, Ffrench-constant RH, Dow JAT. A *Drosophila* systems approach to xenobiotic metabolism. Physiological Genomics. 2007;**30**:223-231

[25] Drnevich JM, Reedy MM, Ruedi EA, Rodriguez-Zas S, Hughes KA. Quantitative evolutionary genomics: Differential gene expression and male reproductive success in *Drosophila melanogaster*. Proceedings of the Royal Society of London. Series B: Biological Sciences. 2004;**271**:2267-2273

[26] Li HM, Buczkowski G, Mittapalli O, Xie J, Wu J, Westerman R, Schemerhorn BL, Murdock LL, Pittendrigh BR. Transcriptomic profiles of *Drosophila melanogaster* third instar larval midgut and responses to oxidative stress. Journal of Insect Molecular Biology. 2008;**17**:325-339

[27] Mizutani T. Toxicity of xanthene food dyes by inhibition of human drug-metabolizing enzymes in a noncompetitive manner. Journal of Environmental and Public Health. 2009; **2009**:1-9

[28] Willoughby L, Chung H, Lumb C, Robin C, Batterham P, Daborn PJ. A comparison of *Drosophila melanogaster* detoxification gene induction responses for six insecticides, caffeine and phenobarbital. Insect Biochemistry and Molecular Biology. 2006;**36**:934-942

[29] Jensen LJ, Kuhn M, Stark M, Chaffron S, Creevey C, Muller J, Doerks T, Julien P, Roth A, Simonovic M, Bork P. von Mering C. STRING 8—A global view on proteins and their functional interactions in 630 organisms. Nucleic Acids Research 2009;**37**:412-416

[30] UesugiN,FurumiyaK,MizutaniT.InhibitionmechanismofUDP–Glucuronosyltransferase 1A6 by xanthene food dyes. Journal of Health Science. 2006;**52**:549-557

[31] Batovska DI, Todorova IT. Trends in utilization of the pharmacological potential of chalcones. Current Clinical Pharmacology. 2010;**5**:1-29

[32] Armstrong RN. Glutathione *S*-transferases: Reaction mechanism, structure, and function. Chemical Research in Toxicology. 1997;**4**:131-140

[33] Balabaskaran S, Smith JN. The inhibition of 1,1,1-trichloro-2,2-bis-(p-chlorophenyl) ethane (DDT) dehydrochlorinase and glutathione S-aryl transferase in grass-grub and housefly preparations. The Biochemical Journal. 1970;**117**:989-996

[34] Debnam P, Glanville S, Clark AG. Inhibition of glutathione *S*-transferases from rat liver by basic triphenylmethane dyes. Biochemical Pharmacology. 1993;**45**:1227-1233

[35] Glanville SD, Clark AG. Inhibition of human glutathione *S*-transferases by basic triphenylmethane dyes. Journal of Life Sciences. 1997;**60**:1535-1544

Chapter 12

# Antiviral Immunity in the Fruit Fly, *Drosophila melanogaster*

Wilfredo A. Lopez, Alexis M. Page, Brad L. Ericson, Darby J. Carlson and Kimberly A. Carlson

Additional information is available at the end of the chapter

http://dx.doi.org/10.5772/intechopen.69293

**Abstract**

The fruit fly, *Drosophila melanogaster*, is an extremely useful model to study innate immunity mechanisms. A fundamental understanding of these mechanisms as they relate to various pathogens has come to light over the past 30 years. The discovery of Toll-like receptors and their recognition of shared molecules (pathogen-associated molecular patterns or PAMPs) among pathogenic bacteria were the first detailed set of receptors to be described that act in innate immunity. The immune deficiency pathway (Imd) described in *D. melanogaster* functions in a very similar way to the Toll pathway in recognizing PAMPs primarily from Gram-negative bacteria. The discovery of small interfering RNAs (RNAi) provided a means by which antiviral immunity was accomplished in invertebrates. Another related pathway, the JAK/STAT pathway, functions in a similar manner to the interferon pathways described in vertebrates, also providing antiviral defense. Recently, autophagy was also shown to function as a protective pathway against virus infection in *D. melanogaster*. At least three of these pathways (Imd, JAK/STAT, and RNAi) show signal integration in response to viral infection, demonstrating a coordinated immune response against viral infection. The number of pathways and the integration of them reflect the diversity of pathogens to which innate immune mechanisms must be able to respond. The viral pathogens that infect invertebrates have developed countermeasures to some of these pathways, in particular to RNAi. The evolutionary arms race of pathogen vs. host is ever ongoing.

**Keywords:** antiviral immunity, autophagy, innate immunity, RNAi, virus

# 1. Introduction

## 1.1. The fruit fly, *Drosophila melanogaster*, as a model for innate immunity

Immunity is a vital component in understanding host-pathogen relationships. It is composed of two responses: innate and adaptive. Innate immunity recognizes morphological characteristics of pathogens for immediate antimicrobial and antiviral defense [1]. Adaptive immunity develops during infection to produce immunological memory against pathogens. This memory provides an immediate pathogen-specific defense against future infections of the same pathogen [2]. Most vertebrate organisms utilize both immune responses for pathogen defense. However, the fruit fly, *Drosophila melanogaster,* does not have an adaptive immune response and relies solely on an innate immune response [3]. This provides a powerful model system to better understand the interaction between innate immunity and pathogenic infections.

Innate immunity is composed of various pathways that target bacteria, fungi, and viruses. These pathways include the immune deficiency pathway (Imd), Toll-Dorsal pathway (Toll), Janus kinase/signal transducer and activator of transcription pathway (JAK/STAT), autophagy, and RNA interference (RNAi) [3–6]. The Imd and Toll pathways contribute to the antibacterial and antifungal defense. However, their function in antiviral defense is not fully understood [7, 8]. The JAK/STAT, autophagy, and RNAi pathways contribute to antiviral defense, with RNAi as the main contributor for antiviral defense.

## 1.2. *Drosophila* viruses

As a model organism, *D. melanogaster* is used to study host immunity to pathogen interactions. Most research is focused on the interaction between bacteria, fungi, and the *D. melanogaster* innate immune response, but viruses are a subject of current interest. Populations of *Drosophila* have naturally occurring infections of RNA viruses, such as Nora virus, Sigma virus (DmelSV), Drosophila C virus (DCV), and Drosophila X virus (DXV). In addition, the first naturally occurring DNA virus, Kallithea virus, is found in *D. melanogaster* (**Table 1**) [9–13].

Nora virus is a recently discovered picorna-like *D. melanogaster* virus. The virus is sequenced and has a 12 kilobase (kb), single-stranded, positive-sense RNA genome. Viral particles measure 30 nm in diameter and are non-enveloped [9]. It establishes a persistent infection in natural and laboratory populations of *D. melanogaster* with no known effect of viral load and no display of pathology on the fly. The virus is transmitted horizontally through the fecal-oral route with infection localizing to the intestinal tract [14]. The genome is organized into four open reading frames (ORFs), unlike other picorna-like viruses such as DCV, which has two ORFs [15]. ORF1–3 partially overlaps, suggesting ribosomal frame shifting events during translation. However, an 88 nucleotide region is found between ORF3 and ORF4, suggesting that an independent initiation translation event is occurring [16]. ORF1 encodes a highly charged protein, which is a suppressor of RNAi [17]. ORF2 encodes a picorna-like replicative cassette, which consists of a helicase, protease, and RNA-dependent RNA polymerase [9]. The hypothesized major capsid proteins of Nora virus are products of ORF3 and ORF4 at the 3′ end of the genome. ORF3 encodes VP3, which is crucial for the stability of Nora virus virions [18]. ORF3 is not fully

| **Virus** | **Family** | **Genome nucleic acid** | **Mode of transmission in *D. melanogaster*** | **Effects of infection** |
|---|---|---|---|---|
| Nora virus | *Picornavirales* | (+) ssRNA | Horizontal | No documented pathology, slight effect on longevity [14] |
| Sigma virus (DmelSV) | *Rhabdoviridae* | (−) ssRNA | Vertical | Anoxia sensitivity [10] |
| Drosophila C virus (DCV) | *Dicistroviridae* | (+) ssRNA | Horizontal | Intestinal obstruction [21] and increased female fecundity and reduced developmental timing [73] |
| Drosophila X virus (DXV) | *Birnaviridae* | dsRNA | Horizontal | Anoxia sensitivity [12] |
| Cricket paralysis virus (CrPV) | *Dicistroviridae* | (+) ssRNA | Horizontal | No documented pathology |
| Flock house virus (FHV) | *Nodaviridae* | (+) ssRNA | Horizontal | High mortality [34] |
| Sindbis virus (SINV) | *Togaviridae* | (+) ssRNA | Vertical | No documented pathology |
| Vesicular stomatitis virus(VSV) | *Rhabdoviridae* | (−) ssRNA | Horizontal | Anoxia sensitivity [30, 40] |
| Kallithea virus | *Nudiviridae* | dsDNA | Horizontal | No documented pathology |
| Invertebrate iridescent virus 6 (IIV-6) | *Iridoviridae* | dsDNA | Horizontal | Low mortality rate [31] |

**Table 1.** Characteristics of a set of *Drosophila* viruses.

characterized, but certain aspects of its protein products were predicted using bioinformatics. It has a predicted alpha-helical domain as a key structural motif [9]. ORF4 is processed into three major proteins, VP4A, VP4B, and VP4C. VP4A and VP4B are predicted to form jelly roll folds, which are also found in other capsid proteins of *Picornavirales*. The third protein, VP4C, has a predicted alpha-helical structure and is also a structural component of the virus [16].

Another virus naturally occur in *D. melanogaster* is Sigma virus. Sigma virus belongs to the family *Rhabdoviridae* [10]. It is composed of a negative-sense, single-stranded RNA genome that consists of five genes: *N, P, M, G,* and *L*. The gene *N* is a nucleoprotein, *P* is the polymerase-associated protein, *M* is the matrix protein, *G* is the glycoprotein, and *L* is the polymerase [19]. A sixth gene, *X*, exists between *P* and *M*, but its current function is not fully understood [20]. In natural infections, the virus causes paralysis or death if flies are exposed to $CO_2$. It is passed through vertical transmission through the sperm or eggs and is the only known host-specific pathogen of *D. melanogaster* [10, 20].

Drosophila C virus is in the family *Dicistroviridae* [21]. The virus particle measures 30 nm in dm with a 9264 kb, positive-sense, single-stranded RNA genome [22]. The genome consists of two ORFs separated by 191 nucleotides. ORF1 encodes an RNA-dependent RNA polymerase, helicase domain, and protease domain [15]. Also, an RNAi suppressor, DCV-1A, is encoded at the N-terminus of ORF1. The suppressor binds long dsRNA, which inhibits Dicer-2 (Dcr-2) processing [23]. ORF2 encodes the structural proteins VP0, VP1, VP2, VP3, and VP4. VP0 is a precursor for VP3 and VP4, which combine to form the capsid [24]. The capsid proteins are encoded in a different reading frame and initiated independently from ORF1 [15]. In addition, DCV is a naturally occurring pathogen found within *D. melanogaster* and spread through horizontal transmission by infected flies or contaminated food. Viral infection can be lethal if injected, but naturally infected flies display decreased pathogenicity [11].

Drosophila X virus is a double-stranded RNA virus, which belongs to the family *Birnaviridae*. It was discovered in a study involving *D. melanogaster* and Sigma virus. Like Sigma virus, DXV is pathogenic, induces $CO_2$ sensitivity, and is lethal [12]. The virus displays a non-enveloped capsid and a bi-segmented dsRNA genome. Segment A encodes a polyprotein, which forms the capsid. The capsid consists of VP1, preVP2, VP2, VP3, and VP4. Segment B encodes VP1, an RNA-dependent RNA polymerase [25].

Recently, a DNA virus was discovered in wild populations of *Drosophila*. By using a metagenomic approach, the Kallithea virus was identified. The virus is closely related to *D. innubila* and the beetle *Orcytes rhinoceros* Nudiviruses. In addition, this is the first DNA virus found naturally occurring in *D. melanogaster*. However, the virus has not been characterized in *D. melanogaster* with recent research using other *Drosophila* species [13]. In wild *D. innubila*, Nudivirus infection is associated with greatly reduced survival and offspring production. In wild *D. falleni*, infection resulted in greatly reduced offspring production. Additionally, infection is highly pathogenic and mediated through the fecal-oral route [26]. Further research with naturally occurring *Drosophila* viruses is important because not many of these viruses exist or have been discovered.

### 1.3. Non-*Drosophila* viruses

Laboratory populations of *D. melanogaster* can be experimentally inoculated with RNA viruses, such as Cricket paralysis virus (CrPV), Flock House virus (FHV), Sindbis virus (SINV), and Vesicular stomatitis virus (VSV). Also, the DNA virus, Invertebrate iridescent virus 6 (IIV-6), can be experimentally inoculated into flies (**Table 1**) [27–31]. Artificial infections of *D. melanogaster* allow for a better understanding and novel insights of host-pathogen interactions.

Cricket paralysis virus is a positive-sense, single-stranded RNA virus closely related to DCV. It belongs to the family *Dicistroviridae* and was first discovered in field crickets, *Teleogryllus oceanicus* and *T. commodus* [32]. The crickets displayed rapid paralysis and significant mortality [27]. The viral RNA genome consists of two ORFs, ORF1 and 2. To initiate translation, each ORF requires an internal ribosome entry site (IRES) region. ORF1 encodes non-structural replication proteins, and ORF2 encodes structural proteins, which form the viral capsid. In addition, this virus encodes a suppressor of RNAi, CrPV-1A, which binds to Argonaute-2 (AGO2) inhibiting RNA-induced silencing complex (RISC) activity [32].

Flock house virus contains two positive-sense, single-stranded RNAs within a non-enveloped virion. This virus belongs to the *Nodaviridae* family and was first discovered in the grass grub, *Costelytra zealandica* [28, 33]. Viral inoculation kills *D. melanogaster,* and the virus propagates in *D. melanogaster* cell lines [34]. The bipartite genome consists of *RNA1* and *RNA2*. *RNA1* encodes protein A, an RNA-dependent RNA polymerase, whereas *RNA2* encodes the precursor protein for production of the mature capsid protein. For viral replication, both RNAs must be present within the cell or replication will not occur [35]. A subgenomic RNA, *RNA3,* is produced by *RNA1* and encodes an RNAi suppressor protein B2. The protein binds viral dsRNA to protect it from cleavage by Dcr-2 and to inhibit loading of viral siRNAs into the RISC complex [34, 36, 37].

Sindbis virus is a single-stranded, positive-sense RNA virus, belongs to the *Togaviridae* family, and is transmitted vertically in *Drosophila*. In other hosts, it is transmitted horizontally. The viral genome mimics cellular mRNA as the viral mRNA possesses a 5′ methylguanylate cap and a 3′ poly(A) tail. The 5′ region encodes nonstructural proteins, and the 3′ region encodes viral structural proteins [38]. Most Sindbis virus research with invertebrates is conducted with mosquitoes because they are a natural vector for SINV. However, *D. melanogaster* S2 (Schneider 2) cells are successfully infected establishing an additional invertebrate model system to examine the host-pathogen interaction with SINV [29].

Vesicular stomatitis virus is a single-stranded, negative-sense RNA virus that belongs to the Rhabdoviridae family [30]. It belongs to the same family as Sigma virus, which naturally occurs in *Drosophila* [39]. The genome is composed of the structural proteins (G, N, and M), the minor protein (NS), the partially glycosylated G precursor ($G_1$), and the L chain. Insects infected with VSV become paralyzed after exposure to $CO_2$. However, VSV has no observable pathogenic effects in infected insect cells [30, 40].

Invertebrate iridescent virus 6, also known as Chilo iridescent virus, is a large and complex double-stranded DNA virus that belongs to the *Iridoviridae* family. The virus is composed of a capsid, an intermediate lipid layer, and a viral genome composed of linear double-stranded DNA [41]. The viral genome size is approximately 212.5 kb, circular, and encodes 211 ORFs along both strands [31]. Several important ORFs encode a DNA-dependent RNA polymerase II, a helicase, and major capsid proteins [42]. IIV-6 has a broad host range and can be used to experimentally infect *D. melanogaster*. Infections in *D. melanogaster* produce high and stable viral titers exhibiting a low mortality rate [31]. Artificial infections of *D. melanogaster* are important because they provide a valuable model of understanding interactions between virus and host immunity.

## 2. RNA interference (RNAi) and the immune response

### 2.1. Antiviral RNAi in *D. melanogaster*

RNAi is the major antiviral immune response pathway for *D. melanogaster* (**Figure 1**). The general pathway occurs in two steps, initiation and execution. To initiate RNAi, dsRNA must be introduced, such as with viral infection. If dsRNAs are greater than 23 bp in length, it is

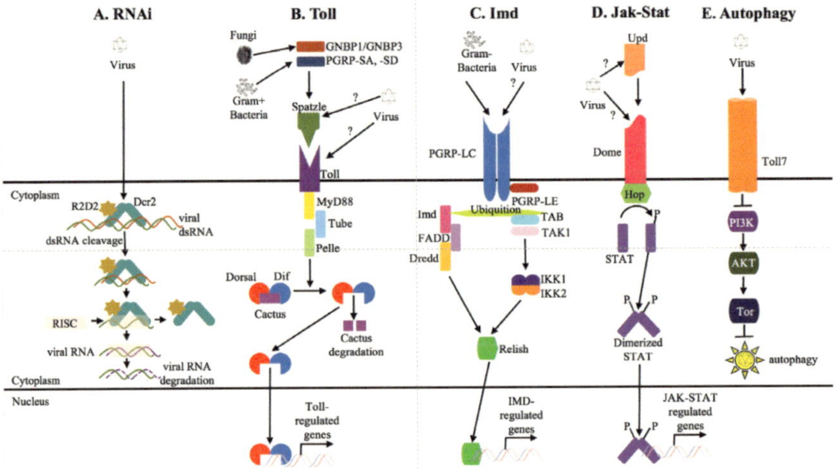


**Figure 1.** The major virus defense pathways of the fruit fly, *Drosophila melanogaster*. (A) RNA interference (RNAi) is the primary defense mechanism against viruses in invertebrate species. Virus replication results in the production of dsRNA replication intermediates that activate the pathway. R2D2 has two binding sites for dsRNA and in conjunction with the RNaseIII-like enzyme, Dicer-2 (Dcr2), will cleave large dsRNAs into small interfering RNAs (siRNA). The Dcr2/R2D2 siRNA complex subsequently interacts with Argonaut-2 protein, a key component of the RNA-induced silencing complex (RISC), and transfers the siRNA component to it. The siRNA acts by targeting viral RNA via base pairing, allowing the targeted viral RNA to be degraded by the nuclease action of RISC. (B) The Toll pathway is activated primarily by pathogen-associated molecular patterns (PAMPs) associated with fungi and Gram-positive cell wall components. The PAMPs are recognized by cytoplasmic receptors, such as Gram-negative bacteria binding proteins (GNBP-1/-3) and peptidoglycan recognition proteins (PGRP-SA, -SD). These receptors are referred to collectively as pattern recognition receptors (PRRs). Once the PRRs are engaged by their specific PAMPs, they now activate proteases that cleave full-length Spatzle to an active form that now can be bound by the Toll receptor. With virus activation of this pathway, it is unclear whether virions can directly interact with Toll or must also activate Spatzle. The binding of the Spatzle ligand to the Toll receptor results in signal transduction through the cytoplasmic adaptor protein, MyD88. This ultimately leads to the proteolytic degradation of Cactus, the inhibitor of the NF-κB-like transcription factors Dorsal and Dif. With the degradation of Cactus, the Dorsal-Dif heterodimer is now able to be transported to the nucleus where it acts to activate the transcription of Toll-regulated genes. (C) The Imd pathway is activated by PAMPs from Gram-negative bacteria and potentially directly by virions. A transmembrane peptidoglycan receptor protein (PGRP-LC) binds the PAMPs and transduces a signal to the cytoplasmic adaptor proteins Imd and FADD, which results in the activation of the caspase-8 like protease, Dredd. Dredd cleaves the NF-κB-like transcription factor, Relish, which removes an IκB-like C-terminal domain that masks a nuclear localization signal. In addition, Dredd also cleaves Imd, which now allows it to become ubiquitinated. This attracts the Tab2/Tak1 complex that activates the IKK1/IKK2 proteins via phosphorylation. These activated kinases now phosphorylate Relish at multiple sites, especially S528 and S529, which are essential to RNA polymerase II recruitment to Imd-regulated genes. (D) The JAK/STAT pathway is activated by the interaction of the ligand unpaired (Upd) with the receptor Dome. In the *Drosophila* immune response, it appears that Upd3, secreted by activated hemocytes, is the preferred ligand for Dome. Most likely, virions are detected by these cells, which in turn secrete Upd3, although direct interaction of virions with Dome has not been ruled out. Once Dome has engaged Upd, it activates, via signal transduction, the Janus kinase Hop, which now is capable of phosphorylating the STAT transcription factors. Phosphorylation of the STAT proteins results in their dimerization and subsequent translocation to the nucleus where they activate the transcription of JAK/STAT regulated genes. (E) Autophagy can also act as a viral defense pathway. In the absence of a ligand for Toll-like receptor 7 (Toll7), the signal transduction pathway involving phosphatidylinositol-3 kinase (PI3K), Akt kinase, and Tor (target of rapamycin) kinase is active and autophagy is inhibited. However, if the Toll-7 receptor is engaged by its ligand, in this case a virion component, this results in the inhibition of PI3K, which ultimately results in the inhibition of Tor, which now relieves inhibition of the autophagy pathway, resulting in the destruction of the cell.

processed into 21–23 bp dsRNA fragments with 3′ overhanging ends by Dcr-1 or Dcr-2 [5]. Dcr-2 produces small interfering RNAs (siRNAs), and Dcr-1 recognizes precursors of micro RNAs (miRNAs). The siRNA products are recruited by AGO2 into the RISC. Once loaded, one of the siRNA strands is degraded in an AGO2-dependent process involving an endoribonuclease, component 3 promoter of RISC (C3PO) [43]. The single strand in the RISC complex is called the guide strand. It acts as a targeting mechanism for locating complementary mRNA. Matching of the guide strand to the targeted mRNA results in either degradation or inhibition of translation. Degradation occurs if the guide strand completely matches the target mRNA. However, inhibition of translation occurs if there is a small mismatching of base pairs (2–3 bp) [5]. Additionally, RNAi is incorporated in two alternative pathways: the miRNA or piwi RNA (piRNA) pathways. In the miRNA pathway, miRNA and Argonaute-1 (AGO1) regulate cellular gene expression through different mechanisms, such as cleavage or translational inhibition [44]. The piRNA pathway is involved as a transposon regulatory control mechanism in *D. melanogaster* testes [45]. However, the siRNA pathway is the major contributor to the RNAi antiviral defense pathway in *D. melanogaster*.

### 2.2. Viral suppression of RNAi

RNAi is an effective antiviral mechanism, but viruses have developed strategies to counteract it using virus-encoded suppressors of RNAi (VSRs). RNAi suppression depends on the mechanism the VSR uses to target RNAi components and can vary with each virus [16]. For example, Nora virus VP1, the protein product of ORF1, can suppress RNAi. It inhibits slicer activity of mature RISC by hindering targeted catalytic cleavage by AGO2 [46]. In FHV, *RNA1* produces a subgenomic *RNA3*, which encodes B2, an RNAi suppressor protein. B2 has dual functions for suppression. It binds to long dsRNA to inhibit siRNA production and to siRNA to prevent siRNA assembly into RISC [47]. In CrPV, the N-terminal region of ORF1 encodes the RNAi suppressor protein, CrPV-1A. It directly interacts with AGO2, which suppresses the catalytic activity of the RISC complex [32]. In the DNA virus IIV-6, *ORF340R* encodes a dsRNA-binding domain (dsRBD), which binds dsRNA. For evasion and suppression, the dsRBD binds to long dsRNA shielding it from Dcr-2 processing and inhibiting siRNA loading into the RISC complex, respectively [48]. Viral suppression of RNAi creates an ongoing arms race between viruses and the RNAi pathway. As the RNAi pathway adapts to evade viral infections, viruses counter adapt to evade viral antagonists, which leads to further adaptions of the RNAi pathway [16]. However, RNAi does not clear all viral infections in *D. melanogaster* suggesting that other alternative antiviral mechanisms must exist.

### 2.3. Vago acts as an RNAi-independent antiviral mechanism

During viral infection of *D. melanogaster*, genes are triggered and expressed. One gene of interest is *Vago*, a 160-amino acid protein, with a signal peptide and eight cysteine residues. The signal peptide contains a single von Willebrand factor type C (VWC) motif. Proteins containing a single VWC domain typically respond to environmental changes and nutritional status, such as viral infection [49]. In *D. melanogaster*, *Vago* functions in response to viral infection [50]. During DCV infection, Vago proteins are important in controlling viral load in the fat body of *D. melanogaster*, which suggests that it may have a tissue-specific role. Also, Vago may act as either an antiviral molecule targeting virions or as a cytokine affecting neighboring cells by triggering an antiviral state [51]. Another gene of interest is *virus-induced RNA 1* (*vir-1*). This gene is a marker of viral regulation that is regulated by the JAK/STAT pathway [52]. A potential mechanism is suggested including both genes. Viral infection triggers the induction of a cytokine, Vago, which activates the JAK/STAT pathway (**Figure 2**). Once activated, virus-related gene expression is induced, which includes *vir-1* [51].

Currently, the pathway for activation of Vago begins with induction of RNAi. First, viral infection is detected by Dcr-2. Dcr-2 is a viral sensor, which activates the RNAi pathway and *Vago* expression for antiviral defense (**Figure 2**). For *Vago*, viral RNA interacts with the DExD/H-box helicase domain on Dcr-2 activating an inducible antiviral response. This domain is located at the carboxy-terminal end of the gene and acts as a cytoplasmic sensor of viral RNA [51]. The DExD/H-box helicase domain also belongs to the same family as the retinoic acid-inducible gene 1-like (RIG-I) receptors in mammals, which function as pattern recognition receptors for

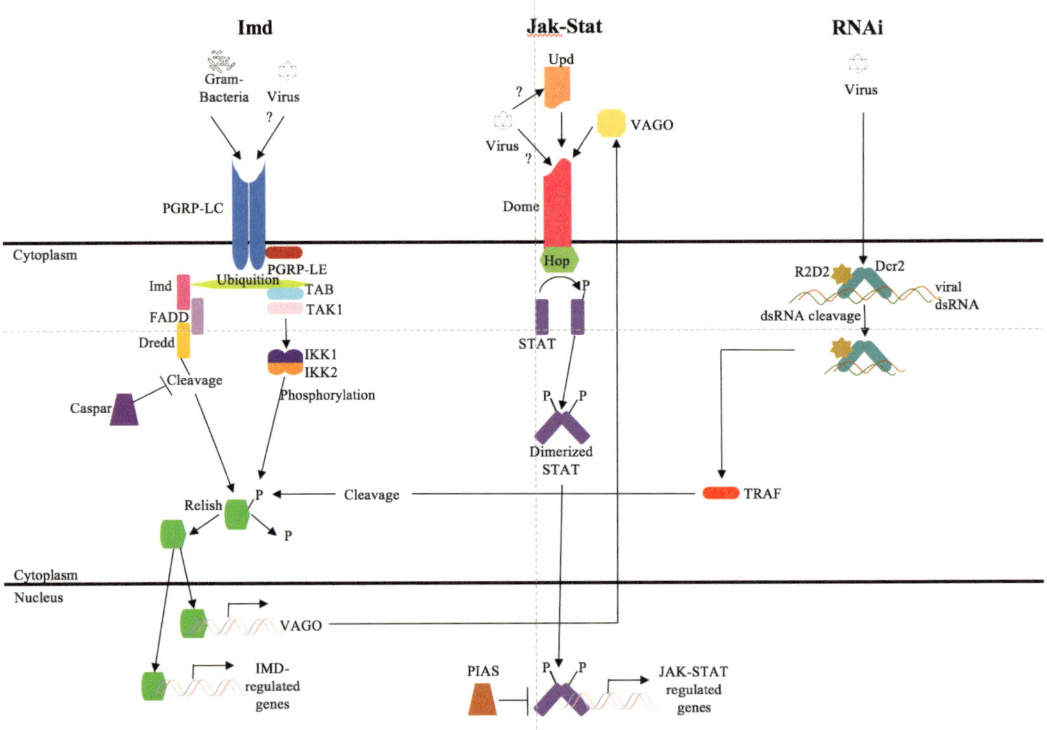


**Figure 2.** Innate immune signaling among several pathways is integrated. The Imd, JAK/STAT, and RNAi virus defense pathways exhibit coordinate expression of anti-viral genes in *Culex* mosquitoes. The RNAi pathway (see **Figure 1**) through the sensing of dsRNA by Dicer-2 activates tumor necrosis factor (TNF) receptor-associated factor (TRAF). TRAF now interacts with the Imd pathway via driving proteolytic cleavage of the N-terminal region of Relish, allowing the C-terminal region of Relish to be transported into the nucleus where it acts as a transcription factor on IMD-regulated genes. One of these genes is *Vago*, which specifies a small secretory cytokine-like molecule. Vago is able to engage the JAK/STAT pathway via the Dome receptor (see **Figure 1**), leading to the expression of JAK/STAT regulated genes.

intracellular dsRNA during viral infection [53]. In addition, other innate immunity pathways are analyzed to determine their role in the induction of *Vago*. However, Toll, Imd, and JAK/STAT were unable to induce *Vago* expression [51]. Currently, the mode of antiviral action of the protein Vago and its role in the RNAi pathway are not fully understood.

Recently, *Vago* was further investigated in the mosquito, *Culex quinquefasciatus*. The orthologue gene, *CxVago*, contributes to antiviral defense during West Nile virus (WNV) infection. In *C. quinquefasciatus*, Dcr-2 is also required for induction and up-regulation of *Cx*Vago. The study suggests that *Cx*Vago is a stable, secreted cytokine that stimulates an antiviral response in insects by activating the JAK/STAT pathway (**Figure 2**). In addition, *Cx*Vago induces expression of the *Culex* orthologue of the *D. melanogaster* gene *vir-1* during viral infection [49]. Another study was not able to establish a relationship between DCV-stimulated *Vago* and induction of *vir-1* in *D. melanogaster* [51]. However, *Vago* may induce vir-1 during viral infection, but in its absence, other unidentified cytokines may also induce vir-1 expression [49].

A mechanism for the activation of *Cx*Vago was proposed (**Figure 2**). First, Dcr-2 senses a viral infection and activates tumor necrosis factor receptor-associated factor (TRAF). This process activates Relish 2 (Rel2) by dephosphorylation, which allows translocation of the molecule from the cytoplasm into the nucleus. Rel2 is a nuclear factor kappa-light-chain-enhancer of activated B cells (NF-κB) transcription factor and induces gene expression of *CxVago* [54]. However, *DmVago* is not induced in *D. melanogaster* by members of the NF-κB family. This may indicate that regulation of *DmVago* occurs through a similar or alternative mechanism [51]. The induction of *Vago* is similar to the RIG-I/TRAF-6/NF-κB-mediated interferon pathway, which is triggered by a viral infection in mammals [54]. Further analysis of the proposed *Cx*Vago pathway in *D. melanogaster* is required to discover the mechanism for antiviral defense. Vago and its associated pathway might be a simplistic interferon response pathway but requires an in-depth investigation to determine its role in antiviral defense.

## 3. Autophagy

Autophagy was first characterized in yeast following starvation, as a process by which cells can degrade long-lived proteins, organelles, and bulk cytoplasm for recycling [55]. Induction of autophagy is both developmentally and nutritionally regulated. When nutrients are sufficient, class I phosphatidylinositol-3-kinases (PI3Ks) and the target of rapamycin (TOR) complex act as inhibitors of autophagy. However, under starvation conditions, class III PI3Ks act to stimulate the production of autophagy-related proteins and induce the autophagy pathway [55] (**Figure 1**).

Following induction, a double-membrane vesicle, the autophagosome, is formed that can sequester cytoplasmic components. Sequestering of the cytoplasmic components is highly regulated by GTPases, phosphatidylinositol kinases, and other various phosphatases. The autophagosome then fuses with the lysosome for the breakdown of the membrane and its contents [56]. In addition, induction of autophagy can occur as an antiviral response during viral infection.

### 3.1. Antiviral autophagy

Autophagy also plays a direct antiviral role against vesicular stomatitis virus (VSV). *D. melanogaster* has homologs of 11 yeast autophagy-related genes and is confirmed for autophagy during development or starvation [57]. *D. melanogaster* encodes nine Toll receptors. Eight of the Toll receptors are not fully understood but may have roles in innate immunity. Activation of the autophagy pathway requires the interaction of Toll receptor 7, which detects VSV G protein. Once G protein is detected, two toll-7 receptors dimerize transmitting a signal through their toll-interleukin-1 receptor (TFR-1) domain [6]. The signal transduction is regulated by the Tor kinase, which leads to the induction of autophagy [56]. Autophagy can be induced under starvation conditions or high stress (i.e., viral infection) conditions. This becomes apparent when *D. melanogaster* S2 cells are infected with VSV and monitored using fluorescent microscopy for autophagy. Cells with mutant autophagy genes have a significantly higher viral titer than those that contain wild-type autophagy genes [57]. This indicates that autophagy not only plays a critical role in recycling of organelles and proteins during times of starvation, but it may also have an antiviral role as well.

## 4. Other antiviral response pathways

The Toll pathway controls the dorsal-ventral patterning within the *D. melanogaster* embryo and is activated during fungal and Gram-positive bacterial infections (**Figure 1**). During fungal and bacterial infection, pathogen recognition proteins (PRRs) recognize common molecules from each pathogen called pathogen-associated molecular patterns (PAMPs). Fungi are detected by their glucans by PRR glucan-binding protein 3 (GNBP3). Gram-positive bacteria are detected by their cell wall components that contain lysine-containing peptidoglycan. Recognition requires a combination of different proteins, including peptidoglycan recognition proteins (PGRP)-SA, PRGP-SD, and GNBP1 [58]. After recognition, the protein creates a complex, which activates the Toll pathway. PGRP-SD is not involved in the complex but is required for detection of certain strains of Gram-positive bacteria. Activation of Toll initiates a protease cascade activating Spätzle (Spz) [59, 60]. Spz is a protein ligand of the Toll receptor. Once activated, Spz induces conformational changes within the receptor to facilitate the recruitment of *Drosophila* Myd88, Tube, and Pelle, a protein kinase. This leads to the phosphorylation and degradation of Cactus and NF-κB-like transcription factors, which allows Dif (Dorsal-related immunity factor) to translocate to the nucleus. Dif mediates Toll-dependent gene expression of certain antimicrobial peptides (AMPs) [61]. There are seven specific AMPs identified in *D. melanogaster:* Drosomycin, Metchnikowin, Diptericin, Drosocin, Cecropin, Defensin, and Attacin [62]. Cecropin, Diptericin, Drosocin, Attacin, and Defensin are involved during bacterial infection, whereas Drosomycin and Cecropin are involved during fungal infection. Metchnikowin is involved in both forms of infection [63, 64]. These peptides are secreted into the hemolymph for antibacterial and antifungal defense.

Recently, Toll was found to elicit an antiviral response (**Figure 1**). A *Dif*$^1$ fly mutant, which did not have a functional Toll pathway, developed higher DXV viral titers and higher mortality

http://dx.doi.org/10.5772/intechopen.69293

when compared to wild-type flies. A gain-of-function Toll mutant, *Tl1°b*, developed a reduced DXV viral titer [65]. These results indicate that Toll may be involved in reducing viral replication of DXV and potentially other viral pathogens and warrants further characterization.

Another pathway involved in antibacterial defense is Imd. Imd has a similar mechanism as Toll but targets Gram-negative bacteria (**Figure 1**). The PAMPs for Gram-negative bacteria are diaminopimelic-containing peptidoglycan (DAP-type PGN), which are recognized by the PRRs, PGRP-LC, and PGRP-LE. This triggers the Imd intracellular signaling cascade [58]. The signaling cascade activates an NF-kB-like factor, Relish (Rel). The Rel domain of Relish translocates to the nucleus, binds to the kB site, and induces transcription of AMPs, regulating expression [3]. Imd and Toll share the same target genes but are activated by different pathogens. In addition, Toll and Imd interact with each other to regulate a coordinated and effective immune response.

The Imd pathway is implicated in an antiviral response in *D. melanogaster* (**Figure 1**). Loss-of-function mutant flies were created for different Imd pathways genes, such as *Rel* and *PGRP-LC*. These flies displayed increased sensitivity to CrPV and had higher viral loads than the controls [8]. The results indicate that Imd signaling may be involved in antiviral innate immune responses during CrPV infection and requires further research.

The JAK/STAT pathway is also involved in the *D. melanogaster* immune response (**Figure 1**). This pathway contributes to a systemic immune response, antiviral response, and regeneration of gut epithelium [52, 66, 67]. JAK/STAT consists of cytokine-like molecules Unpaired (Upd) and the Upd receptor Domeless (Dome), Hopscotch (Hop), the *D. melanogaster* homolog of vertebrate JAK, the signal transducer and activator of transcription protein at 92E (STAT92E), and suppressors of cytokine signaling (SOC3S6E) [4, 68–72]. For activation of the JAK/STAT pathway, Upd binds to Dome. This binding causes Hop to phosphorylate itself and the cytoplasmic tail of Dome [68, 72]. Phosphorylation of Dome allows for the binding of STAT92E proteins. STAT92E is phosphorylated, dimerized, and translocated to the nucleus where it binds and activates transcription. SOCS36E is a negative regulator of the JAK/STAT pathway. It inhibits activation by binding the JAK complex, preventing autophosphorylation [69, 71]. JAK/STAT is also implicated in antibacterial and/or antifungal defense, but its role in antiviral defense needs further investigation.

## 5. Conclusion

Viral pathogens infect all organisms, including insects. For successful infection, viruses must be able to replicate and evade host immunity. *D. melanogaster* must rely on innate immunity to combat infection. Viral infections are easily controlled and can develop a persistent infection with no apparent pathogenesis. However, this regulation of infection is poorly understood. An uncharacterized antiviral mechanism must exist, which may include *Vago*, but further research is needed. A better understanding of antiviral immunity is important because many of the factors and pathways are conserved among species. Further research with viruses, especially new viruses, will help promote a better understanding of host immunity to pathogen interactions.

## Acknowledgements

The financial support was provided by grants from the National Center for Research Resources (NCRR; 5P20RR016469) and the National Institute of General Medical Science (NIGMS; 8P20GM103427), a component of the National Institutes of Health (NIH). This publication's contents are the sole responsibility of the authors and do not necessarily represent the official views of the NIH or NIGMS.

## Author details

Wilfredo A. Lopez*, Alexis M. Page, Brad L. Ericson, Darby J. Carlson and Kimberly A. Carlson

*Address all correspondence to: carlsonka1@unk.edu

Department of Biology, University of Nebraska at Kearney, Kearney, Nebraska, USA

## References

[1] Iwasaki A, Medzhitov R. Control of adaptive immunity by the innate immune system. Nature Immunology. 2015;**16**(4):343-353. DOI: 10.1038/ni.3123

[2] Fearon DT, Locksley RM. The instructive role of innate immunity in the acquired immune response. Science. 1996;**27**(5258):50. DOI: 10.1126/science.272.5258.50

[3] De Gregorio E, Spellman PT, Tzou P, Rubin GM, Lemaitre B. The Toll and Imd pathways are the major regulators of the immune response in *Drosophila*. The EMBO Journal. 2002;**21**(11):2568-2579. DOI: 10.1093/emboj/21.11.2568

[4] Baeg GH, Zhou R, Perrimon N. Genome-wide RNAi analysis of JAK/STAT signaling components in *Drosophila*. Genes & Development. 2005;**19**(16):1861-1870. DOI: 10.1101/gad.1320705

[5] Zambon RA, Vakharia VN, Wu LP. RNAi is an antiviral immune response against a dsRNA virus in *Drosophila melanogaster*. Cellular Microbiology. 2006;**8**(5):880-889. DOI: 10.1111/j.1462-5822.2006.00688.x

[6] Nakamoto M, Moy RH, Xu J, Bambina S, Yasunaga A, Shelly SS, et al. Virus recognition by Toll-7 activates antiviral autophagy in *Drosophila*. Immunity. 2012;**36**(4):658-667. DOI: 10.1016/j.immuni.2012.03.003

[7] Ferreira ÁG, Naylor H, Esteves SS, Pais IS, Martins NE, Teixeira L. The Toll-Dorsal pathway is required for resistance to viral oral infection in *Drosophila*. PLoS Pathogens. 2014;**10**(12):e1004507. DOI: 10.1371/journal.ppat.1004507

[8] Costa A, Jan E, Sarnow P, Schneider D. The Imd pathway is involved in antiviral immune responses in *Drosophila*. PLoS One. 2009;**4**(10):e743. DOI: 10.1371/journal.pone.0007436

[9] Habayeb MS, Ekengren SK, Hultmark D. Nora virus, a persistent virus in *Drosophila*, defines a new picorna-like virus family. Journal of General Virology. 2006;**87**(10):3045-3051. DOI: 10.1099/vir.0.81997-0

[10] Carpenter JA, Obbard DJ, Maside X, Jiggins FM. The recent spread of a vertically transmitted virus through populations of *Drosophila melanogaster*. Molecular Ecology. 2007;**16**(18):3947-3954. DOI: 10.1111/j.1365-294X.2007.03460.x

[11] Jousset FX, Bergoin M, Revet B. Characterization of the *Drosophila* C virus. Journal of General Virology. 1977;**34**(2):269-283. DOI: 10.1099/0022-1317-34-2-269

[12] Teninges D, Ohanessian A, Richard-Molard C, Contamine D. Isolation and biological properties of *Drosophila* X virus. Journal of General Virology. 1979;**42**(2):241-254. DOI: 10.1111/j.1742-4658.2006.05333.*x*

[13] Webster CL, Waldron FM, Robertson S, Crowson D, Ferrari G, Quintana JF. et al. The discovery, distribution, and evolution of viruses associated with *Drosophila melanogaster*. PLoS Biology. 2015;**13**(7). DOI: 10.1371/journal.pbio.1002210

[14] Habayeb MS, Cantera R, Casanova G, Ekström JO, Albright S, Hultmark D. The *Drosophila* Nora virus is an enteric virus, transmitted via feces. Journal of Invertebrate Pathology. 2009;**101**(1):29-33. DOI: 10.1016/j.jip.2009.02.003

[15] Johnson KN, Christian PD. The novel genome organization of the insect picorna-like virus *Drosophila* C virus suggests this virus belongs to a previously undescribed virus family. Journal of General Virology. 1998;**79**(1):191-203. DOI: 10.1099/0022-1317-79-1-191

[16] Ekström JO, Habayeb MS, Srivastava V, Kieselbach T, Wingsle G, Hultmark D. *Drosophila* Nora virus capsid proteins differ from those of other picorna-like viruses. Virus Research. 2011;**160**(1):51-58. DOI: 10.1016/j.virusres.2011.05.006

[17] van Mierlo JT, Overheul GJ, Obadia B, van Cleef KW, Webster CL, Saleh MC. et al. Novel *Drosophila* viruses encode host-specific suppressors of RNAi. PLoS Pathogens. 2014;**10**(7):e1004256. DOI: 10.1371/journal.ppat.1004256

[18] Sadanandan SA, Ekström JO, Jonna VR, Hofer A, Hultmark D. VP3 is crucial for the stability of Nora virus virions. Virus Research. 2016;**223**:20-27. DOI: 10.1016/j.virusres.2016.06.011

[19] Teninges D, Bras F, Dezélée S. Genome organization of the sigma rhabdovirus: Six genes and a gene overlap. Virology. 1993;**193**(2):1018-1023. DOI: 10.1006/viro.1993.1219

[20] Longdon B, Obbard DJ, Jiggins FM. Sigma viruses from three species of *Drosophila* form a major new clade in the rhabdovirus phylogeny. Proceedings of the Royal Society B: Biological Sciences. 2009;**277**(1678):35-44. DOI: 10.1098/rspb.2009.1472

[21] Chtarbanova S, Lamiable O, Lee KZ, Galiana D, Troxler L, Meignin C. et al. *Drosophila* C virus systemic infection leads to intestinal obstruction. Journal of Virology. 2014;**88**(24):14057-14069. DOI: 10.1128/JVI.02320-14

[22] Zhu F, Ding H, Zhu B. Transcriptional profiling of *Drosophila* S2 cells in early response to *Drosophila* C virus. Virology Journal. 2013;**10**(1):1-9. DOI: 10.1186/1743-422X-10-210

[23] van Rij RP, Saleh MC, Berry B, Foo C, Houk A, Antoniewski C. et al. The RNA silencing endonuclease Argonaute 2 mediates specific antiviral immunity in *Drosophila melanogaster*. Genes & Development. 2006;**20**(21):2985-2995. DOI: 10.1101/gad.1482006

[24] Reavy B, Moore NF. Cell-free translation of *Drosophila* C virus RNA: Identification of a virus protease activity involved in capsid protein synthesis and further studies on *in vitro* processing of Cricket paralysis virus specified proteins. Archives of Virology. 1983;**76**(2):101-115

[25] Chung HK, Kordyban S, Cameron L, Dobos P. Sequence analysis of the bicistronic *Drosophila* X virus genome segment A and its encode polypeptides. Virology. 1996;**225**(2):359-368. DOI: 10.1006/viro.1996.0610

[26] Unckless RL. A DNA virus of *Drosophila*. PLoS One. 2011;**6**(10):e26564. DOI: 10.1371/journal.pone.0026564

[27] Reinganum C, O'Loughlin GT, Hogan TW. A nonoccluded virus of the field crickets *Teleogryllus oceanicus* and *T. commodus* (Orthoptera: Gryllidae). Journal of Invertebrate Pathology. 1970;**16**(2):214-220. DOI: 10.1016/0022-2011(70)90062-5

[28] Dasgupta R, Free HM, Zietlow SL, Paskewitz SM, Aksoy S, Shi L. et al. Replication of flock house virus in three genera of medically important insects. Journal of Medical Entomology. 2007;**44**(1):102-110. DOI: 10.1093/jmedent/41.5.102

[29] Mudiganti U, Hernandez R, Ferreira D, Brown DT. Sindbis virus infection of two model insect cell systems—A comparative study. Virus Research. 2006;**122**(1):28-34. DOI: 10.1016/j.virusres.2006.06.007

[30] Blondel D, Dezelee S, Wyers F. Vesicular stomatitis virus growth in *Drosophila melanogaster* cells. II. Modifications of viral protein phosphorylation. Journal of General Virology. 1983;**64**(8):1793-1799. DOI: 10.1099/0022-1317-64-8-1793

[31] Bronkhorst AW, van Cleef KW, Vodovar N, İnce İA, Blanc H, Vlak JM. et al. The DNA virus Invertebrate iridescent virus 6 is a target of the *Drosophila* RNAi machinery. Proceedings of the National Academy of Sciences of the United States. 2012;**109**(51):e3604-e3613. DOI: 10.1073/pnas.1207213109

[32] Nayak A, Berry B, Tassetto M, Kunitomi M, Acevedo A, Deng C. et al. Cricket paralysis virus antagonizes Argonaute 2 to modulate antiviral defense in *Drosophila*. Nature Structural & Molecular Biology. 2010;**17**(5):547-554. DOI: 10.1038/nsmb.1810

[33] Dearing SC, Scotti PD, Wigley PJ, Dhana SD. A small RNA virus isolated from the grass grub, *Costelytra zealandica* (Coleoptera: Scarabaeidae). New Zealand Journal of Zoology. 1980;**7**(2):267-269. DOI: 10.1080/03014223.1980.10423785

[34] Jovel J, Schneemann A. Molecular characterization of *Drosophila* cells persistently infected with Flock House virus. Virology. 2011;**419**(1):43-53. DOI: 10.1016/j.virol.2011.08.002

[35] Miller DJ, Schwartz MD, Ahlquist P. Flock house virus RNA replicates on outer mitochondrial membranes in *Drosophila* cells. Journal of Virology. 2001;**75**(23):11664-11676. DOI: 10.1128/JVI.75.23.11664-11676.2001

[36] Galiana-Arnoux D, Dostert C, Schneemann A, Hoffmann JA, Imler JL. Essential function in vivo for Dicer-2 in host defense against RNA viruses in *Drosophila*. Nature Immunology.2006;**7**:590-597. DOI: 10.1038/ni1335

[37] Li H, Li WX, Ding SW. Induction and suppression of RNA silencing by an animal virus. Science. 2002;**296**(5571):1319-1321. DOI: 10.1038/nmicrobiol.2016.250

[38] Frolov I, Hardy R, Rice CM. Cis-acting RNA elements at the 5'end of Sindbis virus genome RNA regulate minus-and plus-strand RNA synthesis.RNA. 2001;**7**(11):1638-1651. DOI: 10.1017.S135583820101010X

[39] Mudd JA, Leavitt RW, Kingsbury DT, Holland JJ. Natural selection of mutants of vesicular stomatitis virus by cultured cells of *Drosophila melanogaster*. Journal of General Virology. 1973;**20**(3):341-351. DOI: 10.1099/0022-1317-20-3-341

[40] Wyers FR, Richard-Molard C, Blondel D, Dezelee S. Vesicular stomatitis virus growth in *Drosophila melanogaster* cells: G protein deficiency. Journal of Virology. 1980;**33**(1):411-422

[41] Jakob NJ, Müller K, Bahr U, Darai G. Analysis of the first complete DNA sequence of an invertebrate iridovirus: Coding strategy of the genome of Chilo iridescent virus. Virology. 2001;**286**(1):182-196. DOI: 10.1006/viro.2001.0963

[42] Eaton HE, Metcalf J, Penny E, Tcherepanov V, Upton C, Brunetti CR. Comparative genomic analysis of the family Iridoviridae: Re-annotating and defining the core set of iridovirus genes. Virology Journal. 2007;**4**(1):1. DOI: 10.1186/1743-422X-4-11

[43] Liu Y, Ye X, Jiang F, Liang C, Chen D, Peng J. et al. C3PO, an endoribonuclease that promotes RNAi by facilitating RISC activation. Science. 2009;**325**(5941):750-753. DOI: 10.1126/science.1176325

[44] Okamura K, Ishizuka A, Siomi H, Siomi MC. Distinct roles for Argonaute proteins in small RNA-directed RNA cleavage pathways. Genes & Development. 2004;**18**(14):1655-1666. DOI: 10.1101/gad.1210204

[45] Olivieri D, Sykora MM, Sachidanandam R, Mechtler K, Brennecke J. An *in vivo* RNAi assay identifies major genetic and cellular requirements for primary piRNA biogenesis in *Drosophila*. The EMBO Journal. 2010;**29**(19):3301-3317. DOI: 10.1038/emboj.2010.212

[46] Van Mierlo JT, Bronkhorst AW, Overheul GJ, Sadanandan SA, Ekström JO, Heestermans M. et al. Convergent evolution of argonaute-2 slicer antagonism in two distinct insect RNA viruses. PLoS Pathogens. 2012;**8**(8):e1002872. DOI: 10.1371/journal.ppat.1002872

[47] Aliyari R, Wu Q, Li HW, Wang XH, Li F, Green LD. et al. Mechanism of induction and suppression of antiviral immunity directed by virus-derived small RNAs in *Drosophila*. Cell Host & Microbe. 2008;**4**(4):387-397. DOI: 10.1016/j.chom.2008.09.001

[48] Bronkhorst AW, van Cleef KW, Venselaar H, van Rij RP. A dsRNA-binding protein of a complex invertebrate DNA virus suppresses the *Drosophila* RNAi response. Nucleic Acids Research. 2014;**42**(19):12237-12248. DOI: 10.1093/nar/gku910

[49] Paradkar PN, Trinidad L, Voysey R, Duchemin JB, Walker PJ. Secreted Vago restricts West Nile virus infection in *Culex* mosquito cells by activating the JAK/STAT pathway. Proceedings of the National Academy of Sciences of the United States of America. 2012;**109**(46):18915-18920. DOI: 10.1073/pnas.1205231109

[50] Cordes EJ, Licking-Murray KD, Carlson KA. Differential gene expression related to Nora virus infection of *Drosophila melanogaster*. Virus Research. 2013;**175**(2):95-100. DOI: 10.1016/j.virusres.2013.03.021

[51] Deddouche S, Matt N, Budd A, Mueller S, Kemp C, Galiana-Arnoux D. et al. The DExD/H-box helicase Dicer-2 mediates the induction of antiviral activity in *Drosophila*. Nature Immunology. 2008;**9**(12):1425-1432. DOI: 10.1038/ni.1664

[52] Dostert C, Jouanguy E, Irving P, Troxler L, Galiana-Arnoux D, Hetru C. et al. The JAK/STAT signaling pathway is required but not sufficient for the antiviral response of *Drosophila*. Nature Immunology. 2005;**6**(9):946-953. DOI: 10.1038/ni1237

[53] Yoneyama M, Kikuchi M, Matsumoto K, Imaizumi T, Miyagishi M, Taira K. et al. Shared and unique functions of the DExD/H-box helicases RIG-I, MDA5, and LGP2 in antiviral innate immunity. The Journal of Immunology. 2005;**175**(5):2851-2858. DOI: 10.4049/jimmunol.175.5.2851

[54] Paradkar PN, Duchemin JB, Voysey R, Walker PJ. Dicer-2-dependent activation of *Culex* Vago occurs via the TRAF-Rel2 signaling pathway. PLoS Neglected Tropical Diseases. 2014;**8**(4):e2823. DOI: 10.1371/journal.pntd.0002823

[55] Berry DL, Baehrecke EH. Growth arrest and autophagy are required for salivary gland cell degradation in *Drosophila*. Cell. 2007;**131**(6):1137-1148. DOI: 10.1016/j.cell.2007.10.048

[56] Klionsky DJ, Emr SD. Autophagy as a regulated pathway of cellular degradation. Science. 2000;**290**(5497):1717-1721. DOI: 10.1126/science.290.5497.1717

[57] Shelly S, Lukinova N, Bambina S, Berman A, Cherry S. Autophagy is an essential component of *Drosophila* immunity against vesicular stomatitis virus. Immunity. 2009;**30**(4):588-598. DOI: 10.1016/j.immuni.2009.02.009

[58] Chamy LE, Leclerc V, Caldelari I, Reichhart J-M. Danger signal and PAMP sensing define binary signaling pathways upstream of Toll. Nature Immunology. 2008;**9**(10):1165-1170. DOI: 10.1038/ni.1643

[59] Gottar M, Gobert V, Matskevich AA, Reichhart JM, Wang C, Butt TM. et al. Dual detection of fungal infections in *Drosophila* via recognition of glucans and sensing of virulence factors. Cell. 2006;**127**(7):1425-1437. DOI: 10.1016/j.cell.2006.10.046

[60] Jang IH, Chosa N, Kim SH, Nam HJ, Lemaitre B, Ochiai M. et al. A Spätzle-processing enzyme required for toll signaling activation in *Drosophila* innate immunity. Developmental Cell. 2006;**10**(1):45-55. DOI: 10.1016/j.devcel.2005.11.013

[61] Kanoh H, Kuraishi T, Tong LL, Watanabe R, Nagata S, Kurata S. *Ex vivo* genome-wide RNAi screening of the *Drosophila* Toll signaling pathway elicited by a larva-derived tissue extract. Biochemical and Biophysical Research Communications. 2015;**467**(2):400-406. DOI: 10.1016/j.bbrc.2015.09.138

[62] Rutschmann S, Jung AC, Hetru C, Reichhart JM, Hoffmann JA, Ferrandon D. The Rel protein DIF mediates the antifungal but not the antibacterial host defense in *Drosophila*. Immunity. 2000;**12**(5):569-580. DOI: 10.1016/S1074-7613(00)80208-3

[63] Lemaitre B, Nicolas E, Michaut L, Reichhart JM, Hoffmann JA. The dorsoventral regulatory gene cassette spätzle/Toll/cactus controls the potent antifungal response in *Drosophila* adults. Cell. 1996;**86**(6):973-983. DOI: 10.1016/S0092-8674(00)80172-5

[64] Lemaitre B, Reichart JM, Hoffmann JA. *Drosophila* host defense: Differential induction of antimicrobial peptide genes after infection by various classes of microorganisms. Proceedings of the National Academy of Sciences of the United States of America. 1997;**94**(26):14614-14619

[65] Zambon RA, Nandakumar M, Vakharia VN, Wu LP. The Toll pathway is important for an antiviral response in *Drosophila*. Proceedings of the National Academy of Sciences of the United States of America. 2005;**102**(20):7257-7262. DOI: 10.1073/pnas.0409181102

[66] Yang H, Kronhamn J, Ekström JO, Korkut GG, Hultmark D. JAK/STAT signaling in *Drosophila* muscles controls the cellular immune response against parasitoid infection. EMBO Reports. 2015;**16**(12):1664-1672. DOI: 10.15252/embr.201540277

[67] Jiang H, Patel PH, Kohlmaier A, Grenley MO, McEwen DG, Edgar BA. Cytokine/Jak/Stat signaling mediates regeneration and homeostasis in the *Drosophila* midgut. Cell. 2009;**137**:1343-1355. DOI: 10.1016/j.cell.2009.05.014

[68] Binari R, Perrimon N. Stripe-specific regulation of pair-rule genes by hopscotch, a putative Jak family tyrosine kinase in *Drosophila*. Genes & Development. 1994;**8**(3):300-312. DOI: 10.1101/gad.8.3.300

[69] Yan R, Small S, Desplan C, Dearolf CR, Darnell JE. Identification of a Stat gene that functions in *Drosophila* development. Cell. 1996;**84**(3):421-430. DOI: 10.1016/S0092-8674(00)81287-8

[70] Harrison DA, McCoon PE, Binari R, Gilman M, Perrimon N. *Drosophila* unpaired encodes a secreted protein that activates the JAK signaling pathway. Genes & Development. 1998;**12**(20):3252-3263. DOI: 10.1101/gad.12.20.3252

[71] Zeidler MP, Bach EA, Perrimon N. The roles of the *Drosophila* JAK/STAT pathway. Oncogene. 2000;**19**(21):2598-2606. DOI: 10.1038/sj.onc.1203482

[72] Brown S, Hu N, Hombriá JC. Identification of the first invertebrate interleukin JAK/STAT receptor, the *Drosophila* gene domeless. Current Biology. 2001;**21**:1700-1705. DOI: 10.1016/S0960-9822(01)00524-3

[73] Thomas-Orillard M. Modifications of mean ovariole number, fresh weight of adult females and developmental time in *Drosophila melanogaster* induced by *Drosophila* C virus. Genetics. 1984;**107**(4):635-644